Laboratory Manual

Inquiry into Life

Fourteenth Edition

Sylvia S. Mader

Connect
Learn
Succeed™

Senior Vice President, Products & Markets: *Kurt L. Strand*
Vice President, General Manager, Products & Markets: *Marty Lange*
Vice President, Content Production & Technology Services: *Kimberly Meriwether David*
Managing Director: *Michael S. Hackett*
Director, Biology: *Lynn Breithaupt*
Brand Manager: *Eric Weber*
Director of Development: *Rose Koos*
Developmental Editor: *Liz Recker*
Marketing Manager: *Chris Loewenberg*
Director, Content Production: *Terri Schiesl*
Project Manager: *Sherry Kane*
Senior Buyer: *Sandy Ludovissy*
Senior Designer: *Laurie B. Janssen*
Cover Image: *Christian Beirle González, Getty Images*
Senior Content Licensing Specialist: *Lori Hancock*
Photo Research: *Evelyn Jo Hebert*
Compositor and Art Studio: *Electronic Publishing Services Inc., NYC*
Typeface: *11/13 Utopia Std*
Printer: *Quad/Graphics*

Contents

Preface

To the Instructor

The laboratory exercises in this manual are coordinated with *Inquiry into Life,* a general biology text that covers the entire field of biology. The text emphasizes how we can apply biological knowledge to our own lives and to the biological world in general.

Although each laboratory is referenced to the appropriate chapter(s) in *Inquiry,* this manual may also be used in coordination with other general biology texts. In addition, this laboratory manual can be adapted to a variety of course orientations and designs. There are a sufficient number of laboratories and exercises within each lab to tailor the laboratory experience as desired. Then, too, many exercises may be performed as demonstrations rather than as student activities, thereby shortening the time required to cover a particular concept.

The Exercises

All exercises have been tested for student interest, preparation time, estimated time of completion, and feasibility. The following features are particularly appreciated by adopters:

Integrated opening: Each laboratory begins with a list of Learning Outcomes organized according to the major sections of the laboratory. The major sections of the laboratory are numbered on the opening page and in the laboratory text material. This organization will help students better understand the goals of each laboratory session.

Self-contained content: Each laboratory contains all the background information necessary to understand the concepts being studied and to answer the questions asked. This feature will reduce student frustration and increase learning.

Scientific process: All laboratories stress the scientific process, and many opportunities are given for students to gain an appreciation of the scientific method. The first laboratory of this edition explicitly explains the steps of the scientific method and gives students an opportunity to use them. I particularly recommend this laboratory because it utilizes the pillbug, a living subject.

Student activities: A color bar is used to designate each student activity. Some student exercises are Observations and some are Experimental Procedures. An icon (shaped like a clock) appears whenever a procedure requires a period of time before results can be viewed. Sequentially numbered steps guide students as they perform each activity.

Live materials: Although students work with living material during some part of almost all laboratories, the exercises are designed to be completed within one laboratory session. This facilitates the use of the manual in multiple-section courses.

Virtual labs: New to this edition, virtual labs currently available on the *Inquiry into Life* website **www.mhhe.com/maderinquiry14** are announced and described whenever the announcement seems appropriate. Instructors can use the virtual labs as separate assignments or integrate them into the laboratory experience. In a few instances I have revised the instructions for a virtual lab to create an exercise that stresses the scientific method. These exercises have become part of the laboratory itself. Instructors should still feel free to use the virtual labs as they see fit.

Laboratory safety: Throughout the laboratories, safety precautions specific to an activity are highlighted and identified by a caution symbol.

Improvements in This Edition

All laboratories have been revised to

1. Improve the Introduction so that it becomes an integral part of the laboratory experience. For example, in Laboratory 1, pillbug anatomy is more clearly described in the Introduction.
2. Improve the Laboratory Review so that the questions better reflect the Learning Outcomes. The following laboratories have been extensively revised:

Laboratory 1 Scientific Method: The laboratory was rewritten to improve the flow and better emphasize the scientific method.

Laboratory 7 Cellular Respiration: The laboratory now begins with an introduction that includes the function of the mitochondrion, and better stresses the role of the mitochondrion in cellular respiration.

Laboratory 10 Reproduction in Flowering Plants: The laboratory was rewritten to better introduce students to the life cycle of plants in general and the specific life cycle of flowering plants before considering pollination and seed production in particular.

Laboratory 13 Chemical Aspects of Digestion: The laboratory now follows a dissection lab that includes the digestive system, allowing a better correlation between the experiments in this lab and the digestive tract of humans.

Laboratory 16 Homeostasis: The laboratory was rewritten to introduce more hands-on activities, including blood pressure and lung volume measurements and comparative urine analysis to diagnose particular illnesses.

Laboratory 17 Nervous System and Senses: The laboratory was reorganized by changing 17.1 to the Central Nervous System. This new section now includes both the brain and the spinal cord. Coverage of the cerebral lobes was added to better allow students to associate the various senses to particular lobes.

Laboratory 19 Development: The laboratory now centers around the cellular, the tissue, and the organ stages of development as they pertain to human development. The sea star, the frog, and the chick allow an examination of these stages before the extraembryonic membranes and fetal development are discussed.

Laboratory 20 Patterns of Inheritance: The laboratory was rewritten to streamline the experiments to provide students with the opportunity to do hands-on or simulated genetic experiments as desired.

Laboratory 21 Human Genetics: The laboratory was reorganized and revised to include problems concerning genetic disorders and a genetic counseling section that considers chromosomal anomalies and pedigree analysis.

Laboratory 24 Microbiology: The laboratory is improved by the addition of a virtual lab experience that gives the classification of bacteria an evolutionary emphasis and asks students to complete an evolutionary tree based on available data.

Laboratory 27 Introduction to Invertebrates: The laboratory was rewritten to better stress a hands-on examination of living hydras and planarians, vinegar eels, and rotifers.

Laboratory 29 The Vertebrates: The portion of the laboratory dealing with vertebrate anatomy was rewritten to remove discrepancies and to better allow students to come to conclusions about the comparative anatomy of vertebrates.

Customized Editions

The 31 laboratories in this manual are now available as individual "lab separates," so instructors can custom-tailor the manual to their particular course needs.

Laboratory Resource Guide

The *Laboratory Resource Guide,* an essential aid for instructors and laboratory assistants, free to adopters of the *Inquiry into Life Laboratory Manual,* is online at **www.mhhe.com/maderinquiry14.** The answers to the Laboratory Review questions are in the Resource Guide.

To the Student

Special care has been taken in preparing the *Inquiry into Life Laboratory Manual* to enable you to **enjoy** the laboratory experience as you **learn** from it. The instructions and discussions are written clearly so that you can understand the material while working through it. Student aids are designed to help you focus on important aspects of each exercise. Student learning aids are carefully integrated throughout this manual:

The Learning Outcomes set the goals of each laboratory session and help you review the material for a laboratory practical or any other kind of exam. The major sections of each laboratory are numbered, and the Learning Outcomes are grouped according to these topics. This system allows you to study the chapter in terms of the outcomes presented.

The Introduction to the laboratory reviews much of the necessary background information required for comprehending the work you will be doing during the laboratory session. Also, specific information, provided before each of the Observations and Experimental Procedures, will assist you in being engaged while you do these activities.

The Observations and Experimental Procedures require your active participation. Space is provided in Tables for you to record the results and conclusions of observations and experiments. Space is also provided for you to answer questions pertaining to these activities.

Observation: An activity in which models, slides, and preserved or live organisms are observed to achieve a learning outcome.

Experimental Procedure: An activity in which a series of steps uses laboratory equipment to gather data and come to a conclusion.

At the end of an observation or experimental procedure you are often asked to formulate explanations or conclusions. You should be sure you are truly writing an explanation or conclusion and not just restating the observations made. To do so, you will need to synthesize information from a variety of sources, including the following:

1. Your experimental results and/or the results of other groups in the class. If your data are different from those of other groups in your class, do not erase your answer; add the other groups' answers in parentheses.
2. Your knowledge of underlying principles. Obtain this information from the laboratory Introduction or the appropriate section of the laboratory and from the corresponding chapter of your text.
3. Your understanding of how the experiment was conducted and/or the materials used. (If by chance the results seem inappropriate, it's possible the ingredients were contaminated or you misunderstood the directions. If this occurs, consult with other students and your instructor to see if you should repeat the experiment.)

Color Bars, Time Icon, and Safety Boxes

Observations are identified by the color bar shown to the left. Whenever you see this color you know that the activity will require you to make careful observations and answer questions about these activities.

Experimental procedures are identified by the color bar shown to the left. Whenever you see this color you know that the activity will require you to use laboratory equipment to perform an experiment and answer questions about this experiment.

A time icon is used to designate when time is needed for a reaction to occur. You may be asked to start these activities at the beginning of the laboratory, proceed to other activities, and return to these when the designated time is up.

A safety icon throughout the manual alerts you to any specific activity that requires a cautionary approach. Read these boxes and follow the advice given in the box and/or your instructor when performing the activity.

The Laboratory Review is a set of questions covering the day's work. Do all the review questions as an aid to understanding the laboratory. Your instructor may require you to hand in these questions for credit.

The Appendices at the end of the book provide useful information on preparing a laboratory report and the metric system. Also, the tree of life appendix describes the major groups of organisms and an evolutionary tree shows how they are related through the process of evolution. Practical Examination Answer Sheets are available on the *Inquiry into Life* website when and if they are needed.

Laboratory Preparation

It will be very helpful to you to read the entire laboratory chapter before coming to lab. **Study** the introductory material and the Observations and Experimental Procedures so you know ahead of time what you will be doing that week and how it correlates with the lecture material. If necessary, to obtain a better understanding, read the corresponding chapter in your text. If your text is *Inquiry into Life,* by Sylvia S. Mader and Michael Windelspecht, the "text chapter reference" column in the table of contents at the beginning of the *Laboratory Manual* lists the corresponding chapter in the text.

Student Feedback

If you have any suggestions for how this laboratory manual could be improved, you can send your comments to

The McGraw-Hill Companies
Product Development—General Biology
501 Bell St.
Dubuque, Iowa 52001

Acknowledgments

We gratefully acknowledge the following reviewers for their assistance in the development of this lab manual:

Pebble Barbero
Eastfield College

Darrell D. Barnes
Northwest Mississippi Community College

Sandra Bobick
Community College of Allegheny County

Sandra G. Devenny
Delaware County Community College

Michael J. Farabee
Estrella Mountain Community College

Jill Feinstein
Richland Community College

Brian Jeffrey Ferguson
Lewis and Clark Community College

Jose Flores
Eastfield College

Melissa Greene
Northwest Mississippi Community College

Cynthia Kincer
Wytheville Community College

Mary Delores McCright
Texarkana College

Syed Naqvi
South Louisiana Community College, Lafayette

Angel Nickens
Northwest Mississippi Community College

Bernadette A. Nowicki
Morton College

Benjamin Oubre
South Louisiana Community College

Gregory A. Owens
Richland Community College

Snehlata Pandey
Hampton University

Pamela Riddell
Macomb Community College

Chris Robichaux
South Louisiana Community College

Robin Robison
Northwest Mississippi Community College

Lisa Strong
Northwest Mississippi Community College

Mike Taylor
Santiago Canyon College

Artenzia Young-Seigler
Tennessee State University

McGraw-Hill LabSmart™

Based on the same world-class super-adaptive technology as LearnSmart, McGraw-Hill LabSmart is a must-see, outcomes-based lab simulation. It assesses a student's knowledge and adaptively corrects deficiencies, allowing the student to learn faster and retain more knowledge with greater success.

First, a student's knowledge is adaptively leveled on core learning outcomes: Questioning reveals knowledge deficiencies that are corrected by the delivery of content that is conditional on a student's response.

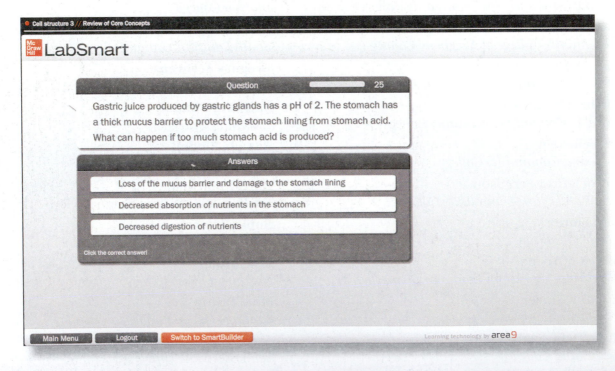

THE Virtual Lab Experience!

Then, a simulated lab experience requires the student to think and act like a scientist: Recording, interpreting, and analyzing data using simulated equipment found in labs and clinics. The student is allowed to make mistakes—a powerful part of the learning experience!

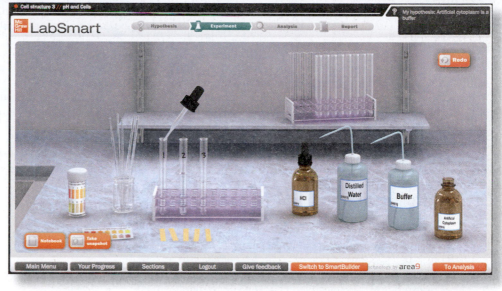

A virtual coach provides subtle hints when needed; asks questions about the student's choices; and allows the student to reflect upon and correct those mistakes.

Whether your need is to overcome the logistical challenges of a traditional lab, provide better lab prep, improve student performance, or make your online experience one that rivals the real world, LabSmart accomplishes it all.

Learn more at www.mhlabsmart.com

Laboratory Safety

The following is a list of practices required for safety purposes in the biology laboratory and in outdoor activities. Following rules of lab safety and using common sense throughout the course will enhance your learning experience by increasing your confidence in your ability to safely use chemicals and equipment. Pay particular attention to oral and written safety instructions given by the instructor. If you do not understand a procedure, ask the instructor, rather than a fellow student, for clarification. Be aware of your school's policy regarding accident liability and any medical care needed as a result of a laboratory or outdoor accident.

The following rules of laboratory safety should become a habit:

1. Wear safety glasses or goggles during exercises in which glassware and chemical reagents are handled, or when dangerous fumes may be present, creating possible hazards to eyes or contact lenses.
2. Assume that all reagents are poisonous and act accordingly. Read the labels on chemical bottles for safety precautions and know the nature of the chemical you are using. If chemicals come into contact with skin, wash immediately with water.
3. **DO NOT**
   a. ingest any reagents.
   b. eat, drink, or smoke in the laboratory. Toxic material may be present, and some chemicals are flammable.
   c. carry reagent bottles around the room.
   d. pipette anything by mouth.
   e. put chemicals in the sink or trash unless instructed to do so.
   f. pour chemicals back into containers unless instructed to do so.
   g. operate any equipment until you are instructed in its use.
   h. dispose of biological or chemical wastes in regular classroom trash receptacles.
4. **DO**
   a. note the location of emergency equipment such as a first aid kit, eyewash bottle, fire extinguisher, switch for ceiling showers, fire blanket(s), sand bucket, and telephone (911).
   b. be familiar with the experiments you will be doing before coming to the laboratory. This will increase your understanding, enjoyment, and safety during exercises. Confusion is dangerous. Completely follow the procedure set forth by the instructor.
   c. keep your work area neat, clean, and organized. Before beginning, remove everything from your work area except the lab manual, pen, and equipment used for the experiment. Wash hands and desk area, including desk top and edge, before and after each experiment. Use clean glassware at the beginning of each exercise, and wash glassware at the end of each exercise or before leaving the laboratory.
   d. wear clothing that, if damaged, would not be a serious loss, or use aprons or laboratory coats, since chemicals may damage fabrics.
   e. wear shoes as protection against broken glass or spillage that may not have been adequately cleaned up.
   f. handle hot glassware with a test tube clamp or tongs. Use caution when using heat, especially when heating chemicals. Do not leave a flame unattended; do not light a Bunsen burner near a gas tank or cylinder; do not move a lit Bunsen burner; do keep long hair and loose clothing well away from the flame; do make certain gas jets are off when the Bunsen burner is not in use. Use proper ventilation and hoods when instructed.
   g. read chemical bottle labels; be aware of the hazards of all chemicals used. Know the safety precautions for each.
   h. stopper all reagent bottles when not in use. Immediately wash reagents off yourself and your clothing if they spill on you, and immediately inform the instructor. If you accidentally get any reagent in your mouth, rinse the mouth thoroughly, and immediately inform your instructor.
   i. use extra care and wear disposable gloves when working with glass tubing and when using dissection equipment (scalpels, knives, or razor blades), whether cutting or assisting.
   j. administer first aid immediately to clean, sterilize, and cover any scrapes, cuts, and burns where the skin is broken and/or where there may be bleeding. Wear bandages over open skin wounds.
   k. report all accidents to the instructor immediately, and ask your instructor for assistance in cleaning up broken glassware and spills.
   l. report to the instructor any condition that appears unsafe or hazardous.
   m. use caution during any outdoor activities. Watch for snakes, poisonous insects or spiders, stinging insects, poison oak, poison ivy, etc. Be careful near water.
   n. wash your hands thoroughly after handling any preserved biological specimens.

I understand the safety rules as presented above. I agree to follow them and all other instructions given by the instructor.

Name: _____   Date: _____

Laboratory Class and Time: _____

x

1
Scientific Method

Introduction

This laboratory will provide you with an opportunity to use the scientific method in the same manner as scientists. Today your subject will be the pillbug, *Armadillidium vulgare*, a type of crustacean that lives on land.

Pillbugs have an exoskeleton consisting of overlapping "armored" plates that make them look like little armadillos. As pillbugs grow, they molt (shed the exoskeleton) four or five times during a lifetime. A pillbug can roll up into such a tight ball that its legs and head are no longer visible, earning it the nickname "roly-poly." They have three body parts: head, thorax, and abdomen. The head bears compound eyes and two pairs of antennae. The thorax bears pairs of walking legs; gills are located at the top of the first five pairs. The gills must be kept slightly moist, which explains why pillbugs are usually found in damp places. The final pair of appendages, the uropods, which are sensory and defensive in function, project from the abdomen of the animal.

Pillbugs are commonly found in damp leaf litter, under rocks, and in basements or crawl spaces under houses. Following an inactive winter, pillbugs mate in

Pillbugs on leaf

*The garden snail, *Helix aspersa,* or the earthworm, *Lumbricus terrestris,* can be substituted as desired.

the spring. Several weeks later, the eggs hatch and remain for six weeks in a brood pouch on the underside of the female's body. Once they leave the pouch, they eat primarily dead organic matter, including decaying leaves. Therefore, they are easy to find and to maintain in a moist terrarium with leaf litter, rocks, and wood chips. You are encouraged to collect some for your experiment. Since they live in the same locations as snakes, be careful when collecting them.

1.1 Using the Scientific Method

Some scientists work alone but often scientists belong to a community of scientists who are working together to study some aspect of the natural world. For example, many scientists from different institutions work together to study the AIDS virus (Fig. 1.1).

Figure 1.1 Scientists work together.
Robert Gallo and his colleagues do research on how viruses, such as the AIDS virus, invade humans.

> You will share your study of pillbugs with the other members of the class.

Even though the methodology can vary, scientists often use the **scientific method** (Fig. 1.2) when doing research. The scientific method involves these steps:

Making observations. Observations help scientists begin their study of a particular topic.

> To learn about pillbugs you will visually observe one. You could also do a Google search of the Web or talk to someone who has worked with pillbugs for a long time.

Why does the scientific method begin with observations? _____

_____

Formulating a hypothesis. Based on their observations, scientists come to a tentative decision, called a hypothesis, about their topic. Formulating hypotheses helps scientists decide how an experiment will be conducted.

> Based on your observations you might hypothesize that a pillbug will be attracted to fruit juice.

Now you know what you will actually do. What is the benefit of formulating a hypothesis? _____

_____

Testing the hypothesis. Scientists make further observations or perform experiments in order to test the hypothesis.

> **Virtual Lab Mealworm Behavior**   A virtual lab called Mealworm Behavior is available on the *Inquiry into Life* website **www.mhhe.com/maderinquiry14**. This virtual lab demonstrates how investigators conducted an experiment similar to the one you will be doing.

> You could decide to expose a pillbug to fruit juice and observe its reaction.

A well-designed experiment must have a **negative control**—that is, a sample or event—that is not exposed to the testing procedure. If the negative control and the test sample produce the same results, either the procedure is flawed or the hypothesis is false.

> Water can substitute for fruit juice and be the control in your experiment.

Scientists call the results of their experiments the **data**. It is very important for scientists to keep accurate records of all their data.

You will record your data in a table that can be easily examined by another person.

When another person repeats the same experiment, and the data is the same, both experiments have merit. Why must a scientist keep a complete record of an experiment? _____

Coming to a conclusion. Scientists come to a conclusion as to whether their data support or do not support the hypothesis.

If a pillbug is attracted to fruit juice, your hypothesis is supported. If the pillbug is not attracted to fruit juice, your hypothesis is not supported.

A scientist never says that a hypothesis has been proven true because, after all, some future knowledge might have a bearing on the experiment. What is the purpose of the conclusion? _____

_____

Developing a scientific theory. A *theory* in science is an encompassing conclusion based on many individual conclusions in the same field. For example, the gene theory states that organisms inherit coded information that controls their anatomy, physiology, and behavior. It takes many years for scientists to develop a theory and, therefore, we will not be developing any theories today. How is a scientific theory different from a conclusion? _____

_____

Figure 1.2   Flow diagram for the scientific method.
Often, scientists use this methodology to come to conclusions and develop theories about the natural world. The return arrow shows that scientists often choose to retest the same hypothesis or test a related hypothesis before arriving at a conclusion.

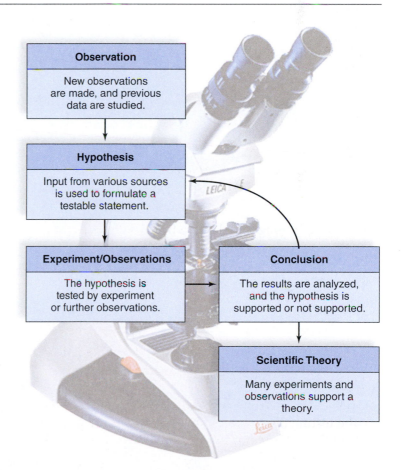

Observation
New observations are made, and previous data are studied.

Hypothesis
Input from various sources is used to formulate a testable statement.

Experiment/Observations
The hypothesis is tested by experiment or further observations.

Conclusion
The results are analyzed, and the hypothesis is supported or not supported.

Scientific Theory
Many experiments and observations support a theory.

1.2 Observing a Pillbug

Wash your hands before and after handling pillbugs. Please handle them carefully so they are not crushed. When touched, they roll up into a ball or "pill" shape as a defense mechanism. They will soon recover if left alone.

Observation: Pillbug's External Anatomy

Obtain a pillbug that has been numbered with white correction fluid or tape tags. Put the pillbug in a small glass or plastic dish to keep it contained.

1. Examine the exterior of the pillbug with the unaided eye and with a magnifying lens or dissecting microscope.

   - How can you recognize the head end of a pillbug? _____

   - How many segments and pairs of walking legs are in the thorax? _____

   - The abdomen ends in uropods, appendages with a sensory and defense function. (Females have leaflike growths at the base of some legs where developing eggs and embryos are held in pouches.)

2. In the following space, draw an outline of your pillbug (at least 7 cm long). Label the head, thorax, abdomen, antennae, eyes, uropods, and one of the seven pairs of legs.

3. Draw a pillbug rolled into a ball.

1. Watch a pillbug's underside as the pillbug moves up a transparent surface, such as the side of a graduated cylinder or beaker.

   a. Describe the action of the feet and any other motion you see. _____

   b. Allow a pillbug to crawl on your hand. Describe how it feels. _____

   c. Does a pillbug have the ability to move directly forward? _____

   d. Do you see evidence of mouthparts on the underside of the pillbug? _____

2. As you watch the pillbug, identify

   a. the anatomical parts that allow a pillbug to identify and take in food. _____

   b. behaviors that will help the pillbug acquire food. For example, is the ability of the pillbug to move directly forward a help in acquiring food? Explain. _____

   What other behaviors allow a pillbug to acquire food? _____

   c. a behavior that helps a pillbug avoid dangerous situations. _____

   If a pillbug rolls up into a ball, wait a few minutes and it may uncurl itself.

3. Measure the speed of three pillbugs.

   a. Place each pillbug on a metric ruler, and use a stopwatch to measure the number of seconds (sec) it takes for the pillbug to move several centimeters (cm). Quickly record here the number of cm moved and the time in sec.

   pillbug 1 _____

   pillbug 2 _____

   pillbug 3 _____

   b. Knowing that 10 millimeters (mm) are in a cm, convert the number of cm traveled to mm and record this in the first column of Table 1.1. Record the total time taken in the second column.

   c. Use the space above to calculate the speed of each pillbug in mm/sec and record the speed for each pillbug in the last column of Table 1.1.

   d. Average the speed for your three pillbugs. (Since you have already calculated the mm/sec for each pillbug, it is only necessary to take an average of the mm moved.) Record the average speed of pillbug motion in Table 1.1.

   When you conduct the experiment in Section 1.4 you will have to be patient with your pillbug as it moves toward or away from a substance.

Table 1.1 Pillbug Speed			
Pillbug	**Millimeters (mm) Traveled**	**Time (sec)**	**Speed (mm/sec)**
1			
2			
3			
		Average speed:	

1.3 Formulating Hypotheses

You will be testing whether pillbugs are attracted to (move toward and eat), repelled by (move away from), or unresponsive to (don't move away from and do not move toward and eat) the particular substances, which are potential foods. If a pillbug simply rolls into a ball, nothing can be concluded, and you may wish to choose another pillbug or wait a minute or two to check for further response.

1. Choose
   a. two dry substances, such as flour, cornstarch, coffee creamer, or baking soda. Fine sand will serve as a control for dry substances. Record your "dry" choices as 1, 2, and 3 in the first column of Table 1.2.
   b. two liquids, such as milk, orange juice, ketchup, applesauce, or carbonated beverage. Water will serve as a control for liquid substances. Record your "wet" choices as 4, 5, and 6 in the first column of Table 1.2.
2. In the second column of Table 1.2, hypothesize how you expect the pillbug to respond to each substance. Use a plus (+) sign if you hypothesize that the pillbug will move toward and eat the substance; a minus (−) sign if you hypothesize that the pillbug will be repelled by the substance; and a zero (0) if you expect the pillbug to show neither behavior.
3. In the third column of Table 1.2 offer a reason for your hypothesis based on your knowledge of pillbugs from the introduction and your examination of the animal.

Table 1.2 Hypotheses About Pillbug's Response to Potential Foods		
Substance	Hypothesis About Pillbug's Response	Reason for Hypothesis
1		
2		
3 (control)		
4		
5		
6 (control)		

1.4   Performing the Experiment and Coming to a Conclusion

A good experimental design would be to keep your pillbug in a petri dish to test its reaction to the chosen substances. During your experiment, no substance must be put directly on the pillbug, nor can the pillbug be placed directly onto the substances.

Experimental Procedure: Pillbug's Response to Potential Foods

1. Before testing the pillbug's reaction, fill in the first column of Table 1.3. It will look exactly like the first column of Table 1.2.
2. Since pillbugs tend to walk around the edge of a petri dish, you could put the wet or dry substance there; or for the wet substance you could put liquid-soaked cotton in the pillbug's path.
3. Rinse your pillbug between procedures by spritzing it with distilled water from a spray bottle. Then put it on a paper towel to dry it off.
4. Watch the pillbug's response to each substance, and record it in Table 1.3, using +, −, or 0 as before.

Table 1.3   Pillbug's Response to Potential Foods

Substance	Pillbug's Response	Hypothesis Supported?
1		
2		
3 (control)		
4		
5		
6 (control)		

Conclusion

5. Do your results support your hypotheses? Answer yes or no in the last column of Table 1.3.
6. Are there any hypotheses that were not supported by the experimental results (data)? Does this

   difference give you more insight into pillbug behavior? Explain. _____

   _____

   _____

7. **Class Results.** Compare your results with those of other students who tested the same substance. Calculate the proportional response to each potential food (%+, %−, %0) and record your calculations in Table 1.4. As a group, your class can decide what proportion is needed to designate this response as typical. For example, if the pillbugs as a whole were attracted to a substance 70% or more of the time, you can call that response the "typical response."

Table 1.4   Pillbug's Response to Potential Foods: Class Results

Substance	Pillbug's Response			Hypothesis Supported?
1	%+	%−	%0	
2	%+	%−	%0	
3 (control)	%+	%−	%0	
4	%+	%−	%0	
5	%+	%−	%0	
6 (control)	%+	%−	%0	

8. On the basis of the class data do you need to revise your conclusion for any particular pillbug

   response? _____ Scientists prefer to come to conclusions on the basis of many trials.

   Why is this the best methodology? _____

   _____

9. Did the pillbugs respond as expected to the controls, i.e., did not eat them? _____ If they

   did not respond as expected, what can you conclude about your experimental results? _____

   _____

_____ **1.** Which is more comprehensive, a conclusion or a theory?

_____ **2.** What is a tentative explanation of observed phenomena?

_____ **3.** What do you call the information scientists collect when doing experiments and making observations?

_____ **4.** What step in the scientific method follows experiments and observations?

_____ **5.** What do you call a sample that goes through all the steps of an experiment and does not contain the factor being tested?

_____ **6.** Can data prove a hypothesis true? (Yes or No)

Indicate whether statements 7 and 8 are hypotheses, conclusions, or scientific theories:

_____ **7.** The data show that vaccines protect people from disease.

_____ **8.** All living things are made of cells.

_____ **9.** How many body divisions does a pillbug have?

_____ **10.** If a pillbug travels 5 mm in 60 sec, what is its rate of speed?

_____ **11.** What can be concluded if a pillbug curls into a ball?

_____ **12.** Pillbugs that back away from a substance are (attracted to/repelled by) the substance.

Thought Questions

13. What is a scientific theory?

14. Why is it important to use one substance at a time when testing a pillbug's reaction?

2

Metric Measurement and Microscopy

Introduction

This laboratory introduces you to the metric system, which biologists use to indicate the sizes of cells and cell structures. This laboratory also examines the features, functions, and use of the compound light microscope and the stereomicroscope (dissecting microscope). Transmission and scanning electron microscopes are explained, and micrographs produced using these microscopes appear throughout this lab manual. The stereomicroscope and the scanning electron microscope view the surface and/or the three-dimensional structure of an object. The compound light microscope and the transmission electron microscope can view only extremely thin sections of a specimen. If a subject was sectioned lengthwise for viewing, the interior of the projections at the top of the cell, called cilia, would appear in the micrograph. A lengthwise cut through any type of specimen is called a **longitudinal section (ls).** On the other hand, if the subject in Figure 2.1 was sectioned crosswise below the area of the cilia, you would see other portions of the interior of the subject. A crosswise cut through any type of specimen is called a **cross section (cs).**

Figure 2.1   Longitudinal and cross sections.
a. Transparent view of a cell. **b.** A longitudinal section would show the cilia at the top of the cell. **c.** A cross section shows only the interior where the cut is made.

a. The cell   b. Longitudinal   c. Cross section
                        section

2.1 The Metric System

The **metric system** is the standard system of measurement in the sciences, including biology, chemistry, and physics. It has tremendous advantages because all conversions, whether for volume, mass (weight), or length, can be in units of ten. Refer to Appendix B, p. B–1, for an in-depth look at the units of the metric system.

Length

Metric units of length measurement include the **meter (m), centimeter (cm), millimeter (mm), micrometer (μm),** and **nanometer (nm)** (Table 2.1). The prefixes milli- (10^{-3}), micro- (10^{-6}), and nano (10^{-9}) are used with length, weight, and volume.

Table 2.1	Metric Units of Length Measurement			
Unit	**Meters**	**Millimeters**	**Centimeters**	**Relative Size**
Meter (m)	1 m	1,000 mm	100 cm	Largest
Centimeter (cm)	0.01 (10^{-2}) m	10 mm	1 cm	
Millimeter (mm)	0.001 (10^{-3}) m	1.0 mm	0.1 cm	
Micrometer (μm)	0.000001 (10^{-6}) m	0.001 (10^{-3}) mm	0.0001 (10^{-4}) cm	
Nanometer (nm)	0.000000001 (10^{-9}) m	0.000001 (10^{-6}) mm	0.0000001 (10^{-7}) cm	Smallest

Experimental Procedure: Length

1. Obtain a small ruler marked in centimeters and millimeters. How many centimeters are

   represented? _____ One centimeter equals how many millimeters? _____ To express

   the size of small objects, such as cell contents, biologists use even smaller units of the metric system than those on the ruler. These units are the micrometer (μm) and the nanometer (nm).

   According to Table 2.1, 1 μm = _____ mm, and 1 nm = _____ mm.

   Therefore, 1 mm = _____ μm = _____ nm.

2. Measure the diameter of the circle shown below to the nearest millimeter. This circle is

   _____ mm = _____ μm = _____ nm.

For example, to convert mm to μm:

$$\text{_____ mm} \times \frac{1,000\ \mu m}{mm} = \text{_____ } \mu m$$

3. Obtain a meter stick. On one side, find the numbers 1 through 39, which denote inches. One meter equals 39.37 inches; therefore, 1 meter is roughly equivalent to 1 yard. Turn the meter stick over, and observe the metric subdivisions. How many centimeters are in a meter? _____ How many millimeters are in a meter? _____ The prefix *milli* means _____.

4. Use the meter stick and the method shown in Figure 2.2 to measure the length of two long bones from a disarticulated human skeleton. Lay the meter stick flat on the lab table. Place a long bone next to the meter stick between two pieces of cardboard (each about 10 cm × 30 cm), held upright at right angles to the stick. The narrow end of each piece of cardboard should touch the meter stick. The length between the cards is the length of the bone in centimeters. For example, if the bone measures from the 22 cm mark to the 50 cm mark, the length of the bone is _____ cm. If the bone measures from the 22 cm mark to midway between the 50 cm and 51 cm marks, its length is _____ mm, or _____ cm.

5. Record the length of two bones. First bone: _____ cm = _____ mm.
   Second bone: _____ cm = _____ mm.

Figure 2.2   Measurement of a long bone.
How to measure a long bone using a meter stick.

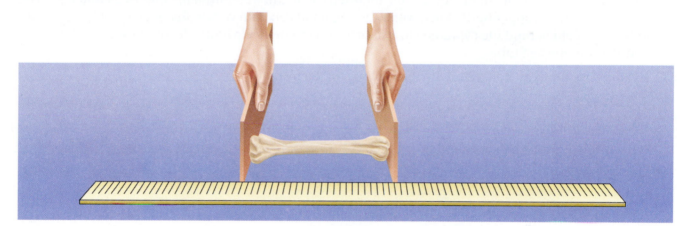

Weight

Two metric units of weight are the **gram (g)** and the **milligram (mg).** A paper clip weighs about 1 g, which equals 1,000 mg. 2 g = _____ mg; 0.2 g = _____ mg; and 2 mg = _____ g.

Experimental Procedure: Weight

1. Use a balance scale to measure the weight of a wooden block small enough to hold in the palm of your hand.
2. Measure the weight of the block to the tenth of a gram. The weight of the wooden block is _____ g = _____ mg.
3. Measure the weight of an item small enough to fit inside the opening of a 50 ml graduated cylinder. The item, a(n) _____, is _____ g = _____ mg.

Volume

Two metric units of volume are the **liter (l)** and the **milliliter (ml).** One liter = 1,000 ml.

Experimental Procedure: Volume

1. Volume measurements can be related to those of length. For example, use a millimeter ruler to measure the wooden block used in the previous Experimental Procedure to get its length, width, and depth.

    length = _____ cm; width = _____ cm; depth = _____ cm

    The volume, or space, occupied by the wooden block can be expressed in cubic centimeters

    (cc or cm^3) by multiplying: length × width × depth = _____ cm^3. For purposes of this

    Experimental Procedure, 1 cubic centimeter equals 1 milliliter; therefore, the wooden block has

    a volume of _____ ml.

2. In the biology laboratory, liquid volume is usually measured directly in liters or milliliters with appropriate measuring devices. For example, use a 50 ml graduated cylinder to add 20 ml of water to a test tube. First, fill the graduated cylinder to the 20 ml mark. To do this properly, you have to make sure that the lowest margin of the water level, or the **meniscus** (Fig. 2.3), is at the 20 ml mark. Place your eye directly parallel to the level of the meniscus, and add water until the meniscus is at the 20 ml mark. (Having a dropper bottle filled with water on hand can help you do this.) A large, blank, white index card held behind the cylinder can also help you see the scale more clearly. Now pour the 20 ml of water into the test tube.

3. Hypothesize how you could find the total volume of the test tube. _____

    _____

    _____

    What is the test tube's total volume? _____

Figure 2.3   Meniscus.
The proper way to view the meniscus.

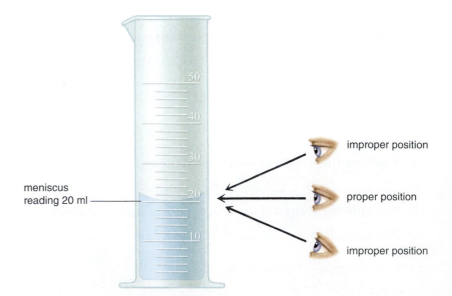

4. Fill a 50 ml graduated cylinder with water to about the 20 ml mark. Hypothesize how you could use this setup to calculate the volume of an object. _____

_____

Now perform the operation you suggested. The object, _____, has a volume of _____ ml.

5. Hypothesize how you could determine how many drops from the pipette of the dropper bottle equal 1 ml. _____

Now perform the operation you suggested. How many drops from the pipette of the dropper bottle equal 1 ml? _____ Some pipettes are graduated and can be filled to a certain level as a way to measure volume directly. Your instructor will demonstrate this. Are pipettes customarily used to measure large or small volumes? _____

Temperature

There are two temperature scales: the **Fahrenheit (F)** and **Celsius (centigrade, C)** scales (Fig. 2.4). Scientists use the Celsius scale.

Experimental Procedure: Temperature

1. Study the two scales in Figure 2.4, and complete the following information:

   a. Water freezes at either _____ °F or _____ °C.

   b. Water boils at either _____ °F or _____ °C.

2. To convert from the Fahrenheit to the Celsius scale, use the following equation:

$$°C = (°F - 32°)/1.8$$
$$or$$
$$°F = (1.8°C) + 32$$

   Human body temperature of 98°F is what temperature on the Celsius scale? _____

3. Record any two of the following temperatures in your lab environment. In each case, allow the end bulb of the Celsius thermometer to remain in or on the sample for one minute.

   Room temperature = _____ °C

   Surface of your skin = _____ °C

   Cold tap water in a 50 ml beaker = _____ °C

   Hot tap water in a 50 ml beaker = _____ °C

   Ice water = _____ °C

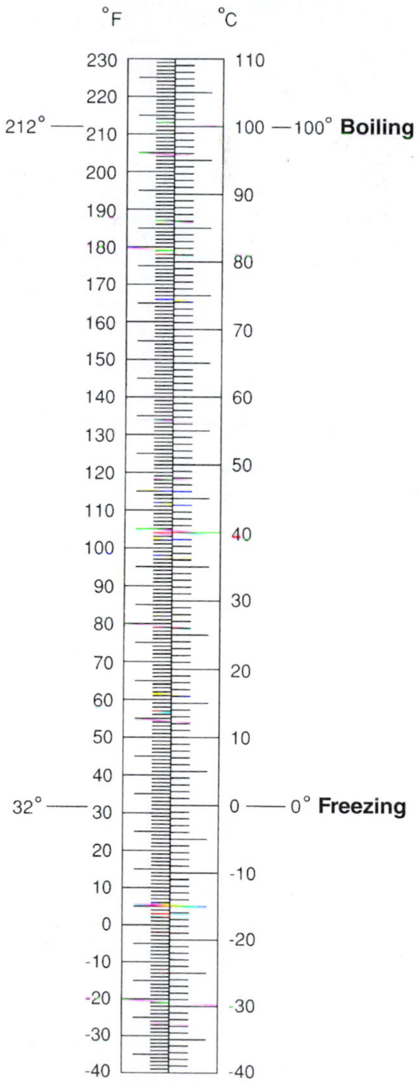

Figure 2.4 Temperature scales.
The Fahrenheit (°F) scale is on the left, and the Celsius (°C) scale is on the right.

2.2 Microscopy

Because biological objects can be very small, we often use a microscope to view them. Many kinds of instruments, ranging from the hand lens to the electron microscope, are effective magnifying devices. A short description of two kinds of light microscopes and two kinds of electron microscopes follows.

Light Microscopes

Light microscopes use light rays passing through lenses to magnify the object. The **stereomicroscope (dissecting microscope)** is designed to study entire objects in three dimensions at low magnification. The **compound light microscope** is used for examining small or thinly sliced sections of objects under higher magnification than that of the stereomicroscope. The term **compound** refers to the use of two sets of lenses: the ocular lenses located near the eyes and the objective lenses located near the object. Illumination is from below, and visible light passes through clear portions but does not pass through opaque portions. To improve contrast, the microscopist uses stains or dyes that bind to cellular structures and absorb light. Photomicrographs, also called light micrographs, are images produced by a compound light microscope (Fig. 2.5*a*).

Figure 2.5   Comparative micrographs of a lymphocyte.
A lymphocyte is a type of white blood cell. **a.** A photomicrograph (light micrograph) of a lymphocyte shows less detail than a **(b).** transmission electron micrograph (TEM). **c.** A scanning electron micrograph (SEM) of a lymphocyte shows the cell surface in three dimensions.

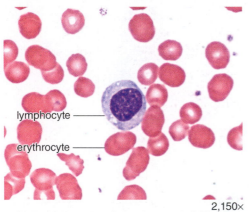

2,150×
a. Photomicrograph or light micrograph (LM)

2,150×
b. Transmission electron micrograph (TEM)

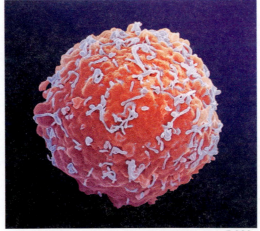

5,000×
c. Scanning electron micrograph (SEM)

Electron Microscopes

Electron microscopes use beams of electrons to magnify the object. The beams are focused on a photographic plate by means of electromagnets. The **transmission electron microscope** is analogous to the compound light microscope. The object is ultra-thinly sliced and treated with heavy metal salts to improve contrast. Figure 2.5*b* is a micrograph produced by this type of microscope. The **scanning electron microscope** is analogous to the dissecting light microscope. It gives an image of the surface and dimensions of an object, as is apparent from the scanning electron micrograph in Figure 2.5*c*.

The micrographs in Figure 2.5 demonstrate that an object is magnified more with an electron microscope than with a compound light microscope. The difference between these two types of microscopes, however, is not simply a matter of magnification; it is also the electron microscope's ability to show detail. The electron microscope has greater resolving power. **Resolution** is the minimum distance between two objects at which they can still be seen, or resolved, as two separate objects. The use of high-energy electrons rather than light gives electron microscopes a much greater resolving power since two objects that are much closer together can still be distinguished as separate points. Table 2.2 lists several other differences between the compound light microscope and the transmission electron microscope.

Table 2.2 Comparison of the Compound Light Microscope and the Transmission Electron Microscope	
Compound Light Microscope	**Transmission Electron Microscope**
1. Glass lenses	1. Electromagnetic lenses
2. Illumination by visible light	2. Illumination due to beam of electrons
3. Resolution $\cong$ 200 nm	3. Resolution $\cong$ 0.1 nm
4. Magnifies to 2,000×	4. Magnifies to 1,000,000×
5. Costs up to tens of thousands of dollars	5. Costs up to hundreds of thousands of dollars

Conclusions: Microscopy

- Which two types of microscopes view the surface of an object? _____
- Which two types of microscopes view objects that have been sliced and treated to improve contrast? _____
- Of the microscopes just mentioned, which one resolves the greater amount of detail?

_____

2.3 Stereomicroscope (Dissecting Microscope)

The **stereomicroscope (dissecting microscope,** Fig. 2.6) allows you to view objects in three dimensions at low magnifications. It is used to study entire small organisms, any object requiring lower magnification, and opaque objects that can be viewed only by reflected light. It is called a stereomicroscope because it produces a three-dimensional image.

Identifying the Parts

After your instructor has explained how to carry a microscope, obtain a stereomicroscope and a separate illuminator, if necessary, from the storage area. Place it securely on the table. Plug in the power cord, and turn on the illuminator. There is a wide variety of stereomicroscope styles, and your instructor will discuss the specific style(s) available to you. Regardless of style, the following features should be present:

1. **Binocular head:** Holds two eyepiece lenses that move to accommodate for the various distances between different individuals' eyes.
2. **Eyepiece lenses:** The two lenses located on the binocular head. What is the magnification of your eyepieces? _____ Some models have one **independent focusing eyepiece** with a knurled knob to allow independent adjustment of each eye. The nonadjustable eyepiece is called the **fixed eyepiece.**
3. **Focusing knob:** A large, black or gray knob located on the arm; used for changing the focus of both eyepieces together.

Figure 2.6 Binocular dissecting microscope (stereomicroscope).
Label this stereomicroscope with the help of the text material.

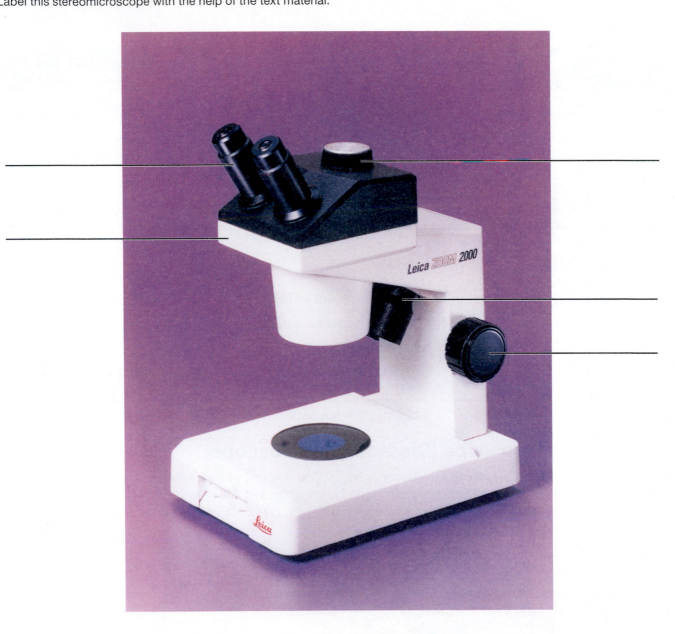

4. **Magnification changing knob:** A knob, often built into the binocular head, used to change magnification in both eyepieces simultaneously. This may be a **zoom** mechanism or a **rotating lens** mechanism of different powers that clicks into place.

5. **Illuminator:** Used to illuminate an object from above; may be built into the microscope or separate.

Locate each of these parts on your stereomicroscope, and label them on Figure 2.6.

Focusing the Stereomicroscope

1. In the center of the stage, place a plastomount that contains small organisms.

2. Adjust the distance between the eyepieces on the binocular head so that they comfortably fit the distance between your eyes. You should be able to see the object with both eyes as one three-dimensional image.

3. Use the focusing knob to bring the object into focus.

4. Does your microscope have an independent focusing eyepiece? _____ If so, use the focusing knob to bring the image in the fixed eyepiece into focus, while keeping the eye at the independent focusing eyepiece closed. Then adjust the independent focusing eyepiece so that the image is clear, while keeping the other eye closed. Is the image inverted? _____

5. Turn the magnification changing knob, and determine the kind of mechanism on your microscope. A zoom mechanism allows continuous viewing while changing the magnification. A rotating lens mechanism blocks the view of the object as the new lenses are rotated. Be sure to click each lens firmly into place. If you do not, the field will be only partially visible. What kind of mechanism is on your microscope? _____

6. Set the magnification changing knob on the lowest magnification. Sketch the object in the following circle as though this represents your entire field of view:

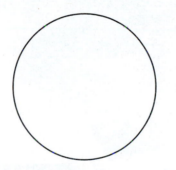

7. Rotate the magnification changing knob to the highest magnification. Draw another circle within the one provided to indicate the reduction of the field of view.

8. Experiment with various objects at various magnifications until you are comfortable with using the stereomicroscope.

2.4 Use of the Compound Light Microscope

As mentioned, the name **compound light microscope** indicates that it uses two sets of lenses and light to view an object. The two sets of lenses are the ocular lenses located near the eyes and the objective lenses located near the object. Illumination is from below, and the light passes through clear portions but does not pass through opaque portions. This microscope is used to examine small or thinly sliced sections of objects under higher magnification than would be possible with the stereomicroscope.

Identifying the Parts

Obtain a compound light microscope from the storage area, and place it securely on the table. *Identify the following parts on your microscope, and label them in Figure 2.7 with the help of the text material.*

Figure 2.7 Compound light microscope.
Compound light microscope with binocular head and mechanical stage. Label this microscope with the help of the text material.

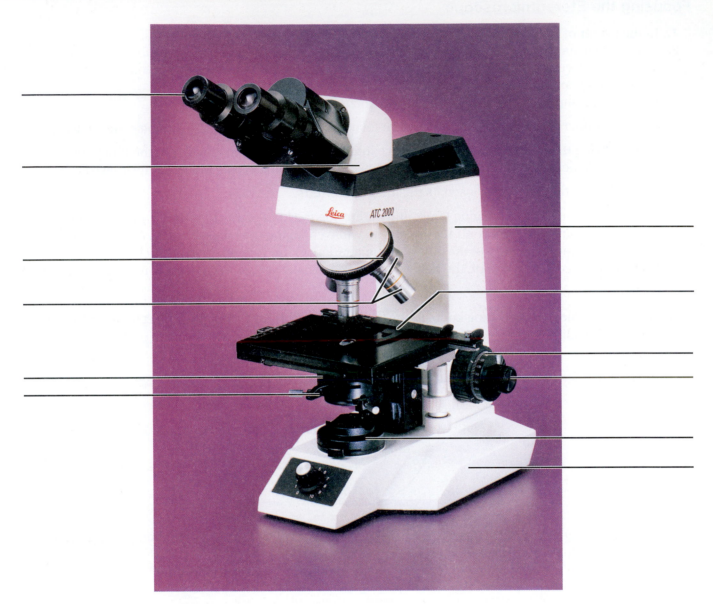

1. **Eyepieces** (ocular lenses): What is the magnifying power of the ocular lenses on your microscope? _____
2. **Viewing head:** Holds the ocular lenses.
3. **Arm:** Supports upper parts and provides carrying handle.
4. **Nosepiece:** Revolving device that holds objectives.
5. **Objectives** (objective lenses):
   a. **Scanning objective:** This is the shortest of the objective lenses and is used to scan the whole slide. The magnification is stamped on the housing of the lens. It is a number followed by an ×. What is the magnifying power of the scanning objective lens on your microscope? _____

b. **Low-power objective:** This lens is longer than the scanning objective lens and is used to view objects in greater detail. What is the magnifying power of the low-power objective lens on your microscope? _____

c. **High-power objective:** If your microscope has three objective lenses, this lens will be the longest. It is used to view an object in even greater detail. What is the magnifying power of the high-power objective lens on your microscope? _____

d. **Oil immersion objective** (on microscopes with four objective lenses): Holds a $95\times$ (to $100\times$) lens and is used in conjunction with immersion oil to view objects with the greatest magnification.

Does your microscope have an oil immersion objective? _____ If this lens is available, your instructor will discuss its use when the lens is needed.

6. **Stage:** Platform that holds and supports microscope slides. A mechanical stage is a movable stage that aids in the accurate positioning of the slide. Does your microscope have a mechanical stage? _____

a. **Stage clips:** Clips that hold a slide in place on the stage.
b. **Mechanical stage control knobs:** Two knobs that control forward/reverse movement and right/left movement, respectively.

7. **Coarse-adjustment knob:** Knob used to bring object into approximate focus; used only with low-power objective.
8. **Fine-adjustment knob:** Knob used to bring object into final focus.
9. **Condenser:** Lens system below the stage used to focus the beam of light on the object being viewed.

a. **Diaphragm or diaphragm control lever:** Lever that controls the amount of light passing through the condenser.

10. **Light source:** An attached lamp that directs a beam of light up through the object.
11. **Base:** The flat surface of the microscope that rests on the table.

Focusing the Compound Light Microscope—Lowest Power

1. Turn the nosepiece so that the *lowest* power objective on your microscope is in straight alignment over the stage.
2. Always begin focusing with the *lowest* power objective on your microscope ($4\times$ [scanning] or $10\times$ [low power]).
3. With the coarse-adjustment knob, lower the stage (or raise the objectives) until it stops.

4. Place a slide of the letter *e* on the stage, and stabilize it with the clips. (If your microscope has a mechanical stage, pinch the spring of the slide arms on the stage, and insert the slide.) Center the *e* as best you can on the stage or use the two control knobs located below the stage (if your microscope has a mechanical stage) to center the *e*.

5. Again, be sure that the lowest-power objective is in place. Then, as you look from the side, decrease the distance between the stage and the tip of the objective lens until the lens comes to an automatic stop or is no closer than 3 mm above the slide.

6. While looking into the eyepiece, rotate the diaphragm (or diaphragm control lever) to give the maximum amount of light.

7. Using the coarse-adjustment knob, slowly increase the distance between the stage and the objective lens until the object—in this case, the letter *e*—comes into view, or focus.

8. Once the object is seen, you may need to adjust the amount of light. To increase or decrease the contrast, rotate the diaphragm slightly.

9. Use the fine-adjustment knob to sharpen the focus if necessary.

10. Practice having both eyes open when looking through the eyepiece, as this greatly reduces eyestrain.

Inversion

Inversion refers to the fact that a microscopic image is upside down and reversed.

Observation: Inversion

1. Draw the letter *e* as it appears on the slide (with the unaided eye, not looking through the eyepiece). _____

2. Draw the letter *e* as it appears when you look through the eyepiece. _____

3. What differences do you notice? _____

4. Move the slide to the right. Which way does the image appear to move? _____

   Explain. _____

Focusing the Compound Light Microscope—Higher Powers

Compound light microscopes are **parfocal;** that is, once the object is in focus with the lowest power, it should also be almost in focus with the higher power.

1. Bring the object into focus under the lowest power by following the instructions in the previous section.

2. Make sure that the letter *e* is centered in the field of the lowest objective.

3. Move to the next higher objective (low power [10×] or high power [40×]) by turning the nosepiece until you hear it click into place. Do not change the focus; parfocal microscope objectives will not "hit" normal slides when changing the focus if the lowest objective is initially in focus. (If you are on low power [10×], proceed to high power [40×] before going on to step 4.)

4. If any adjustment is needed, use only the *fine*-adjustment knob. (*Note:* Always use only the fine-adjustment knob with high power, and do not use the coarse-adjustment knob.)

5. On a drawing of the letter *e*, draw a circle around the portion of the letter that you are now seeing with high-power magnification. _____

6. When you have finished your observations of this slide (or any slide), rotate the nosepiece until the lowest-power objective clicks into place, and then remove the slide.

Total Magnification

Total magnification is calculated by multiplying the magnification of the ocular lens (eyepiece) by the magnification of the objective lens. The magnification of a lens is imprinted on the lens casing.

Calculate total magnification figures for your microscope, and record your findings in Table 2.3.

Table 2.3 Total Magnification			
Objective	**Ocular Lens**	**Objective Lens**	**Total Magnification**
Scanning power (if present)			
Low power			
High power			
Oil immersion (if present)			

Field of View

A microscope's **field of view** is the circle visible through the lenses. The **diameter of field** is the length of the field from one edge to the other.

Low-Power (10×) Diameter of Field

1. Place a clear, plastic ruler across the stage so that the edge of the ruler is visible as a horizontal line along the diameter of the low-power (not scanning) field. Be sure that you are looking at the millimeter side of the ruler.

2. Estimate the number of millimeters, to tenths, that you see along the field: _____ mm. (*Hint:* Start by placing any millimeter marker at the edge of the field.) Convert the observed number

   of millimeters to micrometers: _____ µm. This is the **low-power diameter of field (LPD)** for your microscope in micrometers.

High-Power (40×) Diameter of Field

1. To compute the **high-power diameter of field (HPD),** substitute these data into the formula given:

   a. LPD = low-power diameter of field (in micrometers) = _____

   b. LPM = low-power total magnification (from Table 2.3) = _____

   c. HPM = high-power total magnification (from Table 2.3) = _____

Example: If the diameter of field is about 2 mm, then the LPD is 2,000 µm. Using the LPM and HPM values from Table 2.3, the HPD would be 500 µm.

$$HPD = LPD \times \frac{LPM}{HPM}$$

$$HPD = (\quad\quad) \times \frac{(\quad\quad)}{(\quad\quad)} = \underline{\quad\quad}$$

Conclusions: Total Magnification and Field of View

- Does low power or high power have a larger field of view (one that allows you to see more of the object)? _____

- Which has a smaller field but magnifies to a greater extent? _____

- To locate small objects on a slide, first find them under _____ ; then place them in the center of the field before rotating to _____.

Depth of Field

When viewing an object on a slide under high power, the **depth of field** (Fig. 2.8) is the area—from top to bottom—that comes into focus while slowly focusing up and down with the microscope's fine-adjustment knob.

Observation: Depth of Field

1. Obtain a prepared slide with three or four colored threads mounted together, or prepare a wet-mount slide with three or four crossing threads or hairs of different colors. (Directions for preparing a wet mount are given in Section 2.5.)

2. With low power, find a point where the threads or hairs cross. Slowly focus up and down. Notice that when one thread or hair is in focus, the others seem blurred. Remember, as the stage moves upward (or the objectives move downward), objects on top come into focus first. Determine the order of the threads or hairs, and complete Table 2.4.

3. Switch to high power, and notice that the depth of field is more shallow with high power than with low power. Focusing up and down with the fine-adjustment knob when viewing a slide with high power will give you an idea of the specimen's three-dimensional form. For example, viewing a number of sections from bottom to top allows reconstruction of the three-dimensional structure, as demonstrated in Figure 2.8.

Figure 2.8   Depth of field.
A demonstration of how focusing at depths 1, 2, and 3 would produce three different images (views) that could be used to reconstruct the original three-dimensional structure of the object.

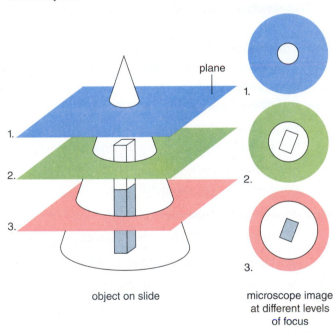

object on slide

microscope image at different levels of focus

Table 2.4 Order of Threads (or Hairs)	
Depth	**Thread (or Hair) Color**
Top	
Middle	
Bottom	

2.5 Microscopic Observations

When a specimen is prepared for observation, the object should always be viewed as a **wet mount.** A wet mount is prepared by placing a drop of liquid on a slide or, if the material is dry, by placing it directly on the slide and adding a drop of water or stain. The mount is then covered with a coverslip, as illustrated in Figure 2.9. Dry the bottom of your slide before placing it on the stage.

Figure 2.9   Preparation of a wet mount.

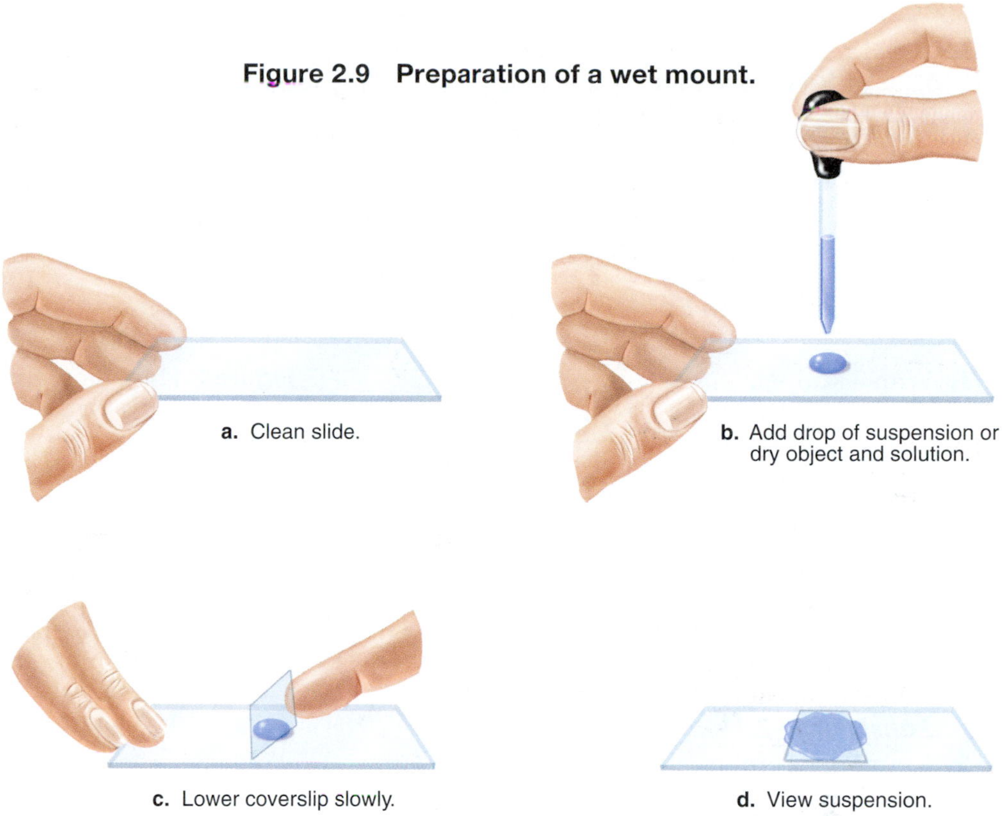

a. Clean slide.

b. Add drop of suspension or dry object and solution.

c. Lower coverslip slowly.

d. View suspension.

Human Epithelial Cells

Epithelial cells (Fig. 2.10) cover the body's surface and line its cavities.

Observation: Human Epithelial Cells

1. Obtain a prepared slide, or make your own as follows:
   a. Obtain a prepackaged flat toothpick (or sanitize one with alcohol or alcohol swabs).
   b. Gently scrape the inside of your cheek with the toothpick, and place the scrapings on a clean, dry slide. Discard used toothpicks in the biohazard waste container provided.
   c. Add a drop of very weak *methylene blue* or *iodine solution,* and cover with a coverslip.
2. Observe under the microscope.
3. Locate the nucleus (the central, round body), the cytoplasm, and the plasma membrane (outer cell boundary). *Label Figure 2.10.*

⚠ **Methylene blue** Avoid ingestion, inhalation, and contact with skin, eyes, and mucous membranes. If any should spill on your skin, wash the area with mild soap and water. Methylene blue will also stain clothing.

Figure 2.10   Cheek epithelial cells.
Label the nucleus, the cytoplasm, and the plasma membrane.

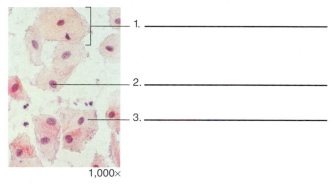

1. _____

2. _____

3. _____

1,000×

4. Because your epithelial slides are biohazardous, they must be disposed of as indicated by your instructor.

Onion Epidermal Cells

Epidermal cells cover the surfaces of plant organs, such as leaves. The bulb of an onion is made up of fleshy leaves.

Observation: Onion Epidermal Cells

1. With a scalpel, strip a small, thin, transparent layer of cells from the inside of a fresh onion leaf.

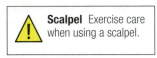
⚠ **Scalpel** Exercise care when using a scalpel.

Figure 2.11 Onion epidermal cells.
Label the cell wall and the nucleus.

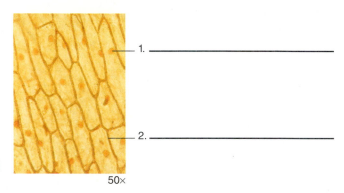

1. _____

2. _____

50×

2. Place it gently on a clean, dry slide, and add a drop of iodine solution (or methylene blue). Cover with a coverslip.
3. Observe under the microscope.
4. Locate the cell wall and the nucleus. *Label Figure 2.11.*
5. Count the number of onion cells that line up end to end in a single line across the diameter of the high-power (40×) field._____

Based on what you learned in Section 2.4 about measuring diameter of field, what is the high-power diameter of field (HPD) in micrometers? _____ μm

Calculate the length of each onion cell:

(HPD ÷ number of cells): _____ μm

6. Note some obvious differences between the human cheek cells and the onion cells, and list them in Table 2.5.

Table 2.5 Differences Between Human Epithelial and Onion Epidermal Cells		
Differences	**Human Epithelial Cells (Cheek)**	**Onion Epidermal Cells**

Euglena

Examination of *Euglena* (a unicellular organism with a flagellum to facilitate movement) will test your ability to observe objects with the microscope, to utilize depth of field, and to control illumination to heighten contrast.

Observation: Euglena

1. Make a wet mount of *Euglena* by using a drop of a *Euglena* culture and adding a drop of Protoslo® (methyl cellulose solution) onto a slide. The Protoslo® slows the organism's swimming.
2. Mix thoroughly with a toothpick, and add a coverslip.
3. Scan the slide for *Euglena:* Start at the upper left-hand corner, and move the slide forward and back as you work across the slide from left to right. The *Euglena* may be at the edge of the slide because they show an aversion to Protoslo®. Use Figure 2.12 to help identify the structural details of *Euglena.*

Figure 2.12   *Euglena.*
Euglena is a unicellular, flagellated organism.

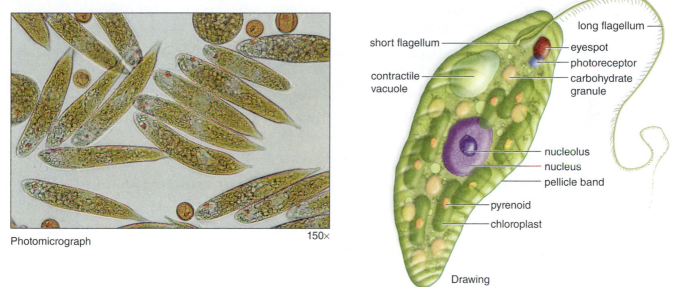

Photomicrograph   150×

Drawing

4. Experiment by using scanning, low-power, and high-power objective lenses; by focusing up and down with the fine-adjustment knob; and by adjusting the light so that it is not too bright.
5. Compare your *Euglena* specimens with Figure 2.12. List the labeled features that you can actually see: _____

_____

Pond Water

Examination of pond water will also test your ability to observe objects with the microscope, to utilize depth of field, and to control illumination to heighten contrast. See Figure 24.17, p. 349, for microorganisms found in pond water.

Laboratory Review 2

_____   **1.** 11 mm equals how many cm?

_____   **2.** 950 mm equals how many m?

_____   **3.** 2.1 liters equals how many ml?

_____   **4.** 122°F equals how many degrees Celsius?

_____   **5.** 4,100 mg equals how many grams?

_____   **6.** Which type of microscope would you use to view a wet mount of *Euglena*?

_____   **7.** What are the ocular lenses?

_____   **8.** Which objective always should be in place both when beginning to use the microscope and also when putting it away?

_____   **9.** A total magnification of 100 requires the use of the $10\times$ ocular lens with which objective?

_____   **10.** Which parts of a microscope regulate the amount of light?

_____   **11.** What word is used to indicate that if the object is in focus at low power it will also be in focus at high power?

_____   **12.** If the thread layers are red, brown, green, from top to bottom, which layer will come into focus first if you are using the microscope properly?

_____   **13.** What adjustment knob is used with high power?

_____   **14.** If a *Euglena* is swimming to the left, which way should you move your slide to keep it in view?

_____   **15.** What type of object do you study with a stereomicroscope?

_____   **16.** Why is a stereomicroscope also called a binocular dissecting microscope?

Thought Questions

17. A virus is 50 nm in size. Which type of microscope should be used to view it? Why?

18. Why is locating an object more difficult if you start with the high-power objective rather than the low-power objective?

19. What advantages does the metric system provide over English units of measure?

20. Which objective (scanning, low power, or high power) allows you to observe the greatest number of cells within the field of view? Explain.

3

Chemical Composition of Cells

Introduction

> **Planning Ahead** To save time, your instructor may have you start the boiling water bath needed for the experiments on page 32 and page 38 at the beginning of the laboratory session.

All living things consist of basic units of matter called **atoms.** Molecules form when atoms bond with one another. Inorganic molecules are often associated with nonliving things, and organic molecules are associated with living organisms. In this laboratory, you will be studying the organic molecules of cells: **proteins, carbohydrates** (monosaccharides, disaccharides, polysaccharides), and **lipids** (i.e., fat).

Large organic molecules form during *condensation synthesis* when smaller molecules bond as water is given off. During *hydrolysis,* bonds are broken as water is added. A fat contains one glycerol and three fatty acids. Proteins and some carbohydrates (called polysaccharides) are **polymers** because they are made up of smaller molecules called **monomers.** Proteins contain a large number of amino acids (the monomer) joined together by a peptide bond. A polysaccharide, such as starch, contains a large number of glucose molecules (the monomer) joined together. Various chemicals will be used in this laboratory to test for the presence of cellular organic molecules. If a color change is observed, the test is said to be *positive* because it indicates that the molecule is present. If the color change is not observed, the test is said to be *negative* because it indicates that the molecule is not present.

3.1 Proteins

Proteins have numerous functions in cells. Antibodies are proteins that combine with pathogens so that the pathogens are destroyed by the body. Transport proteins combine with and move substances from place to place. Hemoglobin transports oxygen throughout the body. Albumin is another transport protein in our blood. Regulatory proteins control cellular metabolism in some way. For example, the hormone insulin regulates the amount of glucose in blood so that cells have a ready supply. Structural proteins include keratin, found in hair, and myosin, found in muscle. **Enzymes** are proteins that speed chemical reactions. A reaction that could take days or weeks to complete can happen within an instant if the correct enzyme is present. Amylase is an enzyme that speeds the breakdown of starch in the mouth and small intestine.

Proteins are made up of **amino acids** joined together. About 20 different common amino acids are found in cells. All amino acids have an acidic group (—COOH) and an amino group (H_2N—). They differ by the **R group** (remainder group) attached to a carbon atom, as shown in Figure 3.1. The R groups have varying sizes, shapes, and chemical activities.

A chain of two or more amino acids is called a **peptide,** and the bond between the amino acids is called a **peptide bond.** A **polypeptide** is a very long chain of amino acids. A protein can contain one or more polypeptide chains. Insulin contains a single chain, while hemoglobin contains four polypeptides. A protein has a particular shape, which is important to its function. The shape comes about because the R groups of the polypeptide chain(s) can interact with one another in various ways.

Figure 3.1 Formation of a dipeptide.
During a dehydration reaction, a dipeptide forms when an amino acid joins with an amino acid as a water molecule is removed. The bond between amino acids is called a peptide bond. During a hydrolysis reaction, water is added and the peptide bond is broken.

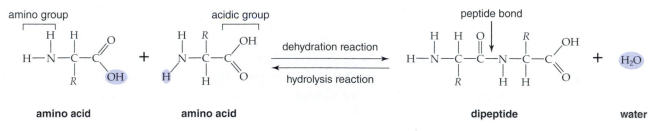

Test for Proteins

Biuret reagent (blue color) contains a strong solution of sodium or potassium hydroxide (NaOH or KOH) and a small amount of dilute copper sulfate ($CuSO_4$) solution. The reagent changes color in the presence of proteins or peptides because the peptide bonds of the protein or peptide chemically combine with the copper ions in biuret reagent (Table 3.1).

Table 3.1 Biuret Test for Protein and Peptides		
	Protein	**Peptides**
Biuret reagent (blue)	Purple	Pinkish-purple

Experimental Procedure: Test for Proteins

With a millimeter ruler and a wax pencil, label and mark four clean test tubes at the 1 cm level. After filling a tube, cover it with Parafilm®, and swirl well to mix. (Do not turn upside down.) The reaction is almost immediate.

> ⚠ **Biuret reagent** Biuret reagent is highly corrosive. Exercise care in using this chemical. If any should spill on your skin, wash the area with mild soap and water. Follow your instructor's directions for its disposal.

*Tube 1
1. Fill to the mark with distilled water, and add about five drops of biuret reagent.
2. Record the final color in Table 3.2.

Tube 2
1. Fill to the mark with albumin solution, and add about five drops of biuret reagent.
2. Record the final color in Table 3.2.

Tube 3
1. Fill to the mark with pepsin solution, and add about five drops of biuret reagent.
2. Record the final color in Table 3.2.

Tube 4
1. Fill to the mark with starch solution, and add about five drops of biuret reagent.
2. Record the final color in Table 3.2.

Table 3.2 Biuret Test for Protein			
Tube	**Contents**	**Final Color**	**Conclusions**
1	Distilled water		
2	Albumin		
3	Pepsin		
4	Starch		

Conclusions: Proteins

- From your test results, conclude if a protein is present or absent and explain. Enter your conclusions in Table 3.2.

- Pepsin is an enzyme. Enzymes are composed of what type of organic molecule? _____

- According to your results, is starch a protein? _____

- Which of the four tubes is the negative control sample? _____ Why? _____

- Why do experimental procedures include control samples? _____

- If your results are not as expected, inform your instructor, who will advise you how to proceed.

*To test a sample for protein, use this procedure. Instead of only water, use a liquefied sample. If protein is present, a pinkish-purple color appears.

3.2 Carbohydrates

Carbohydrates include sugars and molecules that are chains of sugars. **Glucose,** which has only one sugar unit, is a monosaccharide; **maltose,** which has two sugar units, is a disaccharide (Fig. 3.2). Glycogen, starch, and cellulose are polysaccharides, made up of chains of glucose units (Fig. 3.3).

Glucose is used by all organisms as an energy source. Energy is released when glucose is broken down to carbon dioxide and water. This energy is used by the organism to do work. Animals store glucose as glycogen and plants store glucose as starch. Plant cell walls are composed of cellulose.

Figure 3.2 Formation of a disaccharide.

During a dehydration reaction, a disaccharide, such as maltose, forms when a glucose joins with a glucose as a water molecule is removed. During a hydrolysis reaction, the components of water are added, and the bond is broken.

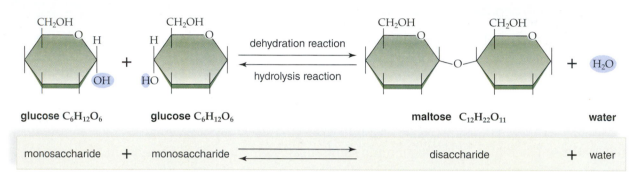

glucose $C_6H_{12}O_6$ glucose $C_6H_{12}O_6$ maltose $C_{12}H_{22}O_{11}$ water

monosaccharide	+	monosaccharide	disaccharide	+	water

Figure 3.3 Starch.

Starch is a polysaccharide composed of many glucose units. **a.** Photomicrograph of starch granules in cells of a potato. **b.** Structure of starch. Starch consists of amylose that is nonbranched and amylopectin that is branched.

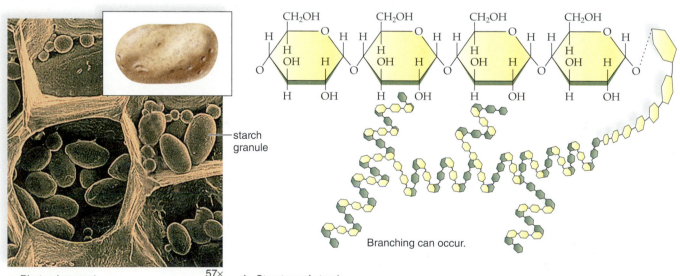

starch granule

Branching can occur.

a. Photomicrograph   57×   b. Structure of starch

Test for Starch

In the presence of starch, iodine solution (yellowish-brown) reacts chemically with starch to form a blue-black color (Table 3.3).

Table 3.3	Iodine Test for Starch
	Starch
Iodine solution	Blue-black

Experimental Procedure: Test for Starch

With a wax pencil, label and mark five clean test tubes at the 1 cm level.

*Tube 1
1. Fill to the 1 cm mark with water, and add five drops of iodine solution.
2. Note the final color change, and record your results in Table 3.4.

Tube 2
1. It is very important to shake the starch suspension well before taking your sample. After shaking, fill this tube to the 1 cm mark with the 1% starch suspension. Add five drops of iodine solution.
2. Note the final color change, and record your results in Table 3.4.

Tube 3
1. Add a few drops of onion juice to the test tube. (Obtain the juice by adding water and crushing a small piece of onion with a mortar and pestle. Clean mortar and pestle after using.) Add five drops of iodine solution.
2. Note the final color change, and record your results in Table 3.4.

Tube 4
1. Add a few drops of potato juice to the test tube. (Obtain the juice by adding water and crushing a small piece of potato with a mortar and pestle. Clean mortar and pestle after using.) Add five drops of iodine solution.
2. Note the final color change, and record your results in Table 3.4.

Tube 5
1. Fill to the 1 cm mark with glucose solution, and add five drops of iodine solution.
2. Note the final color change, and record your results in Table 3.4.

Table 3.4	Iodine (IKI) Test for Starch		
Tube	**Contents**	**Color**	**Conclusions**
1	Water		
2	Starch suspension		
3	Onion juice		
4	Potato juice		
5	Glucose solution		

Conclusions: Starch

- From your test results, draw conclusions about what organic compound is present in each tube. Write these conclusions in Table 3.4.

- Does the potato or the onion store glucose as starch? _____ How do you know? _____

  _____

- If your results are not as expected, offer an explanation. Then inform your instructor, who will advise you how to proceed.

*To test a sample for starch, use this procedure. Instead of only water, use a liquefied sample. If starch is present, a blue-black color appears.

Experimental Procedure: Microscopic Study

Potato

1. With a scalpel, slice a very thin piece of potato. Place it on a microscope slide, add a drop of water and a coverslip, and observe under low power with your compound light microscope. Compare your slide with the photomicrograph of starch granules (see Fig. 3.3*a*). Find the cell wall (large, geometric compartments) and the starch grains (numerous clear, oval-shaped objects).
2. Without removing the coverslip, place two drops of iodine solution onto the microscope slide so that the iodine touches the coverslip. Draw the iodine under the coverslip by placing a small piece of paper towel in contact with the water on the **opposite** side of the coverslip.
3. Microscopically examine the potato again on the side closest to where the iodine solution was applied.

   What is the color of the small, oval bodies? _____

   What is the chemical composition of these oval bodies? _____

Onion

1. Peel a single layer of onion from the bulb. On the inside surface, you will find a thin, transparent layer of onion skin. Peel off a small section of this layer for use on your slide.
2. Add a large drop of iodine solution.
3. Does onion contain starch? _____
4. Are these results consistent with those you recorded for onion juice in Table 3.4? _____

Test for Sugars

> ⚠️ **Benedict's reagent** Benedict's reagent is highly corrosive. Exercise care in using this chemical. If any should spill on your skin, wash the area with mild soap and water. Follow your instructor's directions for disposal of this chemical.

Monosaccharides and some disaccharides will react with **Benedict's reagent** after being heated in a boiling water bath. In this reaction, copper ion (Cu^{2+}) in the Benedict's reagent reacts with part of the sugar molecule, causing a distinctive color change. The color change can range from green to red, and increasing concentrations of sugar will give a continuum of colored products (Table 3.5).

Table 3.5 Benedict's Test for Sugars (Some Typical Reactions)

Chemical	Chemical Category	Benedict's Reagent (After Heating)	
Water	Inorganic	Blue (no change)	
Glucose	Monosaccharide (carbohydrate)	Varies with concentration:	very low—green low—yellow moderate—yellow-orange high—orange very high—orange-red
Maltose	Disaccharide (carbohydrate)	Varies with concentration—see "Glucose"	
Starch	Polysaccharide (carbohydrate)	Blue (no change)	

(This procedure runs for one hour. Prior setup can maximize time efficiency.)

With a wax pencil, label and mark five clean test tubes at the 1 cm level. Save your tubes for comparison with Section 3.4.

*Tube 1   **1.** Fill to the 1 cm mark with water, then add about five drops of Benedict's reagent.

        2. Heat in a boiling water bath for 5 to 10 minutes, note any color change, and record in Table 3.6.

Tube 2   **1.** Fill to the 1 cm mark with glucose solution, then add about five drops of Benedict's reagent.

        2. Heat in a boiling water bath for 5 to 10 minutes, note any color change, and record in Table 3.6.

Tube 3   **1.** Fill to the 1 cm mark with starch suspension, then add about five drops of Benedict's reagent.

        2. Heat in a boiling water bath for 5 to 10 minutes, note any color change, and record in Table 3.6.

Tube 4   **1.** Place a few drops of onion juice in the test tube. (Obtain the juice by adding water and crushing a small piece of onion with a mortar and pestle. Clean mortar and pestle after using.)

        2. Fill to the 1 cm mark with water, then add about five drops of Benedict's reagent.

        3. Heat in a boiling water bath for 5 to 10 minutes, note any color change, and record in Table 3.6.

Tube 5   **1.** Place a few drops of potato juice in the test tube. (Obtain the juice by adding water and crushing a small piece of potato with a mortar and pestle.)

        2. Fill to the 1 cm mark with water, then add about five drops of Benedict's reagent.

        3. Heat in a boiling water bath for 5 to 10 minutes, note any color change, and record in Table 3.6.

Table 3.6 Benedict's Test for Sugars			
Tube	**Contents**	**Color (After Heating)**	**Conclusions**
1	Water		
2	Glucose solution		
3	Starch suspension		
4	Onion juice		
5	Potato juice		

Conclusions: Sugars

- From your test results, conclude what kind of chemical is present. Enter your conclusions in Table 3.6.

- Which tube served as a negative and which as a positive control? _____

*To test a sample for sugars, use this procedure. Instead of only water, use a liquefied sample. If sugar is present, a green to orange-red color appears.

- Compare Table 3.4 with Table 3.6. Sugars are an immediate energy source in cells. In plant cells, glucose (a primary energy molecule) is often stored in the form of starch. Is glucose stored as starch in the potato? _____ Is glucose stored as starch in the onion? _____ Does this explain your results in Table 3.6? _____ Why? _____

_____

3.3 Lipids

Lipids are compounds that are insoluble in water and soluble in solvents, such as alcohol and ether. Lipids include fats, oils, phospholipids, steroids, and cholesterol. Typically, **fat,** such as in the adipose tissue of animals, and **oils,** such as the vegetable oils from plants, are composed of three molecules of fatty acids bonded to one molecule of glycerol (Fig. 3.4). **Phospholipids** have the same structure as fats, except that in place of the third fatty acid there is a phosphate group (a grouping that contains phosphate). **Steroids** are derived from **cholesterol** and, like this molecule, have skeletons of four fused rings of carbon atoms, but they differ by functional groups (attached side chains). Fat, as we know, is long-term stored energy in the human body. Phospholipids are found in the plasma membrane of cells. In recent years, cholesterol, a molecule transported in the blood, has been implicated in causing cardiovascular disease. Regardless, steroids are very important compounds in the body; for example, the sex hormones are steroids.

Figure 3.4   Formation of a fat.
During a dehydration reaction, a fat molecule forms when glycerol joins with three fatty acids as three water molecules are removed. During a hydrolysis reaction, water is added, and the bonds are broken between glycerol and the three fatty acids.

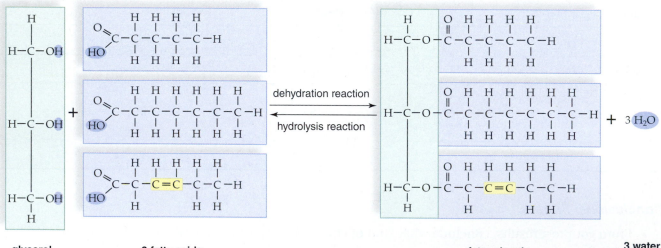

glycerol   3 fatty acids   fat molecule   3 water molecules

Test for Fat

Fats and oils do not evaporate from brown paper or loose-leaf paper; instead, they leave an oily spot.

Experimental Procedure: Paper Test for Fat

*1. Place a small drop of water on a square of brown paper or loose-leaf paper. Describe the immediate effect. _____

2. Place a small drop of vegetable oil on a square of the paper. Describe the immediate effect.

_____

3. Wait at least 15 minutes for the paper to dry. Evaluate which substance penetrates the paper and which is subject to evaporation. Record your observations and conclusions in Table 3.7. Save the paper for comparison use with Section 3.4.

Table 3.7	Paper Test for Fat	
Sample	Observations	Conclusions
Water spot		
Oil spot		

Emulsification of Oil

Some molecules are **polar,** meaning that they have charged groups or atoms, and some are **nonpolar,** meaning that they have no charged groups or atoms. A water molecule is polar, and therefore, water is a good solvent for other polar molecules. When the charged ends of water molecules interact with the charged groups of polar molecules, these polar molecules disperse in water.

Water is not a good solvent for nonpolar molecules, such as fats. A fat has no polar groups to interact with water molecules. An **emulsifier,** however, can cause a fat to disperse in water. An emulsifier contains molecules with both polar and nonpolar ends. When the nonpolar ends interact with the fat and the polar ends interact with the water molecules, the fat disperses in water, and an **emulsion** results (Fig. 3.5).

Bile salts (emulsifiers found in bile produced by the liver) are used in the digestive tract. Today milk, such as 1% milk, has been homogenized so that fat droplets do not congregate and rise to the top of the container. Homogenization requires the addition of natural emulsifiers such as phospholipids—the phosphate part of the molecule is polar and the lipid portion is nonpolar.

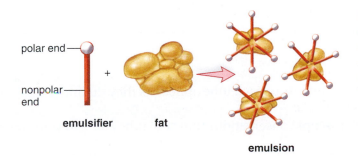

polar end
nonpolar end
emulsifier   fat
emulsion

Figure 3.5   Emulsification.
An emulsifier contains molecules with both a polar and a nonpolar end. The nonpolar ends are attracted to the nonpolar fat, and the polar ends are attracted to the water. This causes droplets of fat molecules to disperse.

*To test an unknown for fat, use this procedure. Do not liquefy the sample; place it as is on a piece of brown paper or loose-leaf paper over the sink, in order to facilitate cleanup. If lipids are present, an oily spot appears.

Experimental Procedure: Emulsification of Lipids

With a wax pencil, label two clean test tubes 1 and 2. Mark tube 1 at the 3 cm and 4 cm levels. Mark tube 2 at the 2 cm, 3 cm, and 4 cm levels.

Tube 1
1. Fill to the 3 cm mark with water and to the 4 cm mark with vegetable oil. Shake.
2. Observe for the initial dispersal of oil, followed by rapid separation into two layers.

   Is vegetable oil soluble in water? _____

3. Let the tube settle for 5 minutes. Label a microscope slide as 1.
4. Use a dropper to remove a sample of the solution that is just below the layer of oil. Place the drop on the slide, add a coverslip, and examine with the low power of your compound light microscope.
5. Record your observations in Table 3.8.

Tube 2
1. Fill to the 2 cm mark with water, to the 3 cm mark with vegetable oil, and to the 4 cm mark with the available emulsifier (Tween® or bile salts). Shake.
2. Describe how the distribution of oil in tube 2 compares with the distribution in tube 1.

   _____

   _____

3. Let the tube settle for 5 minutes. Label a microscope slide as 2.
4. Use a different dropper to remove a sample of the solution that is just below the layer of oil. Place the drop on the slide, add a coverslip, and examine with the low power of your compound light microscope.
5. Record your observations in Table 3.8.

Tube 3
1. Fill to the 1 cm mark with milk and to the 2 cm mark with water. Shake well.
2. Use a different dropper to remove a sample of the solution. Place a drop on a slide, add a coverslip, and examine with the low power of your compound light microscope.
3. Record your observations in Table 3.8.

Table 3.8	Emulsification		
Tube	**Contents**	**Observations**	**Conclusions**
1	Oil Water		
2	Oil Water Emulsifier		
3	Milk Water		

Conclusions: Emulsification

- From your observations, conclude why the contents of each tube appear as they do under the microscope. Record your conclusions in Table 3.8.
- Explain the correlation between your macroscopic observations (how the tubes look to your unaided eye) and your microscopic observations. _____

_____

Adipose Tissue

Adipose tissue in animals such as humans stores droplets of fat. Adipose tissue is found beneath the skin, where it helps insulate and keep the body warm. It also forms a protective cushion around various internal organs.

Observation: Adipose Tissue

1. Obtain a slide of adipose tissue, and view it under the microscope at high power. Refer to Figure 3.6 for help in identifying the structures.
2. Notice how the fat droplets push the cytoplasm to the edges of the cells. The cytoplasm of cells is largely water. Explain why fat does not disperse in cytoplasm.

_____

_____

_____

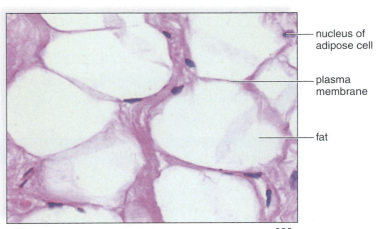

Photomicrograph   200×

Figure 3.6   Adipose tissue.
The cells are so full of fat that the nucleus is pushed to one side.

3.4   Testing Foods and Unknowns

It is common for us to associate the term *organic* with the foods we eat, including carbohydrate foods (Fig. 3.7), protein foods (Fig. 3.8), and lipid foods (Fig. 3.9). Though we may recognize foods as being organic, often we are not aware of what specific types of compounds are found in what we eat. In the following Experimental Procedure, you will use the same tests you used previously to determine the composition of everyday foods and unknowns.

Figure 3.7   Carbohydrate foods.

Figure 3.8   Protein foods.

Figure 3.9   Lipid foods.

Your instructor will provide you with several everyday foods including unknowns, and your task is to

1. State how you will test substances for protein (page 29), carbohydrates (pages 31 and 33), and fat (page 35).

_____

_____

_____

_____

2. Have your instructor okay your procedures, and then conduct the necessary tests.
3. Record your results as positive (+) or negative (–) in Table 3.9.

Table 3.9 Testing Foods and Unknowns

Sample Name	Protein (Biuret)	Starch (Iodine)	Sugar (Benedict's)	Fat (Brown or loose-leaf paper)
Unknown A				
Unknown B				

Conclusions: Testing Foods and Unknowns

- What foods tested positive for only one of the organic compounds? _____
  Explain. _____

- What foods tested positive for more than one of the organic compounds? _____

- What type of carbohydrate would give a positive iodine test but a negative Benedict's test? _____

_____

_____ 1. What type of bond joins amino acids to make peptides?

_____ 2. What type of protein speeds chemical reactions?

_____ 3. What group is different between types of amino acids?

_____ 4. If iodine solution turns blue-black, what substance is present?

_____ 5. If Benedict's reagent turns red, what substance is present?

_____ 6. What is the function of starch in plant cells?

_____ 7. Is starch a monosaccharide or a polysaccharide?

_____ 8. What is the function of fat in animal cells?

_____ 9. What molecules are released when fat undergoes a hydrolysis reaction?

_____ 10. What type of organic molecule requires the action of emulsifiers to be successfully digested?

_____ 11. Are fats polar or nonpolar?

_____ 12. If Biuret reagent turns purple, what substance is present?

_____ 13. A student adds iodine solution to egg white and waits for a color change. How long will the student have to wait?

_____ 14. To test whether a sample contains glucose, what reagent should be used?

Thought Questions

15. Why is it necessary to shake a bottle of salad dressing that contains oil and vinegar before adding it to a salad?

16. An unknown sample is tested with both Biuret reagent and Benedict's reagent. Both tests result in a blue color. What has been learned?

17. When starch and water are mixed together as when you make gravy, why doesn't starch react with water to yield monosaccharides?

18. Based on your study of organic molecules in this lab, why do plants benefit from nitrogen fertilizer?

4

Cell Structure and Function

Learning Outcomes

4.1 Prokaryotic versus Eukaryotic Cells
- Distinguish between prokaryotic and eukaryotic cells by description and examples.   42

4.2 Animal Cell and Plant Cell Structure
- Label an animal cell diagram, and state a function for the structures labeled.   43–44
- Label a plant cell diagram, and state a function for the structures labeled.   45
- Use microscopic techniques to observe plant cell structure.   46

4.3 Diffusion
- Define and describe the process of diffusion as affected by the medium.   47
- Predict and observe which substances will or will not diffuse across a plasma membrane.   48–49

4.4 Osmosis: Diffusion of Water across Plasma Membrane
- Explain an osmosis experiment based on a knowledge of diffusion principles.   49–50
- Define isotonic, hypertonic, and hypotonic solutions, and give examples in terms of NaCl concentrations.   50–51
- Predict the effect of different tonicities on animal (e.g., red blood) cells and on plant (e.g., *Elodea*) cells.   51–53

4.5 pH and Cells
- Predict the change in pH before and after the addition of an acid to nonbuffered and buffered solutions.   54–55
- Suggest a method by which it is possible to test the effectiveness of antacid medications.   55

Introduction

The molecules we studied in Laboratory 3 are not alive—the basic units of life are cells. The **cell theory** states that all living things are composed of cells and that cells come only from other cells. While we are accustomed to considering the heart, the liver, or the intestines as enabling the human body to function, it is actually cells that do the work of these organs.

Figure 2.10 shows human cheek epithelial cells as viewed by an ordinary compound light microscope available in general biology laboratories. It shows that the content of a cell, called the **cytoplasm,** is bounded by a **plasma membrane.** The plasma membrane regulates the movement of molecules into and out of the cytoplasm. In this lab, we will study how the passage of water into a cell depends on the difference in concentration of solutes (particles) between the cytoplasm and the surrounding medium or solution. The well-being of cells also depends upon the pH of the solution surrounding them. We will see how a buffer can maintain the pH within a narrow range and how buffers within cells can protect them against damaging pH changes.

> **Planning Ahead** To save time, your instructor may have you start a boiling water bath (page 49) and the potato strip experiment (page 53) at the beginning of the laboratory.

Because a photomicrograph shows only a minimal amount of detail, it is necessary to turn to the electron microscope to study the contents of a cell in greater depth. The models of plant and animal cells available in the laboratory today are based on electron micrographs.

4.1 Prokaryotic versus Eukaryotic Cells

All living cells are classified as either prokaryotic or eukaryotic. One of the basic differences between the two types is that prokaryotic cells do not contain nuclei (*pro* means "before"; *karyote* means "nucleus"), while eukaryotic cells do contain nuclei (*eu* means "true"; *karyote* means "nucleus"). Only bacteria (including cyanobacteria) and archaea are prokaryotes; all other organisms are eukaryotes.

Prokaryotes also don't have the organelles found in eukaryotic cells (Fig. 4.1). **Organelles** are small, membranous bodies, each with a specific structure and function. Prokaryotes do have **cytoplasm,** the material bounded by a plasma membrane and cell wall. The cytoplasm contains ribosomes, small granules that coordinate the synthesis of proteins; thylakoids (only in cyanobacteria) that participate in photosynthesis; and innumerable enzymes. Prokaryotes also have a nucleoid, a region in the bacterial cell interior in which the DNA is physically organized but not enclosed by a membrane.

Figure 4.1 Prokaryotic cell.
Prokaryotic cells lack membrane-bounded organelles, as well as a nucleus. Their DNA is in a nucleoid region.

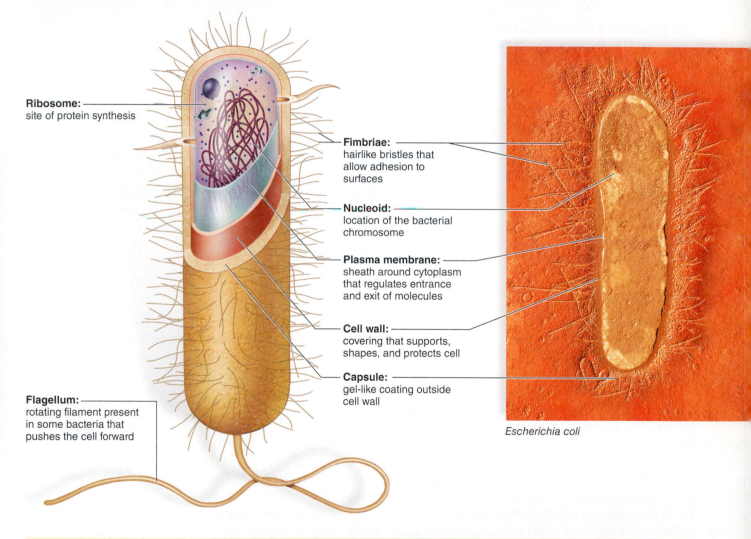

Ribosome:
site of protein synthesis

Fimbriae:
hairlike bristles that
allow adhesion to
surfaces

Nucleoid:
location of the bacterial
chromosome

Plasma membrane:
sheath around cytoplasm
that regulates entrance
and exit of molecules

Cell wall:
covering that supports,
shapes, and protects cell

Capsule:
gel-like coating outside
cell wall

Flagellum:
rotating filament present
in some bacteria that
pushes the cell forward

Escherichia coli

Observation: Prokaryotic/Eukaryotic Cells

Two microscope slides on display will show you the main difference between prokaryotic and eukaryotic cells.

1. Examine a prepared slide of a bacterium. There are no nuclei in these cells.

2. Examine a prepared slide of cuboidal cells from a human kidney (see page 147). Can you make out a nucleus? _____

4.2   Animal Cell and Plant Cell Structure

Table 4.1 lists the structures found in animal and plant cells. The **nucleus** in a eukaryotic cell is bounded by a **nuclear envelope** and contains **nucleoplasm.** The *cytoplasm,* found between the plasma membrane and the nucleus, consists of a background fluid and the organelles, such as the nucleolus, endoplasmic reticulum, Golgi apparatus, vacuoles and vesicles, lysosomes, peroxisome, mitochondrion, and chloroplast.

Table 4.1	**Eukaryotic Structures in Animal Cells and Plant Cells**	
Name	**Composition**	**Function**
Cell wall*	Contains cellulose fibrils	Provides support and protection
Plasma membrane	Phospholipid bilayer with embedded proteins	Outer cell surface that regulates entrance and exit of molecules
Nucleus	Enclosed by nuclear envelope; contains chromatin (threads of DNA and protein)	Storage of genetic information; synthesis of DNA and RNA
Nucleolus	Concentrated area of chromatin	Produces subunits of ribosomes
Ribosome	Protein and RNA in two subunits	Carries out protein synthesis
Endoplasmic reticulum (ER)	Membranous, flattened channels and tubular canals; rough ER and smooth ER	Synthesis and/or modification of proteins and other substances; transport by vesicle formation
Rough ER	Studded with ribosomes	Protein synthesis
Smooth ER	Lacks ribosomes	Synthesis of lipid molecules
Golgi apparatus	Stack of membranous saccules	Processes, packages, and distributes proteins and lipids
Vesicle/vacuole	Membrane-bounded sac; large central vacuole in plant cells*	Stores and transports substances
Lysosome	Vesicle containing hydrolytic enzymes	Digests macromolecules and cell parts
Peroxisome	Vesicle containing specific enzymes	Breaks down fatty acids and converts resulting hydrogen peroxide to water
Mitochondrion	Bounded by double membrane; inner membrane is cristae	Cellular respiration, producing ATP molecules
Chloroplast*	Membranous grana bounded by double membrane	Photosynthesis, producing sugars
Cytoskeleton	Microtubules, intermediate filaments, actin filaments	Maintains cell shape and assists movement of cell parts
Cilia and flagella	Attachments supported by microtubules	Movement of cell
Centrioles** in centrosome	Microtubule-containing, cylindrically shaped organelle in a structure of complex composition.	Centrioles organize microtubules in cilia and flagella; centrosome organizes microtubules in cell

*Plant cells only

**Animal cells only

Study Table 4.1 to determine structures that are unique to plant cells and unique to animal cells, and write them below the examples given.

	Plant Cells	**Animal Cells**
Unique structures:	1. Large central vacuole	1. Small vacuoles
	2. _____	2. _____
	3. _____	

Animal Cell Structure

Label Figure 4.2. With the help of Table 4.1, give a function for each labeled structure.

Structure	Function
Plasma membrane	
Nucleus	
Nucleolus	
Endoplasmic reticulum	
Rough ER	
Smooth ER	
Golgi apparatus	
Vesicle	
Lysosome	
Mitochondrion	
Centrioles in centrosome	
Cytoskeleton	

Figure 4.2 Animal cell structure.

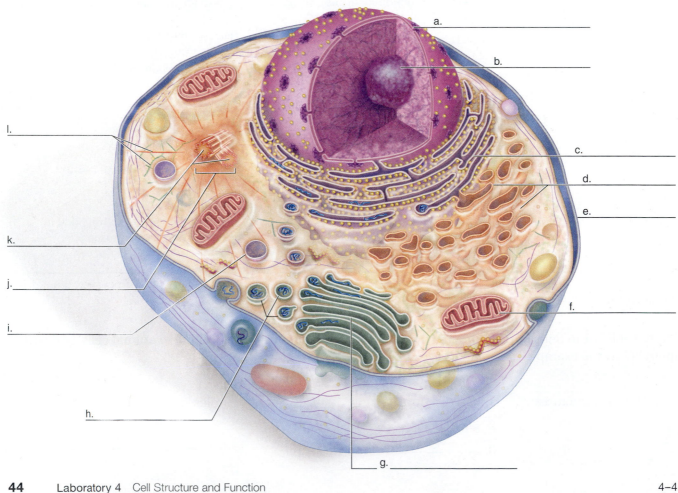

Plant Cell Structure

Label Figure 4.3. With the help of Table 4.1, give a function for each labeled structure unique to plant cells.

Structure	Function
Cell wall	_____
Central vacuole, large	_____
Chloroplast	_____

Figure 4.3   Plant cell structure.

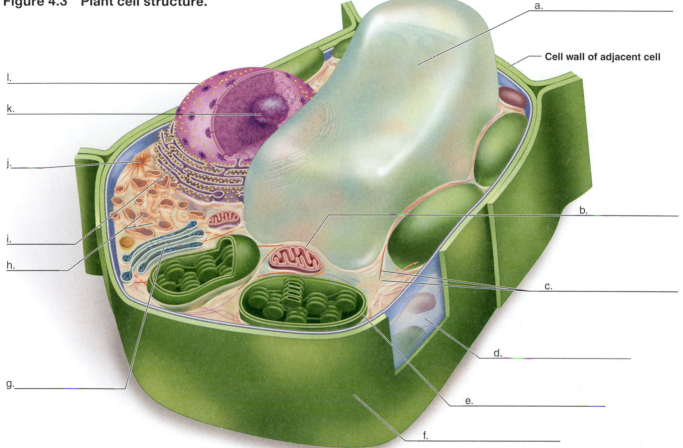

Observation: Plant Cell Structure

1. Prepare a wet mount of a small piece of young *Elodea* leaf in fresh water. *Elodea* is a multicellular, eukaryotic plant found in freshwater ponds and lakes.
2. Have the drop of water ready on your slide so that the leaf does not dry out, even for a few seconds. Take care that the leaf is mounted with its top side up.
3. Using low power bring the leaf surface into focus.
4. Select a cell with numerous chloroplasts for further study, and switch to high power.
5. Carefully focus on the sides of the cell. The chloroplasts appear to be only along the sides of the cell because the large, fluid-filled, membrane-bounded central vacuole pushes the cytoplasm against the cell walls (Fig. 4.4*a*). Then focus on the surface and notice an even distribution of chloroplasts (Fig. 4.4*b*).

Figure 4.4   *Elodea* cell structure.

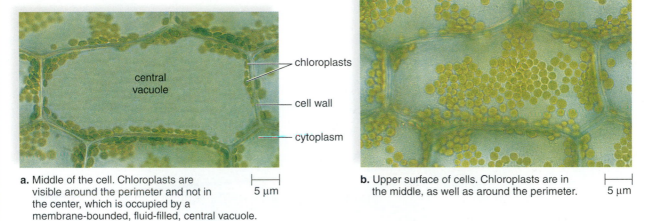

a. Middle of the cell. Chloroplasts are visible around the perimeter and not in the center, which is occupied by a membrane-bounded, fluid-filled, central vacuole.   5 μm

b. Upper surface of cells. Chloroplasts are in the middle, as well as around the perimeter.   5 μm

6. Can you locate the cell nucleus? _____ It may be hidden by the chloroplasts, but when visible, it appears as a faint, grey lump on one side of the cell.

7. Why can't you see the other organelles featured in Figure 4.3?_____
   _____

8. Can you detect movement of chloroplasts in this cell or any other cell? _____ The chloroplasts are not moving under their own power but are being carried by a streaming of the nearly invisible cytoplasm.

9. Save your slide for use later in this laboratory.

4.3 Diffusion

Diffusion is the movement of molecules from a higher to a lower concentration until equilibrium is achieved and the molecules are distributed equally (Fig. 4.5). At equilibrium, molecules may still be moving back and forth, but there is no net movement in any one direction.

Figure 4.5 Process of diffusion.
Diffusion is apparent when dye molecules have equally dispersed.

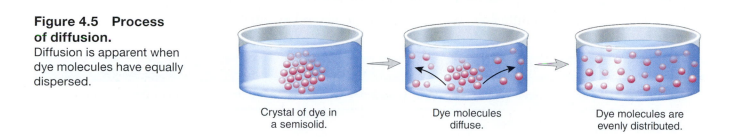

Crystal of dye in a semisolid.

Dye molecules diffuse.

Dye molecules are evenly distributed.

Diffusion is a general phenomenon in the environment. The speed of diffusion is dependent on such factors as the temperature, the size of the molecule, and the type of medium.

Experimental Procedure: Diffusion

Solute Diffusion Through a Semisolid

1. Observe a petri dish containing 1.5% gelatin (or agar) to which potassium permanganate ($KMnO_4$) was added in the center depression at the beginning of the lab.
2. Complete the first four columns of Table 4.2. Your instructor will supply zero time; final time is the current time.

Potassium permanganate ($KMnO_4$) $KMnO_4$ is highly poisonous and is a strong oxidizer. Avoid contact with skin and eyes and with combustible materials. If spillage occurs, wash all surfaces thoroughly. $KMnO_4$ will also stain clothing.

3. Using a ruler placed over the petri dish, measure (in mm) the movement of color from the center of the depression outward in one direction: _____ mm. Add distance moved to Table 4.2.
4. Calculate the speed of diffusion: _____ mm/60 min = mm/hr.
5. Complete Table 4.2.

Solute Diffusion Through a Liquid

1. Add enough water to cover the bottom of a glass petri dish.
2. Place the petri dish over a thin, flat ruler. Position the petri dish directly over a mm measurement line.
3. With tweezers, add a crystal of potassium permanganate ($KMnO_4$) directly over the mm measurement line. Note how far the dye moves in 10 min _____.
4. Complete all columns of Table 4.2 except the last one.
5. Multiply the length of time and the distance moved by 6 to calculate the speed of diffusion: _____ mm/hr. Record in Table 4.2.

Solute Diffusion Through Air

1. Measure the distance from a spot designated by your instructor to your laboratory work area today. Record this distance under distance moved in Table 4.2.
2. Record time zero in Table 4.2 when a perfume or similar substance is released into the air.
3. Note the time when you can smell the perfume. Record this as the final time in Table 4.2. Calculate the length of time (min) since the perfume was released, and record it in Table 4.2.
4. Calculate the speed of diffusion: _____ mm/hr. Record in Table 4.2.

Table 4.2 Speed of Diffusion					
Medium	Time Zero	Final Time	Length of Time (min)	Distance Moved (mm)	Speed of Diffusion (mm/hr)
Semisolid					
Liquid					
Air					

Conclusions: Solute Diffusion

- In which experiment was diffusion the fastest? _____
- What accounts for the difference in speed? _____

Solute Diffusion Across the Plasma Membrane

Some molecules can diffuse across a plasma membrane, and some cannot. In general, small, noncharged molecules can cross a membrane by simple diffusion, but large molecules cannot diffuse across a membrane. The dialysis tube membrane in the Experimental Procedure simulates a plasma membrane.

Experimental Procedure: Solute Diffusion Across Plasma Membrane

At the start of the experiment,

1. Cut a piece of dialysis tubing approximately 40 cm (approx. 16 in) long. Soak the tubing in water until it is soft and pliable.
2. Close one end of the dialysis tubing with two knots.
3. Fill the bag halfway with glucose solution.
4. Add 4 full droppers of starch solution to the bag.
5. Hold the open end while you mix the contents of the dialysis bag. Rinse off the outside of the bag with distilled water.
6. Fill a beaker 2/3 full with distilled water.
7. Add droppers of iodine solution (IKI) to the water in the beaker until an amber (tealike) color is apparent.
8. Record the color of the solution in the beaker in Table 4.3.
9. Place the bag in the beaker with the open end hanging over the edge. Secure the open end of the bag to the beaker with a rubber band as shown (Fig. 4.6). Make sure the contents do not spill into the beaker.

Figure 4.6   Placement of dialysis bag in water containing iodine.

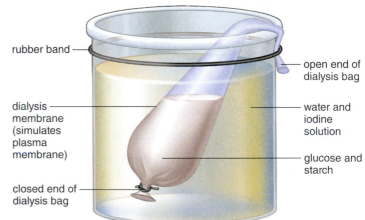

rubber band

open end of dialysis bag

dialysis membrane (simulates plasma membrane)

water and iodine solution

glucose and starch

closed end of dialysis bag

After about 5 minutes, at the end of the experiment,

10. You will note a color change. Record the color of the bag contents in Table 4.3.
11. Mark off a test tube at 1 cm and 3 cm.
12. Draw solution from near the bag and at the bottom of the beaker for testing with Benedict's reagent. Fill the test tube to the first mark with this solution. Add Benedict's reagent to the 3 cm mark. Heat in a boiling water bath for 5 to 10 minutes, observe any color change, and record your results as + or – in Table 4.3. (Optional use of glucose test strip: Dip glucose test strip into beaker. Compare stick with chart provided by instructor.)
13. Remove the dialysis bag from the beaker. Dispose of it and the used Benedict's reagent solution in the manner directed by your instructor.

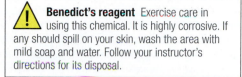

Benedict's reagent Exercise care in using this chemical. It is highly corrosive. If any should spill on your skin, wash the area with mild soap and water. Follow your instructor's directions for its disposal.

Table 4.3 Solute Diffusion Across Plasma Membrane					
At Start of Experiment			**At End of Experiment**		
	Contents	**Color**	**Color**	**Benedict's Test**	**Conclusion**
Bag	Glucose Starch			_____	
Beaker	Water Iodine				

Conclusions: Solute Diffusion Across the Plasma Membrane

- Based on the color change noted in the bag, conclude what solute diffused across the dialysis membrane from the beaker to the bag, and record your conclusion in Table 4.3.
- From the results of the Benedict's test on the beaker contents, conclude what solute diffused across the dialysis membrane from the bag to the beaker, and record your conclusion in Table 4.3.
- Which solute did not diffuse across the dialysis membrane from the bag to the beaker? _____

  How do you know? _____

  _____

  _____

4.4 Osmosis: Diffusion of Water Across Plasma Membrane

Osmosis is the diffusion of water across the plasma membrane of a cell. Just like any other molecule, water follows its concentration gradient and moves from the area of higher concentration to the area of lower concentration.

Experimental Procedure: Osmosis

To demonstrate osmosis, a thistle tube is covered with a membrane at its lower opening and partially filled with 50% corn syrup (starch solution) or a similar substance. The whole apparatus is placed in a beaker containing distilled water (Fig. 4.7). The water concentration in the beaker is 100%. Water molecules can move freely between the thistle tube and the beaker.

Figure 4.7 Osmosis demonstration.
a. A thistle tube, covered at the broad end by a differentially permeable membrane, contains a corn syrup solution. The beaker contains distilled water. **b.** The solute is unable to pass through the membrane, but the water (arrows) passes through in both directions. There is a net movement of water toward the inside of the thistle tube, where there is a lower percentage of water molecules. **c.** Due to the incoming water molecules, the level of the solution rises in the thistle tube.

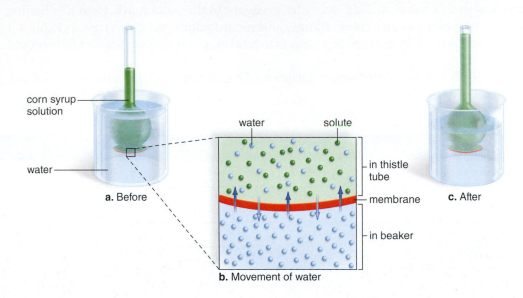

1. Note the level of liquid in the thistle tube, and measure how far it travels in 10 minutes:

   _____ mm

2. Calculate the speed of osmosis under these conditions: _____ mm/hr

Conclusions: Osmosis

- In which direction was there a net movement of water? _____
  Explain what is meant by "net movement" after examining the arrows in Figure 4.7*b*.

  _____

- If the starch molecules in corn syrup moved from the thistle tube to the beaker, would there have
  been a net movement of water into the thistle tube? _____ Why wouldn't large starch molecules
  be able to move across the membrane from the thistle tube to the beaker?

  _____

- Explain why the water level in the thistle tube rose: In terms of solvent concentration, water moved
  from the area of _____ water concentration to the area of _____ water concentration
  across a differentially permeable membrane.

Tonicity In Cells

Tonicity is the relative concentration of solute (particles), and therefore also of solvent (water), outside the cell compared with inside the cell.

- An **isotonic solution** has the same concentration of solute (and therefore of water) as the cell. When cells are placed in an isotonic solution, there is no net movement of water (see Fig. 4.8*a*).

- A **hypertonic solution** has a higher solute (therefore, lower water) concentration than the cell. When cells are placed in a hypertonic solution, water moves out of the cell into the solution (see Fig. 4.8b).
- A **hypotonic solution** has a lower solute (therefore, higher water) concentration than the cell. When cells are placed in a hypotonic solution, water moves from the solution into the cell (see Fig.4.8c).

Animal Cells (Red Blood Cells)

A solution of 0.9% NaCl is isotonic to red blood cells (Fig. 4.8a). A solution greater than 0.9% NaCl is hypertonic to red blood cells. In such a solution, the cells shrivel up, a process called **crenation** (Fig. 4.8b). A solution of less than 0.9% NaCl is hypotonic to red blood cells. In such a solution, the cells swell to bursting, a process called **hemolysis** (Fig. 4.8c).

Figure 4.8 Tonicity and red blood cells.

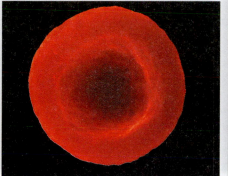
18,000×

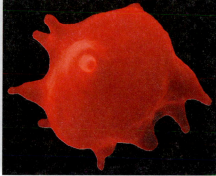
18,000×

18,000×

a. Isotonic solution.
Red blood cell has normal appearance due to no net gain or loss of water.

b. Hypertonic solution.
Red blood cell shrivels due to loss of water.

c. Hypotonic solution.
Red blood cell fills to bursting due to gain of water.

Experimental Procedure: Demonstration of Tonicity in Red Blood Cells

Three stoppered test tubes on display have the following contents:
  Tube 1: 0.9% NaCl plus a few drops of whole sheep blood
  Tube 2: 10% NaCl plus a few drops of whole sheep blood
  Tube 3: 0.9% NaCl plus distilled water and a few drops of whole sheep blood

 Do not remove the stoppers of test tubes during this procedure.

1. In the second column of Table 4.4, record the tonicity of each tube in relation to red blood cells.
2. Hold each tube in front of one of the pages of your Lab Manual. Determine whether you can see the print on the page through the tube. Record your findings in the third column of Table 4.4.
3. Explain in the fourth column of Table 4.4 why you can or cannot see the print.

Table 4.4	Effect of Tonicity on Red Blood Cells		
Tube	Tonicity	Print Visibility	Explanation
1			
2			
3			

Plant Cells

When plant cells are in a hypotonic solution, such as fresh water, the large central vacuole gains water and exerts pressure, called **turgor pressure.** The cytoplasm, including the chloroplasts, is pushed up against the cell wall (Fig. 4.9*a*).

When plant cells are in a hypertonic solution, such as 10% NaCl, the central vacuole loses water, and the cytoplasm, including the chloroplasts, pulls away from the cell wall. This is called **plasmolysis** (Fig. 4.9*b*).

Experimental Procedure: Tonicity in *Elodea* Cells

1. If possible, use the *Elodea* slide you prepared earlier in this laboratory. If not, prepare a new wet mount of a small *Elodea* leaf using fresh water.
2. After several minutes, focus on the surface of the cells, and compare your slide with Figure 4.9*a*.
3. Complete the portion of Table 4.5 that pertains to a hypotonic solution.
4. Prepare a new wet mount of a small *Elodea* leaf using a 10% NaCl solution.
5. After several minutes, focus on the surface of the cells, and compare your slide with Figure 4.9*b*.
6. Complete the portion of Table 4.5 that pertains to a hypertonic solution.

Figure 4.9 *Elodea* cells.

a. Surface view of cells in a hypotonic solution *(above)* and longitudinal section diagram *(below)*. The large central vacuole, filled with water, pushes the cytoplasm, including the chloroplasts, right up against the cell wall. **b.** Surface view of cells in a hypertonic solution *(above)* and longitudinal section diagram *(below)*. When the central vacuole loses water, cytoplasm, including the chloroplasts, piles up in the center of the cell because the cytoplasm has pulled away from the cell wall. (*a:* Magnification ×400)

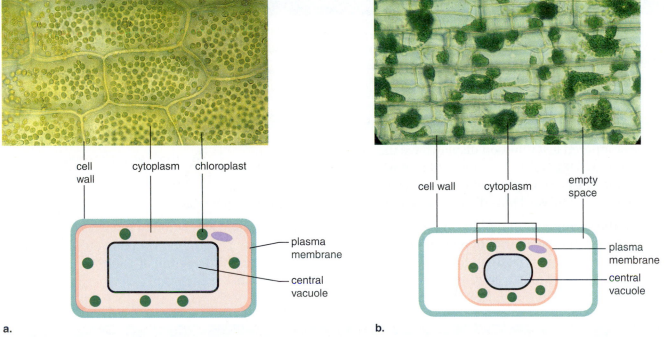

Table 4.5	Effect of Tonicity on *Elodea* Cells	
Tonicity	**Appearance of Cells**	**Due to (Scientific Term)**
Hypotonic		
Hypertonic		

Experimental Procedure: Tonicity in TTw Potato Strips

⏰ (This Experimental Procedure runs for one hour. Prior setup can maximize time efficiency.)

1. Cut two strips of potato, each about 7 cm long and 1.5 cm wide.
2. Label two test tubes 1 and 2. Place one potato strip in each tube.
3. Fill tube 1 with water to cover the potato strip.
4. Fill tube 2 with 10% sodium chloride (NaCl) to cover the potato strip.

⏰ 5. After 1 hour, remove the potato strips from the test tubes and place them on a paper towel. Observe each strip for limpness (water loss) or stiffness (water gain). Which tube has the limp potato strip? _____ Use tonicity to explain why water diffused out of the potato strip in this tube. _____

_____

Which tube has the stiff potato strip? _____ Use tonicity to explain why water diffused into the potato strip in this tube. _____

_____

6. Use this space to create a table to display your results. Give your table a title and columns for tube number and contents, tonicity, results, and explanation.

Conclusions: Tonicity

- In a hypotonic solution, animal cells _____. In red blood cells, this is called _____. In a hypertonic solution, animal cells _____. In red blood cells this is called _____.

- In a hypotonic solution, the central vacuole of *Elodea* cells exerts _____ pressure, and chloroplasts are seen _____. In a hypertonic solution, the central vacuole loses water and _____ occurs. The cytoplasm plus the chloroplasts are seen _____.

- In a hypotonic solution, potato strips _____ water; in a hypertonic solution, potato strips _____ water and become _____.

4.5   pH and Cells

The pH of a solution tells its hydrogen ion concentration [H$^+$]. The **pH scale** ranges from 0 to 14. A pH of 7 is neutral (Fig. 4.10). A pH lower than 7 indicates that the solution is acidic (has more hydrogen ions than hydroxide ions), whereas a pH greater than 7 indicates that the solution is basic (has more hydroxide ions than hydrogen ions). A **buffer** is a system of chemicals that takes up excess hydrogen ions or hydroxide ions, as appropriate.

The concept of pH is important in biology because living organisms are very sensitive to hydrogen ion concentration. For example, in humans the pH of the blood must be maintained at about 7.4 or we become ill. All living things need to maintain the hydrogen ion concentration, or pH, at a constant level.

Why are cells and organisms buffered? _____

_____

Figure 4.10   The pH scale.
The proportionate amount of hydrogen ions (H$^+$) to hydroxide ions (OH$^-$) is indicated by the diagonal line.

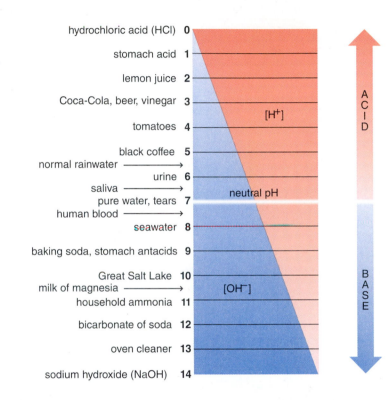

Experimental Procedure: pH and Cells

1. Label three test tubes, and fill them to the halfway mark as follows: tube 1: water; tube 2: buffer (inorganic) solution; and tube 3: simulated cytoplasm (buffered protein solution).
2. Use pH paper to determine the pH of each tube. Dip the end of a stirring rod into the solution, and then touch the stirring rod to a 5 cm strip of pH paper. Read the current pH by matching the color observed with the color code on the pH paper package. Record your results in the "pH Before Acid" column in Table 4.6.
3. Add 0.1 N hydrochloric acid (HCl) dropwise to each tube until you have added 5 drops—shake or swirl after each drop. Use pH paper as in step 2 to determine the new pH of each solution. Record your results in the "pH After Acid" column in Table 4.6.

> ⚠ **Hydrochloric acid (HCl)** used to produce an acid pH is a strong, caustic acid. Exercise care in using this chemical. If any HCl spills on your skin, rinse immediately with clear water. Follow your instructor's directions for disposal of tubes that contain HCl.

Table 4.6 pH and Cells

Tube	Contents	pH Before Acid	pH After Acid	Explanation
1	Water			
2	Buffer			
3	Cytoplasm			

Conclusions: pH and Cells

- Enter your explanations in the last column of Table 4.6.
- Why would you expect cytoplasm to be as effective as the buffer in maintaining pH? _____

_____

Experimental Procedure: Effectiveness of Antacids

This procedure tests the ability of commercial products such as Alka-Seltzer, Rolaids, Tums, or antacid tablets to absorb excess H^+.

1. Use a mortar and pestle to grind up the amount of antacid that is listed as one dose.
2. For each antacid tested, use a 100 ml of phenol red solution diluted to a faint pink to wash the antacid into a 250 ml beaker. Phenol red solution is a pH indicator that turns yellow in an acid and red in a base. Use a stirring rod to get the powder to dissolve.
3. Add and count the number of 0.1 N HCl drops it takes for the solution to turn light yellow.
4. Record your results in Table 4.7.

Table 4.7 Effectiveness of Antacids

Antacid	Drops of Acid Needed to Reach End Point	Evaluation
1		
2		
3		

Conclusions: Effectiveness of Antacids

- Participate with others in concluding which of the antacids tested neutralizes the most acid.

_____

- Did dosage in mg have any effect on the results? _____

_____

- Which of the substances on the label could be a buffer? _____

Laboratory Review 4

_____ **1.** What is the name of the large, often central organelle in eukaryotic cells that contains chromosomes?

_____ **2.** Prokaryotes have what structures necessary for protein synthesis?

_____ **3.** What is the function of the nucleus?

_____ **4.** What is the function of rough endoplasmic reticulum?

_____ **5.** Which organelle carries on intracellular digestion?

_____ **6.** Name a structure present in an animal cell but not in a plant cell.

_____ **7.** Name a structure present in a plant cell but not in an animal cell.

_____ **8.** What term describes the movement of molecules from an area of higher concentration to one of lower concentration?

_____ **9.** What is the name for the movement of water across the plasma membrane?

_____ **10.** In what direction does water move when cells are placed in a hypertonic solution?

_____ **11.** Is 10% NaCl isotonic, hypertonic, or hypotonic to red blood cells?

_____ **12.** What appearance will red blood cells have when they are placed in 9.0% NaCl?

_____ **13.** What scientific term is used to refer to the condition of cells described in question 12?

_____ **14.** What type of molecule prevents extensive changes in the pH of living organisms?

_____ **15.** If acid is added to water, does the pH increase or decrease?

_____ **16.** What is a pH of 7 called?

_____ **17.** Name two features or cellular components that all cells have in common.

Thought Questions

18. The police are trying to determine whether material removed from a crime scene is plant or animal matter. What would you suggest they look for?

19. Your grandmother asks you to fertilize her favorite plant. Without reading the directions on the box, you pour some fertilizer into the pot and then water the plant. The next time you see your grandmother, she tells you the plant died. In terms of osmosis, explain what happened to the plant.

20. Explain what happens to both plant and animal cells when they are placed into a solution that is hypotonic to the interior of the cell. If the two cells meet different fates, explain why.

Mitosis and Meiosis

Introduction

Dividing cells experience nuclear division, cytoplasmic division, and a period of time between divisions called interphase. During **interphase,** the nucleus appears normal, and the cell is performing its usual cellular functions. Also, the cell is increasing all of its components, including such organelles as the mitochondria, ribosomes, and centrioles, if present. DNA replication (making an exact copy of the DNA) occurs toward the end of interphase. Thereafter, the chromosomes, which contain DNA, are duplicated and contain two chromatids held together at a **centromere** (see Table 5.1). These chromatids are called **sister chromatids.**

During nuclear division, called **mitosis,** the new nuclei receive the same number of chromosomes as the parental nucleus. When the cytoplasm divides, a process called **cytokinesis,** two daughter cells are produced. In multicellular organisms, mitosis permits growth and repair of tissues. In eukaryotic, unicellular organisms, mitosis is a form of asexual reproduction. Sexually reproducing organisms utilize another form of nuclear division, called **meiosis.** In animals, meiosis is a part of gametogenesis, the production of gametes (sex cells). The gametes are sperm in male animals and eggs in female animals. As a result of meiosis, the daughter cells have half the number of chromosomes as the parental cell. As we shall see, meiosis contributes to recombination of genetic material and to variation among sexually reproducing organisms.

5.1 The Cell Cycle

As stated in the Introduction, the period of time between cell divisions is known as interphase. Because early investigators noted little visible activity between cell divisions, they dismissed this period of time as a resting state. But when later investigators discovered that DNA replication and chromosome duplication occur during interphase, the **cell cycle** concept was proposed. The cell cycle is divided into the four stages noted in Fig. 5.1. State the events of each stage on the line provided:

G_1 _____

S _____

G_2 _____

M _____

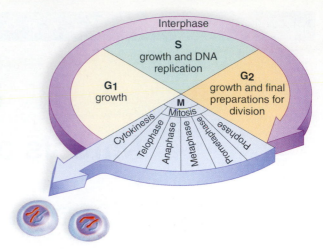

Figure 5.1   The cell cycle.
Immature cells go through a cycle that consists of four stages: G_1, S (for synthesis), G_2, and M (for mitosis). Eventually, some daughter cells "break out" of the cell cycle and become specialized cells.

Explain why the entire process is called the "cell cycle." _____

_____

The Cell Cycle and Cancer

Ordinarily, the cell cycle is tightly regulated by internal (inside the cell) and external (outside the cell) control mechanisms. When these controls are working properly, tumors (cancer) do not occur. Mutations brought about by carcinogens (e.g., radiation, pollution, cigarette smoke) can also cause the control of the cell cycle to falter. Then the cell cycle occurs repeatedly and a tumor develops. The cells of a tumor are disorganized and they do not contribute to the functioning of a tissue or organ. Treatment consists of shrinking or removing the tumor.

Table 5.1	Structures Associated with Mitosis
Structure	**Description**
Nucleus	A large organelle containing the chromosomes and acting as a control center for the cells
Chromosome	Rod-shaped body in the nucleus that is seen during mitosis and meiosis and that contains DNA, and therefore the hereditary units, or genes
Nucleolus	An organelle found inside the nucleus that produces the subunits of ribosomes
Spindle	Microtubule structure that brings about chromosome movement during cell division
Chromatids	The two identical parts of a chromosome following DNA replication
Centromere	A constriction where duplicates (sister chromatids) of a chromosome are held together
Centrosome	The central microtubule-organizing center of cells; consists of granular material; in animal cells, contains two centrioles
Centrioles*	Short, cylindrical organelles in animal cells that contain microtubules and are associated with the formation of the spindle during cell division
Aster*	Short, radiating fibers produced by the centrioles

*Animal cells only

Mitosis

Mitosis is nuclear division that results in two new nuclei, each having the same number of chromosomes as the original nucleus. The **parent cell** is the cell that divides, and the resulting cells are called **daughter cells.**

When cell division is about to begin, chromatin starts to condense and compact to form visible, rodlike sister chromatids held together at the centromere (Fig. 5.2*a*). Consult Table 5.1 and *label the sister chromatids, centromere, and kinetochore in the drawing of a duplicated chromosome in Figure 5.2b.* This illustration represents a chromosome as it would appear just before nuclear division occurs.

Figure 5.2   Duplicated chromosomes.
DNA replication results in a duplicated chromosome that consists of two sister chromatids held together at a centromere.

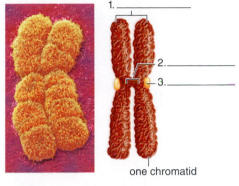

1. _____

2. _____

3. _____

one chromatid

a. SEM   **b.** Drawing

Spindle

Table 5.1 lists the structures that play a role during mitosis. The **spindle** is a structure that appears and brings about an orderly distribution of chromosomes to the daughter cell nuclei. A spindle has fibers that stretch between two poles (ends). Spindle fibers are bundles of microtubules, protein cylinders found in the cytoplasm that can assemble and disassemble. The **centrosome,** the main microtubule-organizing center of the cell, divides before mitosis so that during mitosis, each pole of the spindle has a centrosome. Animal cells contain two **centrioles** in each centrosome and asters, arrays of short microtubules radiating from the poles (see Fig. 5.3). The fact that plant cells lack centrioles suggests that centrioles are not required for spindle formation.

Observation: Animal Mitosis

Animal Mitosis Models

1. Using Figure 5.3 as a guide, identify the phases of animal cell mitosis in models of animal cell mitosis.
2. Each species has its own chromosome number. Counting the number of centromeres tells you the number of chromosomes in the models. What is the number of chromosomes in each of the cells

   in this model series? _____

Whitefish Blastula Slide

The blastula is an early embryonic stage in the development of animals. The **blastomeres** (blastula cells) shown are in different phases of mitosis (see Fig. 5.3).

1. Examine a prepared slide of whitefish blastula cells undergoing mitotic cell division.
2. Try to find a cell in each phase of mitosis. Have a partner or your instructor check your identification.

Mitosis Phases

Figure 5.3   Phases of mitosis in animal and plant cells.
The colors signify that half the chromosomes were inherited from one parent and half were inherited from the other parent.

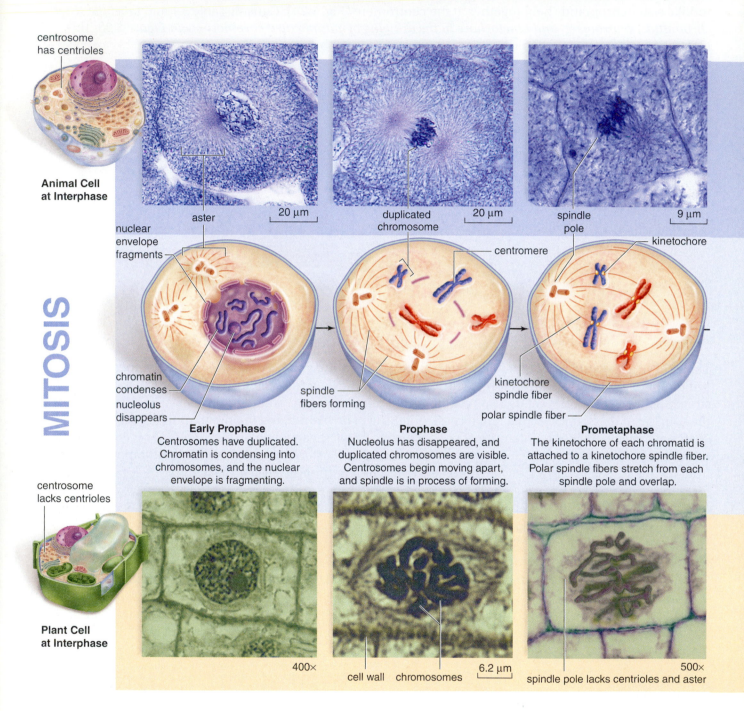

centrosome
has centrioles

**Animal Cell
at Interphase**

aster   20 μm

duplicated   20 μm
chromosome

spindle   9 μm
pole

MITOSIS

nuclear
envelope
fragments

centromere

kinetochore

chromatin
condenses

nucleolus
disappears

spindle
fibers forming

kinetochore
spindle fiber

polar spindle fiber

Early Prophase
Centrosomes have duplicated.
Chromatin is condensing into
chromosomes, and the nuclear
envelope is fragmenting.

Prophase
Nucleolus has disappeared, and
duplicated chromosomes are visible.
Centrosomes begin moving apart,
and spindle is in process of forming.

Prometaphase
The kinetochore of each chromatid is
attached to a kinetochore spindle fiber.
Polar spindle fibers stretch from each
spindle pole and overlap.

centrosome
lacks centrioles

**Plant Cell
at Interphase**

400×

cell wall   chromosomes   6.2 μm

500×

spindle pole lacks centrioles and aster

The phases of mitosis are shown in Figure 5.3. Mitosis is the type of nuclear division that occurs when a person grows larger and when injury heals. Mitosis results in two daughter cells because there is only one round of division, and it keeps the chromosome number constant (same as the parent cell).

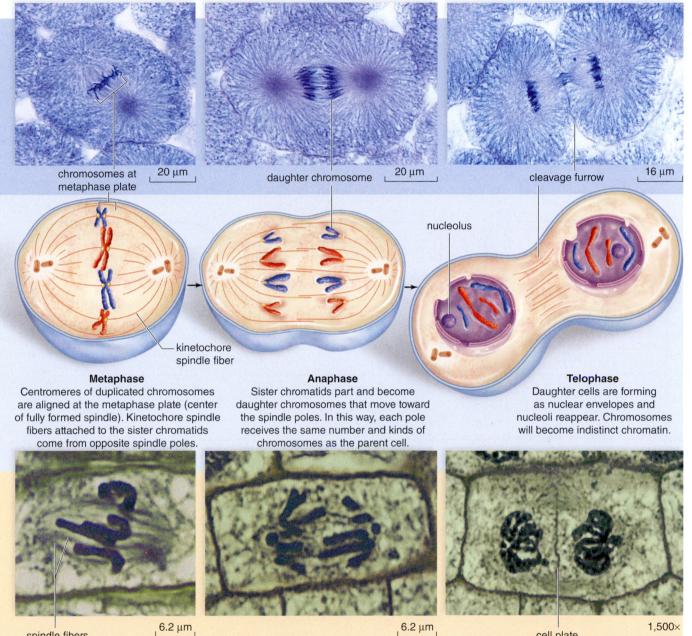

chromosomes at metaphase plate — 20 μm

daughter chromosome — 20 μm

cleavage furrow — 16 μm

nucleolus

kinetochore spindle fiber

Metaphase
Centromeres of duplicated chromosomes are aligned at the metaphase plate (center of fully formed spindle). Kinetochore spindle fibers attached to the sister chromatids come from opposite spindle poles.

Anaphase
Sister chromatids part and become daughter chromosomes that move toward the spindle poles. In this way, each pole receives the same number and kinds of chromosomes as the parent cell.

Telophase
Daughter cells are forming as nuclear envelopes and nucleoli reappear. Chromosomes will become indistinct chromatin.

spindle fibers — 6.2 μm

6.2 μm

cell plate — 1,500×

Plant Mitosis Models

1. Identify the phases of plant cell mitosis using models of plant cell mitosis and Figure 5.3 as a guide.
2. Notice that plant cells do not have centrioles and asters. Plant cells do have centrosomes and this accounts for the formation of a spindle.
3. What is the number of chromosomes in each of the cells in this model series? _____

Onion Root Tip Slide

1. In plants, the root tip contains tissue that is continually dividing and producing new cells. Examine a prepared slide of onion root tip cells (*Allium*) undergoing mitotic cell division. Try to find the phases that correspond to those shown in Figure 5.3.
2. Using high power, focus up and down on a cell in telophase. You may be able to just make out the cell plate, the region where a plasma membrane is forming between the two prospective daughter cells. Later, cell walls appear in this area.
3. In the boxes provided, draw and label the stages of mitosis as observed in the onion root tip slide.

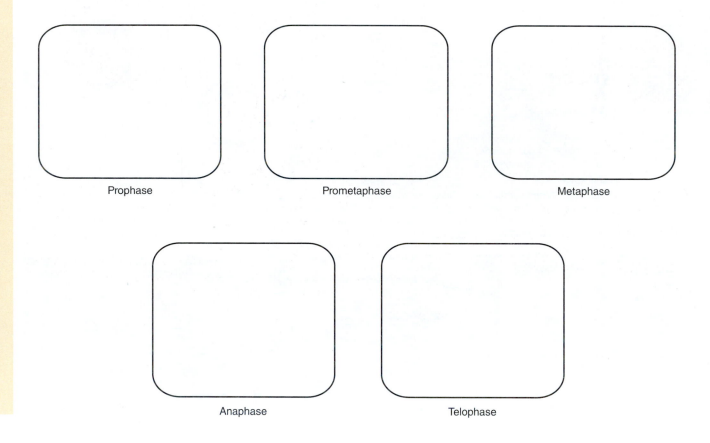

Prophase	Prometaphase	Metaphase
Anaphase	Telophase	

Cytokinesis

Cytokinesis, division of the cytoplasm, usually accompanies mitosis. During cytokinesis, each daughter cell receives a share of the organelles that duplicated during interphase. Cytokinesis begins in anaphase, continues in telophase, and reaches completion by the start of the next interphase.

Cytokinesis in Animal Cells

In animal cells, a **cleavage furrow,** an indentation of the membrane between the daughter nuclei, begins as anaphase draws to a close (Fig. 5.4). The cleavage furrow deepens as a band of actin filaments called the contractile ring slowly constricts the cell, forming two daughter cells.

Were any of the cells of the whitefish blastula slide undergoing cytokinesis? _____

How do you know? _____

_____

_____

> **Virtual Lab   The Cell Cycle and Cancer**   A virtual lab called The Cell Cycle and Cancer is available on the *Inquiry into Life* website **www.mhhe.com/maderinquiry14**. Follow the directions given to review mitosis and compare normal tissues to cancerous tissues.

cleavage furrow

4,000×

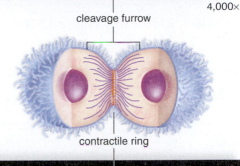

contractile ring

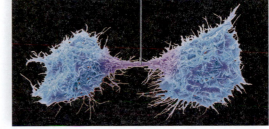

4,000×

Figure 5.4   Cytokinesis in animal cells.
A single cell becomes two cells by a furrowing process. A contractile ring composed of actin filaments gradually gets smaller, and the cleavage furrow pinches the cell into two cells.

Cytokinesis in Plant Cells

After mitosis, the cytoplasm divides by cytokinesis. In plant cells, membrane vesicles derived from the Golgi apparatus migrate to the center of the cell and form a **cell plate** (Fig. 5.5), which is the location of a new plasma membrane for each daughter cell. Later, individual cell walls appear in this area. Were any of the cells of the onion root tip slide undergoing cytokinesis as shown in Figure 5.3, Telophase, p. 61? _____

How do you know? _____

Offer an explanation for why Figure 5.5 is so detailed. _____

_____

Would you hypothesize that the vesicles of the cell plate lay down the new cell wall inside or outside the vesicles? Explain your answer. _____

Figure 5.5   Micrograph showing cytokinesis in plant cells.
During plant cell cytokinesis, vesicles fuse to form a cell plate that separates the daughter nuclei. Later, the cell plate gives rise to a new cell wall.

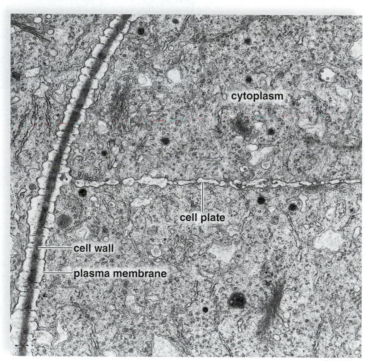

24,000×

Summary of Mitotic Cell Division

1. The nuclei in the daughter cells have the _____ number of chromosomes as the parent cell had.

2. Mitosis is cell division in which the chromosome number _____.

3. If a parent cell has 16 chromosomes, how many chromosomes do the daughter cells have following mitosis? _____

5.2 Meiosis

Meiosis is the type of nuclear division that occurs in the sex organs during **gamete** (egg and sperm) formation. As shown on pages 68–69, meiosis results in four daughter cells because there are two rounds of nuclear division. It also reduces the chromosome number to half that of the parent cell. The parent cell is said to be **diploid (2n)** because it contains **homologues** (homologous chromosomes) that look alike and carry genetic material for the same traits, such as fingers, hairline, or earlobes. Each **gamete** is **haploid (n)** because it contains only one from each pair of homologues.

Experimental Procedure: Meiosis

First, you will build four chromosomes: two pairs of homologues, as in Figure 5.6. In other words the parent cell is 2n = 4.

Building Chromosomes

1. Obtain the following materials: 48 pop beads of one color (e.g., red) and 48 pop beads of another color (e.g., blue) for a total of 96 beads; eight magnetic centromeres.
2. Build a homologous pair of duplicated chromosomes using Figure 5.6a as a guide. Each chromatid will have 16 beads. Be sure to bring the centromeres of the same color together so that they form one duplicated chromosome.
3. Build another homologous pair of duplicated chromosomes using Figure 5.6b as a guide. Each chromatid will have eight beads.
4. Note that your chromosomes are the same as those in Figure 5.6.

Figure 5.6 Two pairs of homologues.
The red chromosomes were inherited from one parent, and the blue chromosomes were inherited from the other parent.

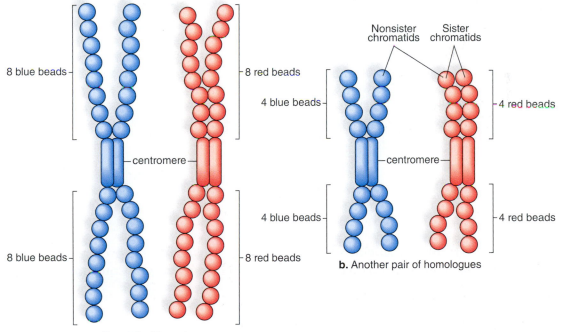

8 blue beads — 8 red beads

Nonsister chromatids   Sister chromatids

4 blue beads — 4 red beads

centromere

8 blue beads — 8 red beads

4 blue beads — 4 red beads

b. Another pair of homologues

a. One pair of homologues

Meiosis I

Meiosis requires two nuclear divisions. The first round is called meiosis I and the second round is called meiosis II. Both meiosis I and meiosis II have stages named the same as in mitosis but the events are different.

Prophase I

5. Put all four of your chromosomes in the center of your work area: this area represents the nucleus. Synapsis, a very important event, occurs during meiosis. To simulate synapsis place the long blue chromosome next to the long red chromosome and the short blue chromosome next to the short red chromosome to show that the homologues pair up as in Figure 5.7 (Prophase I). Now **crossing-over** occurs. During crossing-over an exchange of genetic material occurs between nonsister chromatids. Perform crossing over by switching some blue beads for red beads between the inside chromatids of the homologues.

Genetic Variation Due to Meiosis As a result of crossing-over the genetic material on a chromosome in a gamete can be different from that in the parent cell.

Metaphase I

6. Keep the homologues together and align them at the metaphase plate. Add centrioles to represent the poles of a spindle apparatus.

Anaphase I and Telophase I

7. Separate the homologues, so that each pole of the spindle receives one member from each pair of homologues. This separation causes each pole to receive the haploid number of chromosomes.

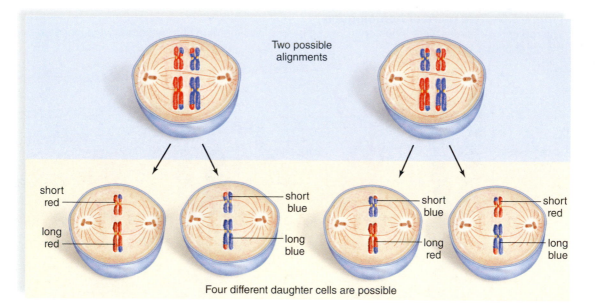

Four different daughter cells are possible

Only two daughter cells result from meiosis I, but two possible alignments of chromosomes can occur during metaphase I. Therefore, four possible combinations of haploid chromosomes are possible in the daughter cells following meiosis I in a 2n=4 parent cell.

Genetic Variation Due to Meiosis As a result of homologue separation all possible combinations of the haploid number of chromosomes can occur among the gametes.

Meiosis II

The second round of nuclear division during meiosis is called **meiosis II** (see Fig. 5.7).

Prophase II

8. Choose one daughter nucleus (see step 7, page 66) to be the parent nucleus undergoing meiosis II.

Metaphase II

9. Move the duplicated chromosomes to the metaphase II metaphase plate, as shown in the art to the right.

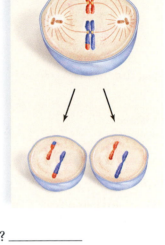

Anaphase II

10. Pull the two magnets of each duplicated chromosome apart. What does this action represent? _____

_____

Telophase II

11. Put the chromosomes—each having one chromatid—at the poles near the centrioles. At the end of telophase, the daughter nuclei reform.

- You chose only one daughter nucleus from meiosis I to be the nucleus that divides. In reality both daughter nuclei go on to divide again.

  Therefore, how many nuclei are usually present when meiosis II is complete? _____

- In this exercise, how many chromosomes were in the parent cell nucleus undergoing meiosis II? _____

- How many chromosomes are in the daughter nuclei? _____ Explain. _____

_____

Summary of Meiosis

1. Meiosis reduces the chromosome number. If the parent cell is 2n=4, the daughter cells are

   n = _____. Without meiosis the chromosome number would double with each generation.

   Instead, when a haploid sperm fertilizes a haploid egg, the new individual is 2n.

2. Sexual reproduction results in offspring that can look very different as represented in this illustration. Why do the puppies born to these parents show variation?

   a. During prophase I, the homologues come together and exchange genetic material. Now the inherited chromosomes will be different from those in the parent cell. This process is

   called_____.

   b. During anaphase I, the homologues separate differently and therefore the daughter cells following telophase I can have different _____ of chromosomes.

   c. During fertilization (see page 73) variant sperm fertilize variant eggs, further helping to ensure that the new individual inherits different _____ of homologues than a parent had.

Meiosis Phases

Figure 5.7   Meiosis I and II in plant cell micrographs and animal cell drawings.
Crossing-over occurs during meiosis I.

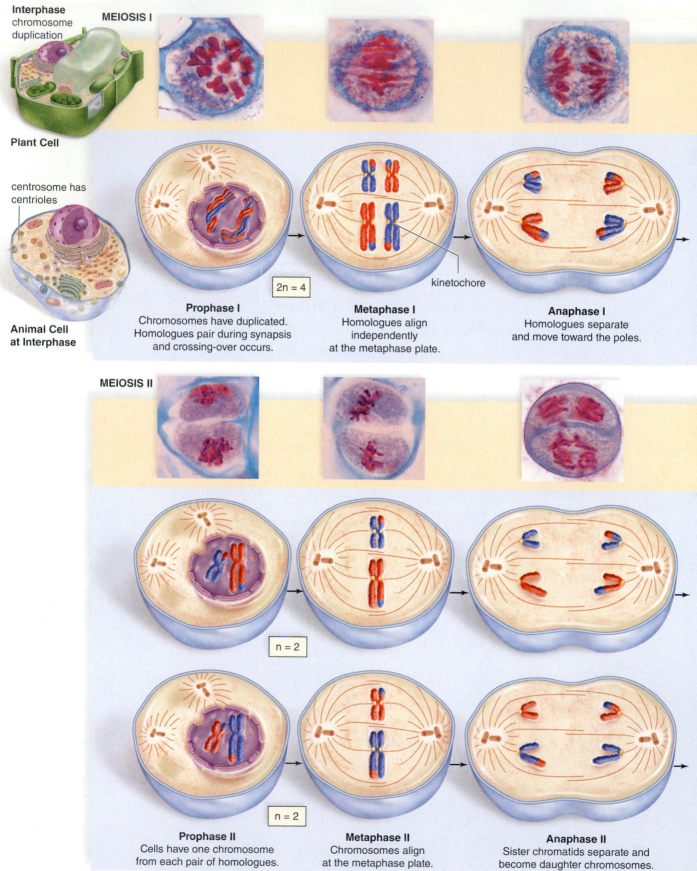

Interphase
chromosome duplication

MEIOSIS I

Plant Cell

centrosome has centrioles

2n = 4

kinetochore

Animal Cell at Interphase

Prophase I
Chromosomes have duplicated.
Homologues pair during synapsis
and crossing-over occurs.

Metaphase I
Homologues align
independently
at the metaphase plate.

Anaphase I
Homologues separate
and move toward the poles.

MEIOSIS II

n = 2

n = 2

Prophase II
Cells have one chromosome
from each pair of homologues.

Metaphase II
Chromosomes align
at the metaphase plate.

Anaphase II
Sister chromatids separate and
become daughter chromosomes.

The phases of meiosis are shown in Figure 5.7. Meiosis is the type of nuclear division that (1) occurs in the sex organs (testes and ovaries); (2) results in four nuclei because there are two rounds of nuclear cell division; and (3) reduces the chromosome number to half that of the parent cell.

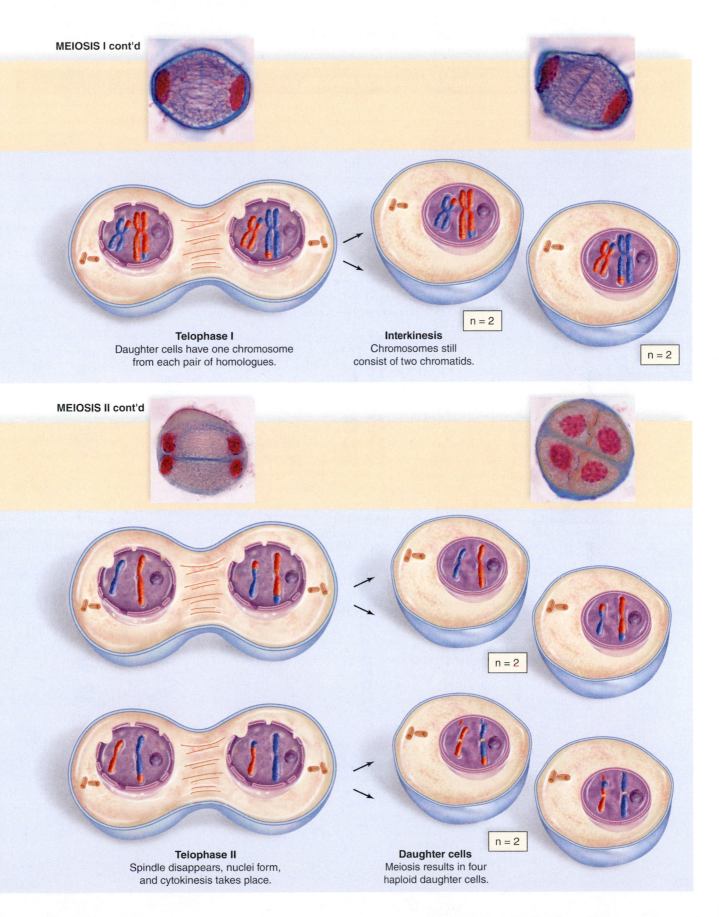

MEIOSIS I cont'd

n = 2

n = 2

Telophase I
Daughter cells have one chromosome from each pair of homologues.

Interkinesis
Chromosomes still consist of two chromatids.

MEIOSIS II cont'd

n = 2

n = 2

Telophase II
Spindle disappears, nuclei form, and cytokinesis takes place.

Daughter cells
Meiosis results in four haploid daughter cells.

5.3 Mitosis versus Meiosis

In comparing mitosis to meiosis it is important to note that meiosis requires two nuclear divisions but mitosis requires only one nuclear division. Therefore mitosis produces two daughter cells and meiosis produces four daughter cells. Following mitosis, the daughter cells are still diploid but following meiosis, the daughter cells are haploid. Figure 5.8 explains why. Fill in Table 5.2 to indicate general

Table 5.2 Differences between Mitosis and Meiosis		
	Mitosis	**Meiosis**
1. Number of divisions		
2. Chromosome number in daughter cells		
3. Number of daughter cells		

Figure 5.8 Meiosis I compared to mitosis.

Compare metaphase I of meiosis I to metaphase of mitosis. Only in metaphase I are the homologues paired at the metaphase plate. Homologues separate during anaphase I, and therefore the daughter cells are haploid. The blue chromosomes were inherited from one parent, and the red chromosomes were inherited from the other parent. The exchange of color between nonsister chromatids represents the crossing-over that occurred during meiosis I.

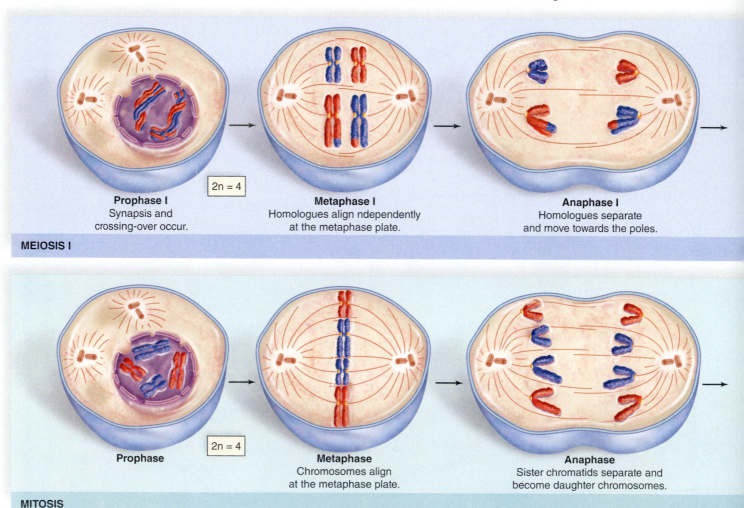

MEIOSIS I

2n = 4

Prophase I
Synapsis and
crossing-over occur.

Metaphase I
Homologues align ndependently
at the metaphase plate.

Anaphase I
Homologues separate
and move towards the poles.

MITOSIS

2n = 4

Prophase

Metaphase
Chromosomes align
at the metaphase plate.

Anaphase
Sister chromatids separate and
become daughter chromosomes.

differences between mitosis and meiosis and complete Table 5.3 to indicate specific differences between mitosis and meiosis I. Mitosis need be compared only with meiosis I because the same events occur during both mitosis and meiosis II except that the cells are diploid during mitosis and haploid during meiosis II.

Table 5.3 Mitosis Compared with Meiosis I	
Mitosis	**Meiosis I**
Prophase: No pairing of chromosomes	Prophase I: _____
Metaphase: Duplicated chromosomes at metaphase plate	Metaphase I: _____
Anaphase: Sister chromatids separate	Anaphase I: _____
Telophase: Chromosomes have one chromatid	Telophase I: _____

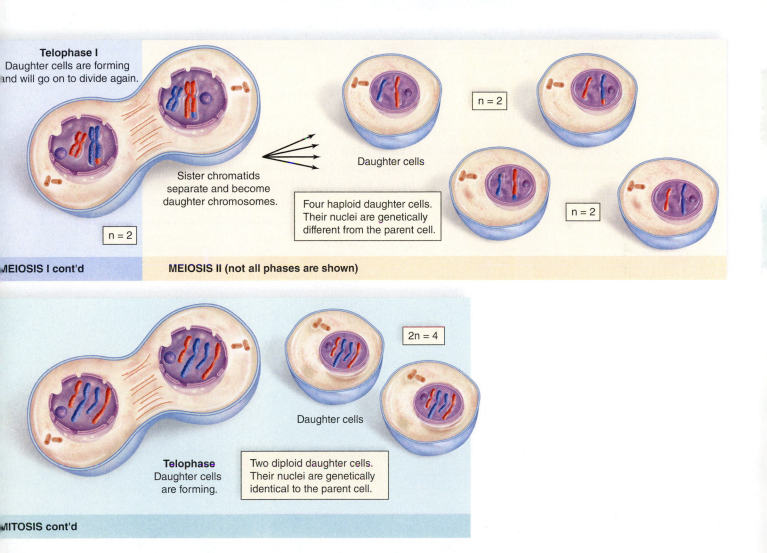

Telophase I
Daughter cells are forming and will go on to divide again.

n = 2

Sister chromatids separate and become daughter chromosomes.

n = 2

Daughter cells

Four haploid daughter cells. Their nuclei are genetically different from the parent cell.

n = 2

MEIOSIS I cont'd   **MEIOSIS II (not all phases are shown)**

2n = 4

Daughter cells

Telophase
Daughter cells are forming.

Two diploid daughter cells. Their nuclei are genetically identical to the parent cell.

MITOSIS cont'd

5.4   Karyotype Anomalies

In a karyotype, the chromosomes of an organism are arranged so that the pairs of chromosomes can be seen (Fig. 5.9). At that time, it is possible to observe any anomalies in chromosome number and structure.

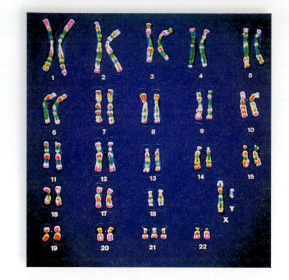

Figure 5.9   Karyotype of a person with Down syndrome.
Note the three number 21 chromosomes.

Chromosome number anomalies usually occur when cells divide during mitosis or during meiosis, signifying that these complex processes do not always occur as expected. The most common anomaly is a change in number following meiosis so that the individual has either 45 chromosomes or 47 chromosomes, instead of 46 chromosomes (the usual number in human beings). An extra or missing chromosome can cause unexpected consequences. For example, a fetus who has a missing X chromosome may fail to develop the appearance of a female and may not have the internal organs of a female. Either sex that has an extra chromosome 21 has the symptoms of Down syndrome (Fig. 5.9).

Chromosome structure anomalies can occur when nuclei divide, particularly if cells have been subject to environmental influences such as radiation or drug intake. Some of the more common structural anomalies are

Deletion: The chromosome is shorter than usual because some portion is missing.
Duplication: The chromosome is longer than usual because some portion is present twice over.
Inversion: The chromosome is normal in length but some portion runs in the opposite direction.
Translocation: Two chromosomes have switched portions and each switched portion is on the wrong chromosome.

Chromosome number or structure anomalies often result in recognized **syndromes,** a collection of symptoms that always occur together. Which syndrome appears depends on the particular anomaly. This topic is studied in more detail in laboratory 10.

5.5 Gametogenesis in Animals

Gametogenesis is the formation of **gametes** (sex cells) in animals. In humans and other mammals, the gametes are sperm and eggs. **Fertilization** occurs when the nucleus of a sperm fuses with the nucleus of an egg.

Gametogenesis in Mammals

Gametogenesis occurs in the testes of males, where **spermatogenesis** produces sperm. Gametogenesis occurs in the ovaries of females, where **oogenesis** produces oocytes (eggs).

A **diploid** (2n) nucleus contains the full number of chromosomes, which is 46 in humans, and a **haploid** (n) nucleus contains half as many, which is 23 in humans. Gametogenesis involves **meiosis,** the process that reduces the chromosome number from 2n to n. In sexually reproducing species, if meiosis did not occur, the chromosome number would double with each generation. Meiosis consists of two divisions: the first meiotic division (meiosis I) and the second meiotic division (meiosis II). Therefore, you would expect four haploid cells at the end of the process. Indeed, there are four sperm as a result of spermatogenesis (Fig. 5.10). However, in females, meiosis I results in a secondary oocyte and one polar body. A **polar body** is a nonfunctioning cell that will disintegrate. A secondary oocyte does not undergo meiosis II unless fertilization (fusion of egg and sperm) occurs. At the completion of oogenesis, there is a single egg. The polar bodies die (Fig. 5.10). Fertilization restores the full chromosome number, which is 46 in humans. Now the individual has two of every type of chromosome.

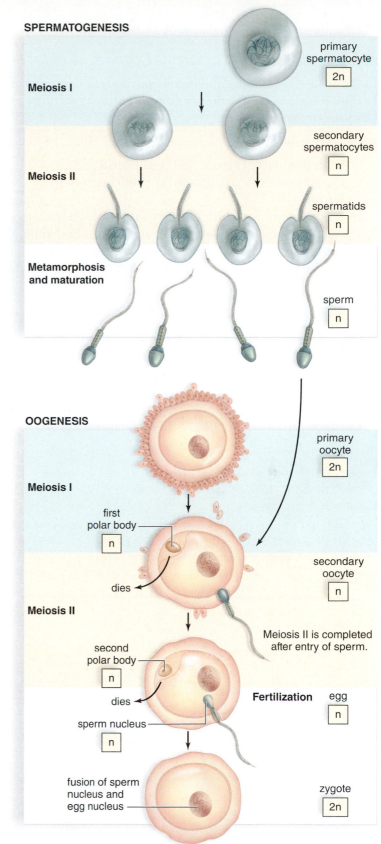

Figure 5.10 Spermatogenesis and oogenesis in mammals. Spermatogenesis produces four viable sperm, whereas oogenesis produces one egg and two polar bodies. The polar bodies die. In humans, both sperm and egg have 23 chromosomes each; therefore, following fertilization, the zygote has 46 chromosomes.

Gametogenesis Models

Examine any available gametogenesis models, and determine the diploid number of the parent cell and the haploid number of a gamete. Remember that counting the number of centromeres tells you the number of chromosomes.

Slide of Ovary

1. With the help of Figure 5.11, examine a prepared slide of an ovary. Under low power, you will see a large number of small, primary follicles near the outer edge. A primary follicle contains a primary oocyte.
2. Find a secondary follicle, and switch to high power. Note the secondary oocyte (egg), surrounded by numerous cells, to one side of the liquid-filled follicle.
3. Also look for a large, fluid-filled vesicular (Graafian) follicle, which contains a mature secondary oocyte to one side. The vesicular follicle will be next to the outer surface of the ovary because this type of follicle releases the egg during ovulation.

4. How many secondary follicles can you find on your slide? _____ How many vesicular follicles can you find? _____ How does this number compare with the number of sperm cells seen in the testis cross section (see Fig. 5.12)? _____

Figure 5.11 Microscopic ovary anatomy.

The stages of follicle and oocyte (egg) development are shown in sequence. Each follicle goes through all the stages. Following ovulation, a follicle becomes the corpus luteum.

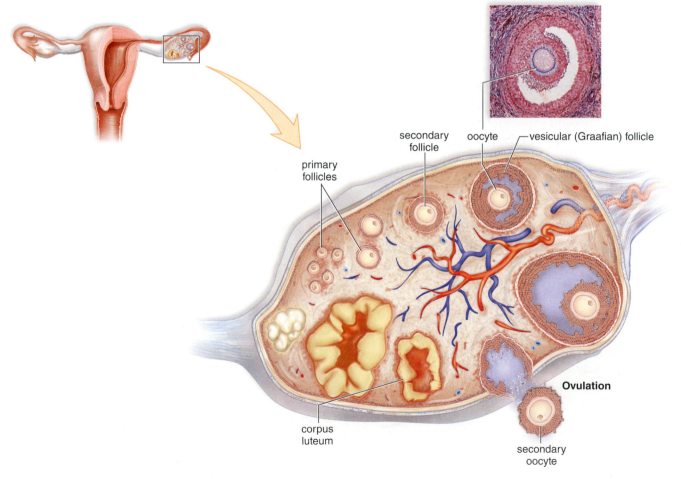

secondary follicle   oocyte   vesicular (Graafian) follicle

primary follicles

corpus luteum

Ovulation

secondary oocyte

Slide of Testis

1. With the help of Figure 5.12, examine a prepared slide of a testis. Under low power, note the many circular structures. These are the **seminiferous tubules,** where sperm formation takes place.
2. Switch to high power, and observe one tubule in particular. Find mature sperm (which look like thin, fine, dark lines) in the middle of the tubule. **Interstitial cells,** which produce the male sex hormone testosterone, are between the tubules.

Figure 5.12 Microscopic testis anatomy.
a. A testis contains many seminiferous tubules. **b.** Scanning electron micrograph of a cross section of the seminiferous tubules, where spermatogenesis occurs. Note the location of interstitial cells in clumps among the seminiferous tubules in this light micrograph.

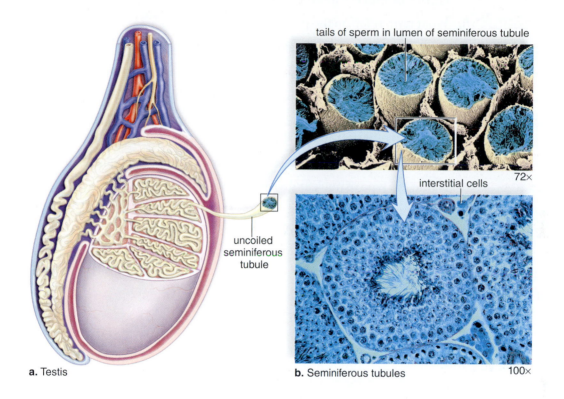

tails of sperm in lumen of seminiferous tubule

interstitial cells

72×

uncoiled seminiferous tubule

a. Testis

b. Seminiferous tubules

100×

Summary of Gametogenesis

1. What is gametogenesis? _____

   In general, how many chromosomes are in a gamete (n or 2n)? _____

2. What is spermatogenesis? _____

   How many chromosomes does a human sperm have (arabic number)? _____

3. What is oogenesis? _____

   How many chromosomes does a human egg have (arabic number)? _____

4. Following fertilization, how many chromosomes does the zygote, the first cell of the new individual, have? _____

Laboratory Review 5

_____ 1. During what stage of the cell cycle does DNA replication occur?

_____ 2. Name the phase of mitosis during which separation of sister chromatids occurs.

_____ 3. By what process does the cytoplasm of a human cell separate?

_____ 4. Name the phase of mitosis when duplicated chromosomes first appear.

_____ 5. What structure forms in plant cells during cytokinesis?

_____ 6. Where in humans would you expect to find meiosis taking place?

_____ 7. If there are 13 pairs of homologues in a primary spermatocyte, how many chromosomes are there in a sperm?

_____ 8. What term refers to the production of an egg?

_____ 9. During which type of gametogenesis would you see polar bodies?

_____ 10. What do you call chromosomes that look alike and carry genes for the same traits?

_____ 11. If homologues are separating, what phase is this?

_____ 12. If the parental cell has 24 chromosomes, how many does each daughter cell have at the completion of meiosis II?

_____ 13. Name the type of nuclear division during which homologues pair.

_____ 14. Name the type of nuclear division described by 2n $\longrightarrow$ 2n.

_____ 15. Does metaphase of mitosis, meiosis I, or meiosis II have the haploid number of chromosomes at the metaphase plate of the spindle?

Thought Questions

16. List four differences when comparing mitosis with meiosis.

17. If the cells of an organism have 12 chromosmes, how does the number of chromosomes differ at the metaphase plate when you compare metaphase of mitosis with metaphase of meiosis II?

18. A student is simulating meiosis I with a pair of homologues that are red-long and yellow-long. Why would you not expect to find both red-long and yellow-long in one resulting daughter cell?

19. With reference to a pair of homologues, describe the change in the two participating nonsister chromatids following crossing-over.

20. What would be the appearance of a cell that completes mitosis without completing cytokinesis?

6

How Enzymes Function

Introduction

The cell carries out many chemical reactions. All the chemical reactions that occur in a cell are collectively called **metabolism.** A possible chemical reaction can be indicated like this:

$$A + B \longrightarrow C + D$$
$$\text{reactants} \qquad \text{products}$$

In all chemical reactions, the **reactants** are molecules that undergo a change, which results in the **products.** The arrow means "produces," as in A + B produces C + D. The number of reactants and products can vary; in the one you are studying today, a single reactant breaks down to two products. All the reactions that occur in a cell have an enzyme. **Enzymes** are organic catalysts that speed metabolic reactions. Because enzymes are specific and speed only one type of reaction, they are given names. In today's laboratory, you will be studying the action of the enzyme **catalase.** The reactants in an enzymatic chemical reaction are called **substrate(s)** (Fig. 6.1).

Figure 6.1   Enzymatic action.
The reaction occurs on the surface of the enzyme at the active site. The enzyme is reusable. **a.** Degradation: substrate is broken down. **b.** Synthesis: substrates are combined.

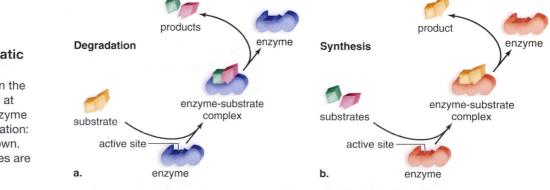

Enzymes are specific because they have a shape that accommodates the shape of their substrates. Enzymatic reactions can be indicated like this:

$$E + S \longrightarrow ES \longrightarrow E + P$$

In this reaction, E = enzyme, ES = enzyme-substrate complex, and P = product.

Two types of enzymatic reactions in cells are shown in Figure 6.1. During degradation reactions, the substrate is broken down to the product(s), and during synthesis reactions, the substrates are joined to form a product. A number of other types of reactions also occur in cells. The location where the enzyme and substrate form an enzyme-substrate complex is called the **active site** because the reaction occurs here. At the end of the reaction, the product is released, and the enzyme can then combine with its substrate again. A cell needs only a small amount of an enzyme because enzymes are used over and over. Some enzymes have turnover rates well in excess of a million product molecules per minute.

> **Planning Ahead** To save time, your instructor may have you start a boiling water bath (page 80) at the beginning of the laboratory.

6.1   Catalase Activity

Catalase is involved in a degradation reaction: Catalase speeds the breakdown of hydrogen peroxide (H_2O_2) in nearly all organisms including bacteria, plants, and animals. A cellular organelle called a peroxisome, which contains catalase, is present in every plant and animal organ. This means that we could use any plant or animal organ as our source of catalase today. Commonly, school laboratories use the potato as a source of catalase because potatoes are easily obtained and cut up.

Catalase performs a useful function in organisms because hydrogen peroxide is harmful to cells. Hydrogen peroxide is a powerful oxidizer that can attack and denature cellular molecules like DNA! Knowing its harmful nature, humans use hydrogen peroxide as a commercial antiseptic to kill germs (Fig. 6.2). In reduced concentration, hydrogen peroxide is a whitening agent used to bleach hair and teeth. Skillful technicians use it to provide oxygen to aquatic plants and fish, but it is also used industrially to clean most anything from tubs to sewage. It's even put in glow sticks, where it reacts with a dye that then emits light.

When catalase speeds the breakdown of hydrogen peroxide, water and oxygen are released.

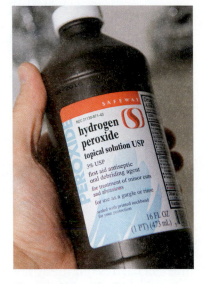

Figure 6.2   Hydrogen peroxide.
Bubbling occurs when you apply hydrogen peroxide to a cut because oxygen is being released when catalase, an enzyme present in the body's cells, degrades hydrogen peroxide.

$$2\,H_2O_2 \xrightarrow{\text{catalase}} 2\,H_2O + O_2$$

hydrogen peroxide   water   oxygen

What is the reactant in this reaction? _____ What is the substrate for

catalase? _____ What are the products in this reaction? _____ and

_____

Bubbling occurs as the reaction proceeds. Why? _____

In the experimental procedure that follows, you will use bubble height to indicate the amount of product per unit time and therefore enzyme activity. Examine Table 6.1 and hypothesize which tube (1, 2, or 3) will

have a greater bubble column height. Explain your answer. _____

_____

_____

With a wax pencil, label and mark three clean test tubes at the 1 cm and 5 cm levels.

Tube 1   **1.** Fill to the first mark with catalase buffered at pH 7.0, the optimum pH for catalase.

2. Fill to the second mark with hydrogen peroxide. Swirl well to mix, and wait at least 20 seconds for bubbling to develop.

3. Measure the height of the bubble column (in mm), and record your results in Table 6.1.

Tube 2   **1.** Fill to the first mark with water.

2. Fill to the second mark with hydrogen peroxide. Swirl well to mix, and wait at least 20 seconds.

3. Measure the height of the bubble column and record your results in Table 6.1.

Tube 3   **1.** Fill to the first mark with catalase.

2. Fill to the second mark with sucrose solution. Swirl well to mix; wait 20 seconds.

3. Measure the height of the bubble column, and record your results in Table 6.1.

Table 6.1	Catalase Activity		
Tube	**Contents**	**Bubble Column Height (mm)**	**Explanation**
1	Catalase Hydrogen peroxide		
2	Water Hydrogen peroxide		
3	Catalase Sucrose solution		

Conclusions: Catalase Activity

- Which tube showed the amount of bubbling you expected? _____ Record your explanation in Table 6.1.

- Which tube is a negative control? _____ If this tube showed bubbling, what could you conclude about your procedure? _____

  Record your explanation in Table 6.1.

- Enzymes are specific; they speed only a reaction that contains their substrate. Which tube exemplifies this characteristic of an enzyme? _____ Record your explanation in Table 6.1.

6.2   Effect of Temperature on Enzyme Activity

The active sites of enzymes increase the likelihood that substrate molecules will find each other and interact. Therefore, enzymes lower the energy of activation (the temperature needed for a reaction to occur). Still, increasing the temperature is expected to increase the likelihood that active sites will be occupied because molecules move about more rapidly as the temperature rises. In this way, a warm temperature increases enzyme activity.

The shape of an enzyme and its active site must be maintained or else they will no longer be functional. A very high temperature, such as the one that causes water to boil, is likely to cause weak bonds of a protein to break; and if this occurs, the enzyme **denatures**—it loses its original shape and the active site will no longer function to bring reactants together. Now enzyme activity plummets.

With this information in mind, examine Table 6.2 below and hypothesize which tube (1, 2, or 3) will have more product per unit time as judged by bubble column height. _____ Explain your answer. _____

_____

Experimental Procedure: Effect of Temperature

With a wax pencil, label and mark three clean test tubes at the 1 cm and 5 cm levels.

1. Fill each tube to the first mark with catalase buffered at pH 7.0, the optimum pH for catalase.
2. Place tube 1 in a refrigerator or cold water bath, tube 2 in an incubator or warm water bath, and tube 3 in a boiling water bath. Complete the second column in Table 6.2. Wait 15 minutes.
3. As soon as you remove the tubes one at a time from the refrigerator, incubator, and boiling water, fill to the second mark with hydrogen peroxide.
4. Swirl well to mix, and wait 20 seconds.
5. Measure the height of the bubble column in each tube, and record your results in Table 6.2. Plot your results in Figure 6.3.

Table 6.2	Effect of Temperature		
Tube	**Temperature °C**	**Bubble Column Height (mm)**	**Explanation**
1 Refrigerator			
2 Incubator			
3 Boiling water			

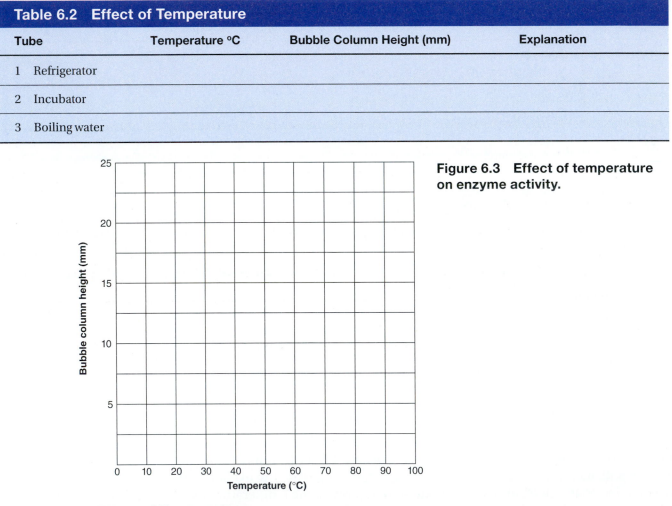

Figure 6.3   Effect of temperature on enzyme activity.

Conclusions: Effect of Temperature

• The bubble column height indicates the degree of enzyme activity. Was your hypothesis supported? _____ Explain in Table 6.2 the degree of enzyme activity per tube.

• What is your conclusion concerning the effect of temperature on enzyme activity? _____

_____

6.3   Effect of Concentration on Enzyme Activity

Consider that if you increase the number of caretakers per number of children, it is more likely that each child will have quality time with a caretaker. So it is if you increase the amount of enzyme per amount of substrate: it is more likely that substrates will find the active site of an enzyme and a reaction will take place. With this in mind, examine Table 6.3 and hypothesize which tube (1, 2, or 3) will have more product per unit time as judged by bubble column height. _____ Explain your answer. _____

_____

Experimental Procedure: Effect of Enzyme Concentration

With a wax pencil, label three clean test tubes.

Tube 1
1. Mark this tube at the 1 cm and 5 cm levels.
2. Fill to the first mark with water and to the second mark with hydrogen peroxide.
3. Swirl well to mix, and wait 20 seconds.
4. Measure the height of the bubble column, and record your results in Table 6.3.

Tube 2
1. Mark this tube at the 1 cm and 5 cm levels.
2. Fill to the first mark with buffered catalase and to the second mark with hydrogen peroxide.
3. Swirl well to mix, and wait 20 seconds.
4. Measure the height of the bubble column, and record your results in Table 6.3.

Tube 3
1. Mark this tube at the 3 cm and 7 cm levels.
2. Fill to the first mark with buffered catalase and to the second mark with hydrogen peroxide.
3. Swirl well to mix, and wait 20 seconds.
4. Measure the height of the bubble column, and record your results in Table 6.3.

Table 6.3   Effect of Enzyme Concentration

Tube	Amount of Enzyme	Bubble Column Height (mm)	Explanation
1	none		
2	1 cm		
3	3 cm		

Conclusions: Effect of Concentration

- The bubble column height indicates the degree of enzyme activity. Was your hypothesis supported? _____ Explain in Table 6.3 the degree of enzyme activity per tube.

- If unlimited time was allotted, would the results be the same in all tubes? _____ Explain why or why not. _____

- Would you expect similar results if the substrate concentration were varied in the same manner as the enzyme concentration? _____ Why or why not? _____

- What is your conclusion concerning the effect of concentration on enzyme activity? _____

_____

6.4 Effect of pH on Enzyme Activity

Each enzyme has a pH at which the speed of the reaction is optimum (occurs best). Any higher or lower pH affects hydrogen bonding and the structure of the enzyme, leading to reduced activity.

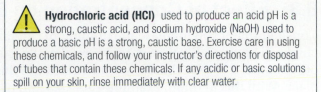

Hydrochloric acid (HCl) used to produce an acid pH is a strong, caustic acid, and sodium hydroxide (NaOH) used to produce a basic pH is a strong, caustic base. Exercise care in using these chemicals, and follow your instructor's directions for disposal of tubes that contain these chemicals. If any acidic or basic solutions spill on your skin, rinse immediately with clear water.

Catalase is an enzyme found in cells where the pH is near 7 (called neutral pH). Other enzymes prefer different pHs. The pancreas secretes a slightly basic (below pH 7) juice into the digestive tract and the stomach wall releases a very acidic digestive juice which can be as low as pH 2. With this information about catalase in mind, examine Table 6.4 and hypothesize which tube (1, 2, or 3) will have more product per unit time as judged by the bubble column height. _____ Explain your answer. _____

_____

Experimental Procedure: Effect of pH

With a wax pencil, label and mark three clean test tubes at the 1 cm, 3 cm, and 7 cm levels. Fill each tube to the 1 cm level with nonbuffered catalase.

Tube 1
1. Fill to the second mark with water adjusted to pH 3 by the addition of HCl. Wait one minute.
2. Fill to the third mark with hydrogen peroxide.
3. Swirl to mix, and wait 20 seconds.
4. Measure the height of the bubble column, and record your results in Table 6.4.

Tube 2
1. Fill to the second mark with water adjusted to pH 7. Wait one minute.
2. Fill to the third mark with hydrogen peroxide.
3. Swirl to mix, and wait 20 seconds.
4. Measure the height of the bubble column, and record your results in Table 6.4.

Tube 3
1. Fill to the second mark with water adjusted to pH 11 by the addition of NaOH. Wait one minute.
2. Fill to the third mark with hydrogen peroxide.
3. Swirl to mix, and wait 20 seconds.
4. Measure the height of the bubble column, and record your results in Table 6.4.

Table 6.4	Effect of pH		
Tube	pH	Bubble Column Height (mm)	Explanation
1	3		
2	7		
3	11		

Plot your results from Table 6.4 here (Fig. 6.4).

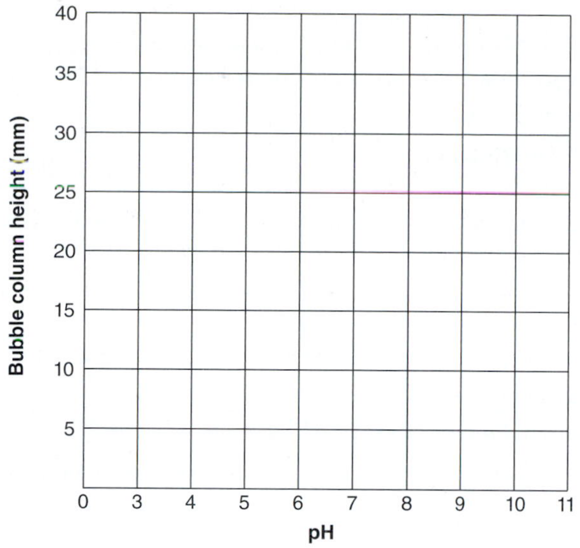

Figure 6.4 Effect of pH on enzyme activity.

Conclusions: Effect of pH

- The amount of bubble column height indicates the degree of enzyme activity. Was your hypothesis supported? _____ Explain in Table 6.4 the degree of enzyme activity per tube.
- What is your conclusion concerning the effect of pH on enzyme activity? _____

6.5 Factors That Affect Enzyme Activity

In Table 6.5, summarize what you have learned about factors that affect the speed of an enzymatic reaction. For example, in general, what type of temperature promotes enzyme activity, and what type inhibits enzyme activity? Answer similarly for enzyme or substrate concentration and pH.

Table 6.5 Factors That Affect Enzyme Activity		
Factors	**Promote Enzyme Activity**	**Inhibit Enzyme Activity**
Enzyme specificity		
Temperature		
Enzyme or substrate concentration		
pH		

Conclusions: Factors That Affect Enzyme Activity

- Why does enzyme specificity promote enzyme activity? _____
- Why does a warm temperature promote enzyme activity? _____
- Why does increasing enzyme concentration promote enzyme activity? _____
- Why does optimum pH promote enzyme activity? _____

Virtual Lab **Enzyme-Controlled Reactions**   A virtual laboratory called Enzyme-Controlled Reactions is available on the *Inquiry into Life* website **www.mhhe.com/maderinquiry14**. After opening this virtual lab, click on the TV screen to watch a video about enzymes.

Experiment 1 Determining an Enzyme's Optimum pH

Hypothesis

Considering that the pH of the medium affects the shape of protein and that the shape of an enzyme is necessary to its enzymatic function, formulate a general hypothesis about the relationship between pH and enzyme function.

Hypothesis: _____.

Steps of the Experiment

In this experiment, it is possible to vary both the amount of substrate and the pH. Which of these factors should you hold

constant during this experiment for determining an enzyme's optimum pH? _____ Knowing how enzymes

work, why might you decide to use the maximum concentration of substrate available? _____

1. Sequentially change the pH to pH 3, pH 5, pH 7, pH 9, and pH 11 by using the down or up arrows beneath the test tubes.

2. Click and drag 8 g of substrate (far right only) to each test tube.

Results

3. Click the computer monitor to see the number of product molecules per minute formed for each test tube. In Table 1 below, sequentially enter your results per pH when the substrate is held constant at 8 g.

Table 1 Product/min/pH at substrate concentration of 8 g				
pH 3	**pH 5**	**pH 7**	**pH 9**	**pH 11**
Product/min				

Did the enzyme perform best at a particular pH? _____ What pH? _____

Use this graph to show your results:

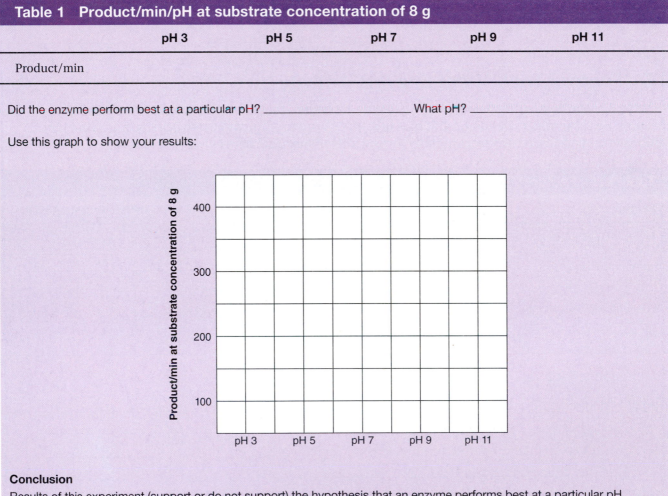

Conclusion

Results of this experiment (support or do not support) the hypothesis that an enzyme performs best at a particular pH. _____

_____ This particular pH is called the **optimum pH.**

Experiment 2 Determining the Optimum Substrate Concentration

Hypothesis

Knowing that enzymes perform best when their active site is always filled with substrate, hypothesize how increasing the substrate concentration could affect the product/min:

Hypothesis: _____

Steps of the Experiment

In this experiment what experimental factor will you hold constant? _____ Which pH would you use and

why? _____

1. Click the reset button and this time hold the pH constant. Note that all test tubes already indicate pH 7.

2. Click and drag substrate concentrations from 0.5 to 8.0 g to the test tubes.

Results

3. Click the computer monitor to see the number of product molecules formed per minute for each test tube. In Table 2 below, sequentially enter your results per amount of substrate when the pH is held constant at pH 7.

Table 2 Product/min at pH 7					
Amount of Substrate	0.5 g	1.0 g	2.0 g	4.0 g	8.0 g
Product/min					

Use this graph to show your results:

Did the amount of product/min continue to increase as substrate concentration increased to 8.0 g? Explain. _____

_____

Why do scientists favor the use of graphs to present their results? _____

_____

Conclusion

Results of this experiment (support or do not support) the hypothesis? _____

Explain your results on the basis of the active site model. _____

Laboratory Review 6

_____ 1. In the representation of a chemical reaction, what does the arrow stand for?

_____ 2. The reactants in an enzymatic reaction are called what?

_____ 3. Where on an enzyme do substrates come together?

_____ 4. If an enzyme is boiled, what happens to the enzyme?

_____ 5. If an enzyme is warmed, what happens to its activity?

_____ 6. A negative control in an experiment for the effect of a warm temperature would lack what substance?

_____ 7. If more enzyme is used, what happens to the amount of product per unit time?

_____ 8. What must be held constant when testing the effect of enzyme concentration on enzyme activity?

_____ 9. What product of the catalase reaction causes bubbling?

_____ 10. What is varied when testing the effect of pH on enzyme activity?

_____ 11. If the pH is unfavorable, what happens to enzyme activity?

_____ 12. What allowed you to measure the amount of catalase activity?

_____ 13. The reaction rate would _____ if another substance competed with catalase for the active site.

Thought Questions

14. Lipase is a digestive enzyme that digests fat droplets in the small intestine. Lipase requires a slightly basic pH, which the presence of $NaHCO_3$ provides. Indicate which of the following test tubes would show digestion following incubation, and explain why the others would not.

    Tube 1 Water, fat droplets _____

    Tube 2 Water, fat droplets, lipase _____

    Tube 3 Water, fat droplets, lipase, $NaHCO_3$ _____

    Tube 4 Water, lipase, $NaHCO_3$ _____

15. In what way does an enzyme speed its reaction? How does this explain why enzymes are specific?

16. As temperature increases, enzymatic activity increases up to a certain point beyond which enzymatic activity decreases. Explain.

17. Why can the same enzyme be used over and over again?

7

Cellular Respiration

Introduction

In this laboratory, you will study **cellular respiration** in germinating soybeans. Cellular respiration is an ATP-generating process that involves the complete breakdown most often of glucose to carbon dioxide and water. In eukaryotes, glucose breakdown begins in the cytoplasm, but is completed in mitochondria. Cellular respiration is aerobic and requires oxygen, and it results in a buildup of ATP (Fig. 7.1). This equation represents cellular respiration:

> **Planning Ahead** You may wish to start the fermentation experiment on page 92 first, to allow time for incubation.

$$C_6H_{12}O_6 + 6\,O_2 \longrightarrow 6\,CO_2 + 6\,H_2O + ATP$$

glucose   oxygen   carbon   water
                                       dioxide

Figure 7.1   Cellular respiration.
Both plant cells and animal cells have mitochondria and carry on cellular respiration, a process that utilizes the equation above.

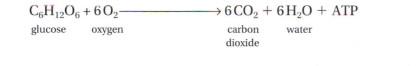

O$_2$ and glucose enter cells, which release H$_2$O and CO$_2$.

Mitochondria use energy from glucose to form ATP from ADP + P.

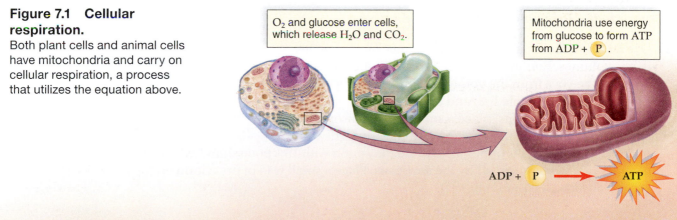

ADP + P   ⟶   ATP

In this laboratory, you will also study **ethanol fermentation,** which is an ATP-generating process. When an organism, such as yeast, breaks down glucose to ethanol and carbon dioxide, only 2 ATP result but the process is anaerobic and does not require oxygen. Fermentation occurs in the cytoplasm and mitochondria are not involved. This reaction represents yeast fermentation:

$$C_6H_{12}O_6 \longrightarrow 2\,CO_2 + 2\,C_2H_5OH + 2\,ATP$$

glucose   carbon   ethanol
                                  dioxide

When animals, such as humans, ferment, they produce lactate instead of ethanol and carbon dioxide.

7.1   Cellular Respiration

When germination occurs and plants begin to grow, cellular respiration can provide them with the ATP they need to produce all the molecules that allow them to grow. Consider, for example, Figure 7.2, which depicts soybean germination.

In the Experimental Procedure that follows, we are going to measure the amount of oxygen uptake as evidence that germinating soybeans are carrying on cellular respiration. The need for oxygen by a germinating soybean will be compared to the need by nongerminating soybeans. State a hypothesis for this experiment here:

Hypothesis: _____

Figure 7.2   Germination of a soybean seed.

Experimental Procedure: Cellular Respiration

1. Obtain a volumeter, an apparatus that measures changes in gas volumes. Remove the three vials from the volumeter. Remove the vials from the volumeter and the stoppers from the vials. Label the vials 1, 2, and 3.

2. Using the same amounts, place a small wad of absorbent cotton in the bottom of each vial. Without getting the sides of the vials wet, use a dropper to saturate the cotton with 15% potassium hydroxide (KOH). The KOH absorbs CO_2 as it is given off by the soybeans. Place a small wad of dry cotton on top of the KOH-soaked absorbent cotton (Fig. 7.3).

> ⚠ **Potassium hydroxide (KOH)** is a strong, caustic base. Exercise care in using this chemical. If any KOH spills on your skin, rinse immediately with clear water. Follow your instructor's directions for disposal of tubes that contain KOH.

Figure 7.3 Vials.
In this experiment, three vials are filled as noted.

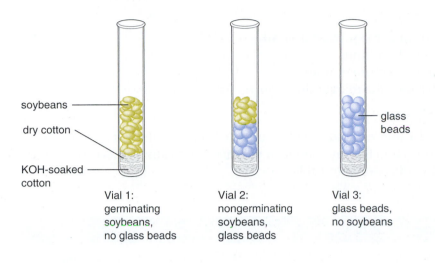

soybeans

dry cotton

KOH-soaked cotton

glass beads

Vial 1:
germinating
soybeans,
no glass beads

Vial 2:
nongerminating
soybeans,
glass beads

Vial 3:
glass beads,
no soybeans

3. Count 25 germinating soybean seeds and add to vial 1. Count 25 dry (nongerminating) soybean seeds and add to vial 2. Estimate the number of glass beads to add to vial 2 so that the volume occupied in vial 2 is approximately the same as in vial 1. Add the glass beads first and then the nongerminating soybeans. Add only glass beads to vial 3 to bring to approximately the same volume.

4. Replace the vials in the volumeter and firmly place the stoppers in the vials (Fig. 7.4). Each stopper has a vent (outlet tube) ending in rubber tubing held shut with a clamp. Remove the clamps. Adjust the graduated side arm until only about 5 mm to 1 cm protrudes through the stopper.

5. Use a Pasteur pipe to inject Brodie manometer fluid (or water colored with vegetable dye and a small amount of detergent) into each graduated side arm so that approximately 1 cm of dye is drawn into each side arm.

Figure 7.4 Volumeter containing three respirometers.
In this experiment, the respirometers are vials filled as per Figure 7.3 with graduated side arms attached. Oxygen uptake is measured by movement of a marker drop in each side arm.

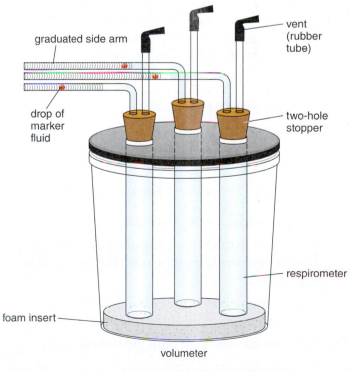

graduated side arm

drop of
marker
fluid

vent
(rubber
tube)

two-hole
stopper

respirometer

foam insert

volumeter

6. After waiting one minute for equilibrium, reattach the pinch clamp to the outlet rubber tube and mark the position of the dye with a wax pencil. The dye should be near the outlet and as O_2 is taken up, the marker drop will move toward the vials.

7. Record the following in Table 7.1 for each vial:
   - the initial position of the marker drop to the nearest 0.01 mL
   - the position of the marker drop after 10 minutes
   - the position of the marker drop after 10 more minutes
   - the net change of position (the distance the marker drop moved from the initial reading)

8. Did the marker drop change in vial 3 (glass beads)? _____ By how much? _____ Enter this number in the "Correction" column of Table 7.1, and use this number to correct the net change you observed in vials 1 and 2. (This is a correction for any change in volume due to atmospheric pressure changes or temperature changes.) This will complete Table 7.1.

Table 7.1 Cellular Respiration							
Vial	Contents	Initial Reading	Reading After 10 Minutes	Reading After 20 Minutes	Net Change	Correction	(Corrected) Net Change
1	Germinating soybeans						
2	Dry (nongerminating) soybeans						
3	Glass beads						

Conclusions: Cellular Respiration

- Do your results support or fail to support the hypothesis? _____
  Explain. _____
- Why was it necessary to absorb the carbon dioxide? _____
  _____
- Explain which respirometer in this experiment was the negative control. _____

7.2 Fermentation

When you study the ability of yeast to ferment sugar, you will need to use a respirometer to measure the amount of CO_2 given off. This experimental procedure shows you how to prepare your **respirometer.**

Experimental Procedure: Respirometer Practice

1. Completely fill a small tube (15 × 125 mm) with *water only* (Fig. 7.5).

2. Invert a large tube (20 × 150 mm) over the small tube, and with your finger or a pencil, push the small tube up into the large tube until the upper lip of the small tube is in contact with the bottom of the large tube.

3. Quickly invert both tubes. Do not permit the small tube to slip away from the bottom of the large tube. A little water will leak out of the small tube and be replaced by an air bubble.

4. Practice this inversion until the bubble in the small tube is as small as you can make it.

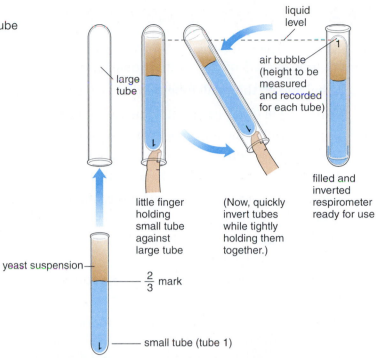

Figure 7.5 Respirometer for fermentation.
Place a small tube inside a large tube. Hold the small tube in place as you rotate the entire apparatus, and an air bubble will form in the small tube.

Ethanol Fermentation

Yeast fermentation to produce ethanol (C_2H_5OH) and carbon dioxide (CO_2) has long been utilized by humans to produce wine and bread. During the production of wine, it is the ethanol that is desired, while bread and other baked goods rise when yeast gives off carbon dioxide (Fig. 7.6). Recently, there has been a great deal of interest in using ethanol produced by yeast fermentation of corn as a substitute for gasoline in cars.

Testing Sugars

In the Experimental Procedure that follows, we are going to test which of several sugars is a better food source for yeast when they ferment. It will be assumed that the ease of sugar fermentation correlates with the amount of carbon dioxide given off within a certain time limit. As background data, observe the structure of the three sugars involved:

Figure 7.6 Products of fermentation.

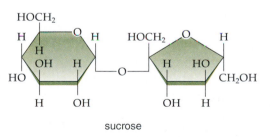

State a hypothesis for your experiment in which you sequence the sugars according to how easily you expect yeast to ferment them:

Hypothesis: _____

Have four large test tubes ready. With a wax pencil, label and mark off a small test tube at the 2/3-full level. Use this tube to mark off three other small tubes at the same level.

1. Label and fill the small tubes as directed, and record the sugar content in Table 7.2.

   Tube 1   Fill to the mark with glucose solution.

   Tube 2   Fill to the mark with fructose solution.

   Tube 3   Fill to the mark with sucrose solution.

   Tube 4   Fill to the mark with distilled water.

2. Resuspend a yeast solution each time, and fill all four tubes to the top with yeast suspension (Fig. 7.5).

3. Slide the large tubes over the small tubes, and invert them in the way you practiced. This will mix the yeast and sugar solutions.

4. Place the respirometers in a tube rack, and measure the initial height of the air space in the rounded bottom of the small tube. Record the height in Table 7.2.

5. Place the respirometers in an incubator or in a warm water bath maintained at 37°C. Note the time, and allow the respirometers to incubate about 20 minutes (incubator) or one hour (water bath). However, watch your respirometers and if they appear to be filling with gas quite rapidly, stop the incubation when appropriate.

6. At the end of the incubation period, measure the final height of the gas bubbles, and record it in Table 7.2. Calculate the net change, and record it in Table 7.2.

Table 7.2   Fermentation by Yeast

Tube	Sugar	Initial Gas Height	Final Gas Height	Net Change	Ease of Fermentation
1					
2					
3					
4					_____

Conclusions: Yeast Fermentation

- From your results, evaluate how the sugars tested compare as an energy source for yeast fermentation. Enter your evaluation in Table 7.2.

- Do your data support or fail to support your hypothesis? _____

  Explain. _____

  _____

- Can your results be correlated with the comparative structure of the sugars? (See page 91.) _____

  Explain. _____

- Explain which respirometer was a negative control. _____

Laboratory Review 7

_____ 1. Both cellular respiration and fermentation begin with what molecule?

_____ 2. What reactant needed for cellular respiration is absent from the fermentation reaction?

_____ 3. What gas do organisms give off when they carry out cellular respiration?

_____ 4. Both fermentation and cellular respiration provide what molecule needed by cells as a source of chemical energy?

_____ 5. Do plant cells or animal cells carry on cellular respiration?

_____ 6. Which process, fermentation or cellular respiration, results in an end product that contains C—H bonds?

_____ 7. Name the device that can measure the amount of gas given off by yeast.

_____ 8. Yeast cells carry out fermentation when they are supplied with what type of molecule?

_____ 9. During the fermentation experiment, the gas bubble got larger. What gas was being produced?

_____ 10. What role was played by KOH in the soybean experiment?

_____ 11. In the cellular respiration experiment, what do you call the tube that contains nongerminating soybeans?

_____ 12. What gas is being taken up when the marker in the side arm of a respirometer moves toward a tube that contains germinating seeds?

Thought Questions

13. Why is it reasonable to hypothesize that glucose would be responsible for the most activity during the fermentation experiment?

14. If you performed the cellular respiration experiment without soaking the cotton with KOH, how would your results change? Why?

15. What role was played by the tube that contained only glass beads in this experiment?

16. If you repeated this experiment using peas instead of soybeans, would you expect approximately the same results? Explain.

Photosynthesis

Introduction

Photosynthesis involves the use of solar energy to produce a carbohydrate:

$$6\ CO_2 + 6\ H_2O \xrightarrow{\text{solar energy}} (C_6H_{12}O_6) + 6\ O_2$$

In this equation, glucose ($C_6H_{12}O_6$) appears as an end product of photosynthesis.

Photosynthesis takes place in chloroplasts (Fig. 8.1). Here membranous thylakoids are stacked in grana surrounded by the stroma.

During the light reactions, pigments within the thylakoid membranes absorb solar energy, water is split, and oxygen is released.

The Calvin cycle reactions occur within the stroma. During these reactions, carbon dioxide (CO_2) is reduced and solar energy is now stored in a carbohydrate (CH_2O).

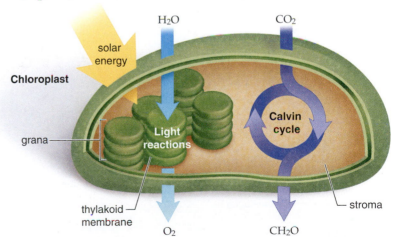

Figure 8.1   Overview of photosynthesis.
Photosynthesis includes the light reactions when energy is collected and O_2 is released and the Calvin cycle reactions when CO_2 is reduced and carbohydrate (CH_2O) is formed.

8.1 Photosynthetic Pigments

Later in this laboratory, we will learn that visible (white) light is composed of different colors of light (see Fig. 8.3). We can hypothesize that leaves contain various pigments that absorb the solar energy of these different colors. Restate this hypothesis here:

Hypothesis: _____

To test this hypothesis, we will use a technique called chromatography to separate the pigments located in leaves. Chromatography separates molecules from each other on the basis of their solubility in particular solvents. The solvents used in the following Experimental Procedure are petroleum ether and acetone, which have no charged groups and are therefore nonpolar. As a nonpolar solvent moves up the chromatography paper, the pigment moves along with it. The more nonpolar a pigment, the more soluble it is in a nonpolar solvent and the faster and farther it proceeds up the chromatography paper.

Experimental Procedure: Photosynthetic Pigments

1. Assemble a **chromatography apparatus** (Fig. 8.2*a*):
   - Obtain a large, dry test tube and a cork with a hook.
   - Attach a strip of precut **chromatography paper** (hold from the top) to the hook and test for fit. The paper should hang straight down and barely touch the bottom of the test tube; trim if necessary.
   - Measure 2 cm from the bottom of the paper; place here a small dot with a pencil (not a pen).
   - With the stopper in place, mark the test tube with a wax pencil 1 cm below where the dot is on the paper.
   - Set the chromatography apparatus in a test tube rack.

Figure 8.2   Paper chromatography.
a. For a chromatography apparatus, the paper must be cut to size and arranged to hang down without touching the sides of a dry tube. **b.** The pigment (chlorophyll) solution is applied to a designated spot. **c.** The chromatogram, which develops after the spotted paper is suspended in the chromatography solution, will show these pigments.

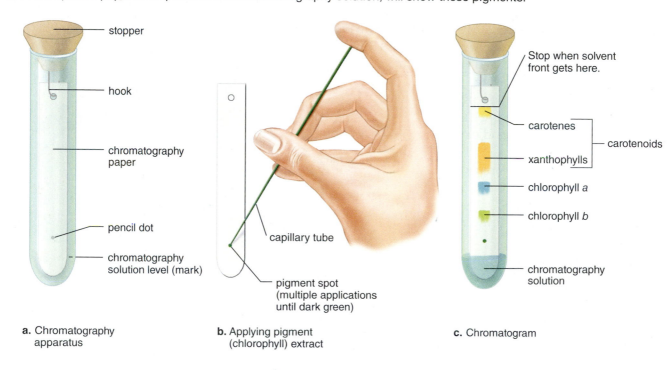

a. Chromatography apparatus

b. Applying pigment (chlorophyll) extract

c. Chromatogram

2. Prepare the chromatography paper (Fig. 8.2b):
   • Remove the chromatography paper from the test tube; place on a paper towel and apply **plant pigments** to the dot on the paper as directed by your instructor. Figure 8.2b shows how to apply pigment extract using a capillary tube.
   • In the **fume hood**, add **chromatography solution** up to the mark you made on the test tube. Place the chromatography paper attached to the hook back in the test tube. The pigment spot should remain above the chromatography solution, but close the chromatography apparatus tightly.

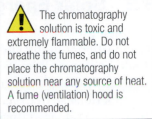

The chromatography solution is toxic and extremely flammable. Do not breathe the fumes, and do not place the chromatography solution near any source of heat. A fume (ventilation) hood is recommended.

3. Develop the chromatogram (Fig. 8.2c):
   • Place the reassembled chromatography apparatus in the test tube rack and allow 10 minutes for the chromatogram to develop, but check frequently so that the solution does not reach the top of the paper.
   • When the solvent has moved to within 1 cm of the upper edge of the paper, remove the paper. Close the empty apparatus tightly. With a pencil, lightly mark the location of the solvent front (where the solvent stopped on the paper) and allow the chromatogram to dry in the fume hood.

4. Read the chromatogram:
   • Compare your chromatogram to that shown in Figure 8.2c. Measure the distance in mm from the dark green pigment spot to the top of each individual pigment band, and record these values in Table 8.1.

   • Measure the distance the solvent moved from the dark green pigment spot to the solvent front and add this value to Table 8.1.

   • Use this formula to calculate the R_f (ratio-factor) values for each pigment, and record these values in Table 8.1:

   $$R_f = \frac{\text{distance moved by pigment}}{\text{distance moved by solvent}}$$

Table 8.1	R_f (Ratio-Factor) Values for Each Pigment	
Pigments	**Distance Moved (mm)**	**R_f Values**
Carotenes		
Xanthophylls		
Chlorophyll *a*		
Chlorophyll *b*		
Solvent		————

Conclusions: Photosynthetic Pigments

• Do your results support the hypothesis that plant leaves contain various photosynthetic pigments? _____ Explain. _____

_____

8.2 Solar Energy

During light reactions of photosynthesis, solar energy is absorbed by the photosynthetic pigments and is transformed into the chemical energy of a carbohydrate (CH_2O). Without solar energy, photosynthesis would be impossible.

Verify that photosynthesis releases oxygen by reviewing the overall equation for photosynthesis on page 95. Release of oxygen from a plant indicates that the light reactions of photosynthesis are occurring. The oxygen released during photosynthesis is taken up by a plant when cellular respiration occurs. This must be taken into account when the rate of photosynthesis is calculated.

Role of White Light

White (sun) light contains different colors of light, as is demonstrated when white light passes through a prism (Fig. 8.3). White light is the best for photosynthesis because it contains all the colors of light.

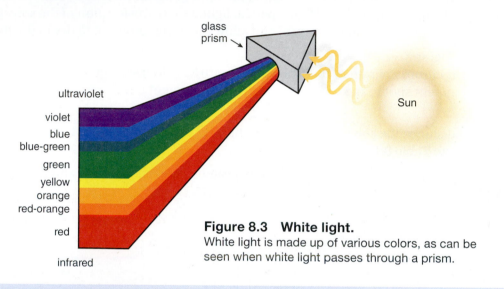

Figure 8.3   White light.
White light is made up of various colors, as can be seen when white light passes through a prism.

Experimental Procedure: White Light

1. Place a generous quantity of duckweed or *Elodea* with the cut end up (make sure the cuts are fresh) in a test tube with a rubber stopper containing a piece of glass tubing, as illustrated in Figure 8.4. When assembled, this is your volumeter for studying the need for light in photosynthesis. (Do not hold the volumeter in your hand, as body heat will also drive the reaction forward.) Your instructor will show you how to fix the volumeter in an upright position.
2. Before stoppering the test tube, add sufficient 3% sodium bicarbonate ($NaHCO_3$) solution so that, when the rubber stopper is inserted into the tube, the solution comes to rest at about ¼ the length of the upright glass tubing. Mark this location on the glass tubing with a wax pencil.
3. Place a beaker of plain water next to the *Elodea* tube to serve as a heat absorber. Place a lamp (150 watt) next to the beaker. The tube, beaker, and lamp should be as close to one another as possible.
4. Turn on the lamp. As soon as the edge of the solution in the tubing begins to move, time the reaction for 10 minutes. Be careful not to bump the tubing or to readjust the stopper, or your readings will be altered. After 10 minutes, mark the edge of the solution, and measure in millimeters the distance the edge moved upward: _____ mm/10 min. This is **net photosynthesis,** a measurement that does not take into account the oxygen that was used up for cellular respiration. Record your results in Table 8.2. Why did the edge move upward? _____

_____

Figure 8.4   Volumeter.
A volumeter apparatus is used
to study the role of light in
photosynthesis.

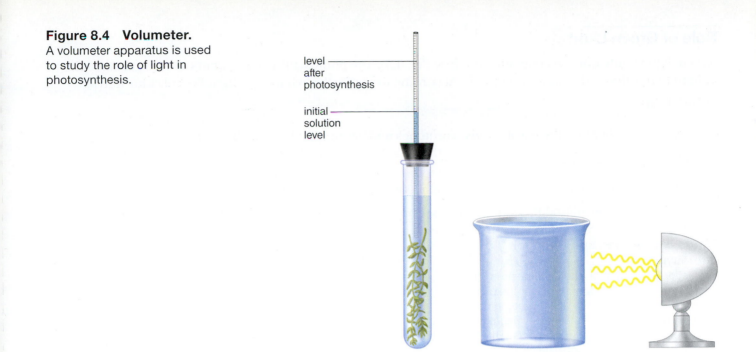

level
after
photosynthesis

initial
solution
level

5. Carefully wrap the tube containing *Elodea* in aluminum foil, and record here the length of time it
takes for the edge of the solution in the tubing to move downward 1 mm: _____. Convert your
measurement to _____ mm/10 min, and record this value for **cellular respiration** in Table 8.2.
(Do not use a minus sign, even though the edge moved downward.) Why does cellular respiration,
which occurs in a plant whether it is light or dark, cause the edge to move downward? _____

_____

6. If the *Elodea* had *not* been respiring in step 4, how far would the edge have moved upward?
_____ mm/10 min. This is **gross photosynthesis** (net photosynthesis + cellular respiration).
Record this number in Table 8.2.
7. Calculate the **rate of photosynthesis** (mm/hr) by multiplying gross photosynthesis (mm/10 min)
by 6 (that is, 10 min × 6 = 60 min = 1 hr): _____ mm/hr. Record this value in Table 8.2.

Table 8.2 Rate of Photosynthesis (White Light)		
	Movement of Edge (mm/10 min)	Rate of Photosynthesis (mm/hr)
Net photosynthesis (white light)		_____
Cellular respiration (no light)		_____
Gross photosynthesis (net + cellular respiration)		

Role of Green Light

Green light is only one part of white light (see Fig. 8.3). The photosynthetic pigments absorb certain colors of light better than other colors (Fig. 8.5). According to Figure 8.5, what color light do the chlorophylls absorb best? _____ Least? _____

What color light do the carotenoids (carotenes and xanthophylls) absorb best? _____ Least? _____

Hypothesize which color light is minimally utilized for photosynthesis. _____
The following Experimental Procedure will test your hypothesis.

Figure 8.5 Action spectrum for photosynthesis.

The action spectrum for photosynthesis is the sum of the absorption spectrums for the pigments chlorophyll *a*, chlorophyll *b*, and carotenoids. The peaks in this diagram represent wavelengths of sunlight absorbed by photosynthetic pigments. The chlorophylls absorb predominantly violet-blue and orange-red light and reflect green light. The carotenoids (carotenes and xanthophylls) absorb mostly blue-green light and reflect yellow-red light.

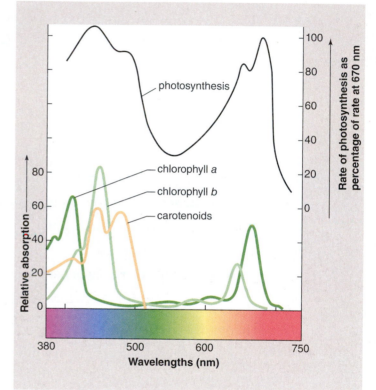

Experimental Procedure: Green Light

1. Add three drops of green dye (or use a green cellophane wrapper) to the beaker of water used in the previous Experimental Procedure until there is a distinctive green color. Remove all previous wax pencil marks from the glass tubing.

2. Record in Table 8.3 your data for gross photosynthesis (mm/10 min) when photosynthesis was using white light (mm/hr) from Table 8.2.

3. Turn on the lamp. Mark the location of the edge of the solution on the glass tubing. As soon as the edge begins to move, time the reaction for 10 minutes. After 10 minutes, mark the edge of the solution, and measure in millimeters the distance the edge moved. Net photosynthesis for green light = _____ mm/10 min.

4. Carefully wrap the tube containing *Elodea* in aluminum foil, and record here the length of time it takes for the edge of the solution in the tubing to recede 1 mm: _____. Convert your

measurement to _____ mm/10 min. As before, this reading shows how much oxygen was used for cellular respiration.

5. Calculate gross photosynthesis for green light (mm/10 min) as you did for white light, and record your data below in Table 8.3. (As before, add your data for net photosynthesis to that for cellular respiration.)

6. Calculate rate of photosynthesis for green light (mm/hr) as you did for white light, and record your data below in Table 8.3. (As before, convert your data for mm/10 min to mm/hr.)

7. Collect the white and green light rates of photosynthesis (mm/hr) from each group in your lab; then average all the rates including your own. In the last column of Table 8.3 record the average for white and green light rates of photosynthesis (mm/hr).

Table 8.3 Rate of Photosynthesis (White and Green Light)		
Gross photosynthesis	**Your Data**	**Class Data**
White (from Table 8.2)	mm/10 min	_____
Green	mm/10 min	_____
Rate of photosynthesis		
White (from Table 8.2)	mm/hr	mm/hr
Green	mm/hr	mm/hr

8. Calculate the rate of photosynthesis (green light) as a percentage of the rate of photosynthesis (white light) using this equation.

$$\text{Percentage} = \frac{\text{rate of photosynthesis (green light)}}{\text{rate of photosynthesis (white light)}} \times 100$$

Percentage based on your own data recorded in Table 8.3 = _____.

Percentage based on class data recorded in Table 8.3 = _____.

Conclusions: Rate of Photosynthesis

- Do your results support the hypothesis that green light is minimally used by a land plant for photosynthesis? _____ Explain, with reference to Figure 8.5. _____

_____

_____

- How does the percentage based on your data differ from that based on class data?

_____

Explain any difference in the percentage. _____

_____

8.3 Carbon Dioxide Uptake

During the Calvin cycle reactions of photosynthesis, the plant takes up carbon dioxide (CO_2) and reduces it to a carbohydrate, such as glucose ($C_6H_{12}O_6$). Therefore, the carbon dioxide in the solution surrounding *Elodea* should disappear as photosynthesis takes place.

Experimental Procedure: Carbon Dioxide Uptake

1. Temporarily remove the *Elodea* from the test tube. Empty the sodium bicarbonate ($NaHCO_3$) solution from the test tube, rinse the test tube thoroughly, and fill with a phenol red solution diluted to a faint pink. (Add more water if the solution is too dark.) Phenol red is a pH indicator that turns yellow in an acid and red in a base.

> ⚠️ **Phenol red** Avoid ingestion, inhalation, and contact with skin, eyes, and mucous membranes. Follow your instructor's directions for disposal of this chemical. Use protective eyewear when performing this experiment.

2. Blow *lightly* on the surface of the solution. Stop blowing as soon as the surface color changes to yellow. Then shake the test tube until the rest of the solution turns yellow.

   Blowing onto the solution adds what gas to the test tube? _____ When carbon dioxide combines with water, it forms carbonic acid; therefore, the solution appears yellow.

3. Thoroughly rinse the *Elodea* with distilled water, return it to the test tube, which now contains a yellow solution, and assemble your volumeter as before.

4. The water in the beaker used to absorb heat should be clear.

5. Turn on the lamp, and wait until the edge of the solution just begins to move upward. Note the time.

   Observe until you note a change in color. How long did it take for the color to change? _____

6. Hypothesize why the solution in the test tube eventually turned red. _____

   _____

7. The carbon cycle includes all the many ways that organisms exchange carbon dioxide with the atmosphere. Figure 8.6 notes the relationship between cellular respiration and photosynthesis. Animals produce carbon dioxide used by plants to carry out photosynthesis. Plants produce the food (and oxygen) that they and animals require to carry out cellular respiration. Therefore, the same carbon atoms pass between animals and plants and between plants and animals (Fig. 8.6).

> 🧪 **Virtual Lab** **Energy in a Cell** A virtual lab called Energy in a Cell is available on the *Inquiry into Life* website **www.mhhe.com/ maderinquiry14**. It will help you test your knowledge of photosynthesis and cellular respiration. Follow the directions given but be aware that in this virtual lab the Celvin cycle reactions are called dark reactions.

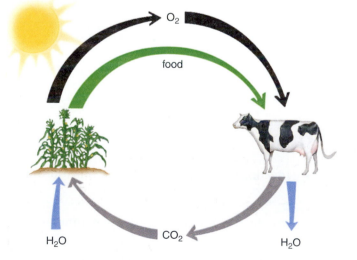

Figure 8.6 Photosynthesis and cellular respiration.
Animals are dependent on plants for a supply of oxygen, and plants are dependent on animals for a supply of carbon dioxide.

8.4   The Light Reactions and the Calvin Cycle Reactions

Review the introduction to this laboratory. Note the overall equation for photosynthesis and that photosynthesis consists of the light reactions and the Calvin cycle reactions. Solar energy is absorbed by photosynthetic pigments during the light reactions and energy is used during the Calvin cycle reactions to reduce carbon dioxide to a carbohydrate.

Light Reactions

1. Photosynthetic pigments

   a. What is the function of the photosynthetic pigments in photosynthesis? _____

   b. How does it benefit a plant to have a variety of photosynthetic pigments? (See Fig. 8.5.) _____

   _____

   c. Green light minimally promotes photosynthesis. Account for this observation with reference to

   the photosynthetic pigments. (See Fig. 8.5.) _____

   d. Does this explain why leaves are green? _____ How so? _____

2. Water and oxygen

   a. What happens to water during the light reactions? _____

   b. What happens to the released oxygen? _____

3. Location of the light reactions

   Fill in the blank: The light reactions take place in the _____ membranes.

4. In your own words, summarize the light reactions based on this laboratory.

Calvin Cycle Reactions

1. Carbon dioxide and carbohydrate

   What happens to carbon dioxide after it is taken up during the Calvin cycle reactions? _____

2. Location of the Calvin cycle reaction

   Fill in the blank: The Calvin cycle reactions take place in the _____.

3. In your own words, summarize the Calvin cycle reactions based on this laboratory.

Light Reactions and the Calvin Cycle Reactions

1. Examine the overall equation for photosynthesis and show that there is a relationship between the light reactions and the Calvin cycle reactions by drawing an arrow between the hydrogen atoms in water and the hydrogen atoms in the carbohydrate.

$$CO_2 + H_2O \longrightarrow (CH_2O) + O_2$$

2. Only because solar energy splits water can hydrogen atoms be used to reduce carbon dioxide. In this sense, solar energy is now stored in the carbohydrate. This energy sustains all the organisms in the biosphere.

_____ 1. Where in a chloroplast do the light reactions of photosynthesis take place?

_____ 2. What procedure did you use to separate plant pigments?

_____ 3. What determines the speed with which a pigment moves up the chromatography paper?

_____ 4. Where do plants ordinarily get the energy they need to carry on photosynthesis?

_____ 5. Green plants do not absorb what color of light?

_____ 6. Blue, red, and green light are all present in what color of light?

_____ 7. Carotenes absorb what color light?

_____ 8. The gas released by plants is used by organisms to do what?

_____ 9. If net photosynthesis is 5 mm/10 min and cellular respiration is .5 mm/10 min, how much is gross photosynthesis?

_____ 10. Phenol red turns what color when carbon dioxide is added?

_____ 11. Chemically speaking, what happens to carbon dioxide during photosynthesis?

_____ 12. What two substances do plants provide for us?

_____ 13. Where, within the chloroplast, does the Calvin cycle take place?

Thought Questions

14. Some plants are colorless. Do you predict that they carry on photosynthesis? Explain.

15. Suppose there were single-celled protozoans (nonphotosynthetic) in the test tube with *Elodea* when you did the white light experiment. Could you still calculate *Elodea's* rate of photosynthesis? Explain.

16. The light reactions of photosynthesis do not directly produce carbohydrates. Why, then, are these reactions required for the photosynthetic process?

17. *Elodea* carries on cellular respiration in the dark and in the light. Explain.

9

Organization of Flowering Plants

Introduction

Despite their great diversity in size and shape, all flowering plants have three vegetative organs that have nothing to do with reproduction: the root, the stem, and the leaf (Fig. 9.1). Roots anchor a plant and absorb water and minerals from the soil. A stem transports substances and supports the leaves so that they are exposed to sunlight. Leaves carry on photosynthesis, and thereby produce the nutrients that sustain a plant and allow it to grow.

Figure 9.1   Organization of plant body.
Roots, stems, and leaves—the vegetative organs of a plant—are shown in this photo of an onion plant.

9.1 External Anatomy of a Flowering Plant

Figure 9.2 shows that a plant has a root system and a shoot system. The **root system** consists of the roots. The **shoot system** consists of the stem and the leaves.

Observation: A Living Plant

Shoot System

What structures are in the shoot system?

_____

_____

_____

The Leaves

Leaves carry on photosynthesis. Which part of a leaf (blade or petiole) is the most expansive part?

_____

_____

The Stem

1. In the **stem**, locate a **node** and an **internode**.
2. Measure the length of the internode in the middle of the stem. Does the internode get larger or smaller toward the apex of the

   stem? _____ Toward the roots? _____
   Based on the fact that a stem elongates as it grows, explain your observation. _____

   _____

   _____

   _____

3. Where is the **terminal bud** (also called the

   shoot tip) of a stem?_____

   _____

   Where are axillary buds? _____

   _____

Root System

Observe the root system of a living plant if the root system is exposed. Does this plant have a strong primary root or many roots of the same size? _____
What structures are in the root system?

_____

_____

_____

Where is the root tip? _____

Figure 9.2 Organization of a plant.
Roots, stems, and leaves are vegetative organs. The flower and fruit are reproductive structures studied in Laboratory 10.

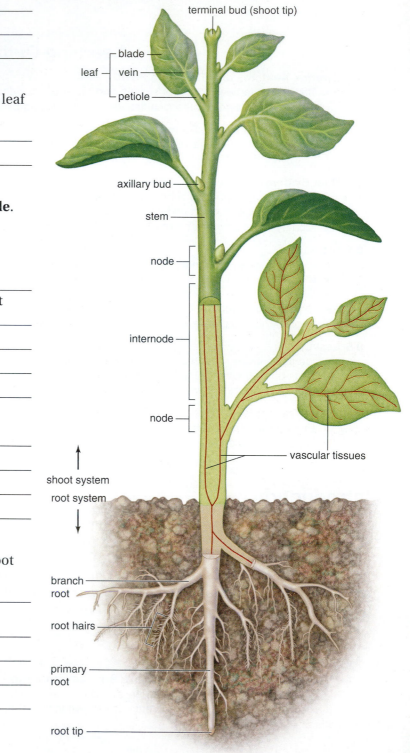

terminal bud (shoot tip)

blade
leaf — vein
petiole

axillary bud

stem

node

internode

node

vascular tissues

shoot system
root system

branch root

root hairs

primary root

root tip

9.2　Major Tissues of Roots, Stems, and Leaves

Unlike humans, flowering plants grow in size their entire life because they have an immature tissue called **meristematic tissue** composed of cells that divide. **Apical meristem** is located at the terminal end of the stem, the branches, and at the root tip and the root branches. When apical meristem cells divide, some of the cells differentiate into the mature tissues of a plant:

Dermal tissue: Forms the outer protective covering of a plant organ

Ground tissue: Fills the interior of a plant organ; photosynthesizes and stores the products of photosynthesis

Vascular tissue: Transports water and sugar, the product of photosynthesis, in a plant and provides support

Note in Table 9.1 that roots, stems, and leaves have all three tissues but they may be given different specific names.

Table 9.1　Mature Tissues of Vegetative Organs

Tissue Type	Roots	Stems	Leaves
1. Dermal tissue (epidermis)	Protect inner tissues Root hairs absorb water and minerals.	Protect inner tissues	Protect inner tissues Cuticle prevents H_2O loss. Stomata carry on gas exchange.
2. Ground tissue	Cortex: Store products of photosynthesis Pith: Store products of photosynthesis	Cortex: Carry on photosynthesis, if green Pith: Store products of photosynthesis	Mesophyll: Photosynthesis
3. Vascular tissue (xylem and phloem)	Vascular cylinder: Transports water and nutrients	Vascular bundle: Transports water and nutrients	Leaf vein: Transports water and nutrients

Observation: Tissues of Roots, Stems, and Leaves

Meristematic tissue. View a slide showing the apical meristem in a shoot tip and another showing the apical meristem in a root tip are on demonstration. Meristematic cells are spherical and stain well when they are dense and have thin cell walls (Fig. 9.3). What is the function of meristematic tissue?

_____

Figure 9.3 Apical meristem.
A shoot tip and a root tip contain meristem tissue, which allows them to grow longer the entire life of a plant.

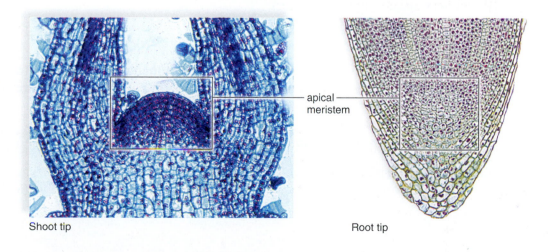

Shoot tip　　　　　　　　　　　　　Root tip

Mature Tissues

1. **Dermal tissue.** In a cross section of a leaf (Fig. 9.4), for example, focus only on the upper or lower epidermis (Fig. 9.4). Epidermal cells tend to be square or rectangular in shape. In a leaf, the epidermis is interrupted by openings called **stomata** (sing., stoma). Later you will have an opportunity to see the epidermis in roots, stems, and leaves. What is a function of epidermis in all three organs (see Table 9.1)?

_____

_____

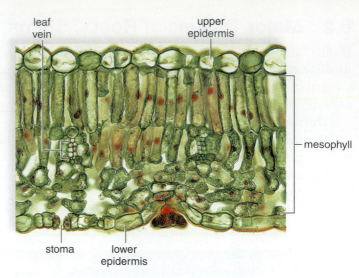

Figure 9.4   Microscopic leaf structure.
Like the stem and root, a leaf contains epidermal tissue, vascular tissue (leaf vein), and ground tissue (mesophyll).

2. **Ground tissue.** The ground tissue fills the space between epidermis in roots, stems, and leaves. In leaves, for example, ground tissue is called mesophyll (see Fig. 9.4). Ground tissue largely contains parenchyma cells and sclerenchyma cells. **Parenchyma cells** can be of different sizes and vary from fairly circular to oval. Those that contain chloroplasts carry on photosynthesis. Those that contain leucoplasts store starches and oils. **Sclerenchyma cells** are usually elongate in shape and have thick walls impregnated with lignin. These dead cells appear hollow and the presence of lignin means that they stain a red color. Sclerenchyma cells are strong and provide support. Which type cell (parenchyma or sclerenchyma) carries on photosynthesis or stores the

products of photosynthesis in a leaf? _____ Which one lends strength to

ground tissue in roots and stems? _____

3. **Vascular tissue.** In a leaf, strands of vascular tissue are called leaf veins (see Fig. 9.4). There are two types of vascular tissue, called xylem and phloem. **Xylem** contains hollow dead cells that transport water. The presence of lignin makes the cell walls strong, stains red, and makes xylem easy to spot. **Phloem** contains thin-walled, smaller living cells that transport sugars in a plant. Phloem is harder to locate than xylem, but it is always found in association with xylem. Which type tissue (xylem or

phloem) transports sugars in a plant? _____

Which type tissue transports water? _____

Monocots Versus Eudicots

Flowering plants are classified into two major groups: **monocots** and **eudicots.** In this laboratory, you will be studying the differences between monocots and eudicots with regard to the roots, stems, and leaves as noted in Figure 9.5.

Experimental Procedure: Monocot Versus Eudicot

1. Examine the live plant again (see Fig. 9.2). The leaf vein pattern—that is, whether the veins run parallel to one another or whether the veins spread out from a central location (called the net

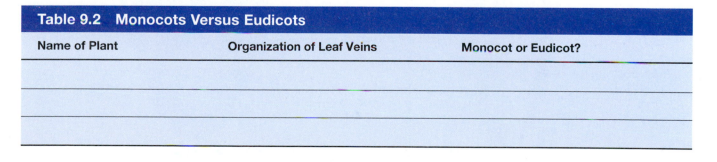

	Seed	Root	Stem	Leaf	Flower
Monocots	One cotyledon in seed	Root xylem and phloem in a ring	Vascular bundles scattered in stem	Leaf veins form a parallel pattern	Flower parts in threes and multiples of three
Eudicots	Two cotyledons in seed	Root phloem between arms of xylem	Vascular bundles in a distinct ring	Leaf veins form a net pattern	Flower parts in fours or fives and their multiples

Figure 9.5   Monocots versus eudicots.
The five features illustrated here are used to distinguish monocots from eudicots.

pattern)—indicates that a plant is either a monocot or a eudicot. Is the plant in Figure 9.2 a monocot or eudicot? _____ The leaf pattern in Figure 9.4 appears to be parallel. Is this the leaf of a monocot or eudicot? _____

2. Observe any other available plants, and note in Table 9.2 the name of the plant and whether it is a monocot or a eudicot based on leaf vein pattern.

3. Aside from leaf vein pattern, other external features indicate whether a plant is a monocot or a eudicot. For example, open a peanut if available. The two halves you see are cotyledons. Is the plant that produced the peanut a monocot or eudicot? _____

If available, examine a flower. If a flower has three petals or six petals, or any multiple of three, is the plant that produced this flower a monocot or a eudicot? _____

In this laboratory you will have the opportunity to examine the cross sections of roots and stems microscopically; the arrangement of vascular tissue roots and stems also indicates whether a plant is a monocot or a eudicot.

Table 9.2 Monocots Versus Eudicots		
Name of Plant	**Organization of Leaf Veins**	**Monocot or Eudicot?**

9.3　Root System

The **root system** anchors the plant in the soil, absorbs water and minerals from the soil, and stores the products of photosynthesis received from the leaves.

Anatomy of a Root Tip

Primary growth of a plant increases its length. Note the location of the root apical meristem in Figure 9.6. As primary growth occurs, root cells enter zones that correspond to various stages of differentiation and specialization.

Figure 9.6　Eudicot root tip.
a. In longitudinal section, the root cap is followed by the zone of cell division, zone of elongation, and zone of maturation. **b.** Micrograph.

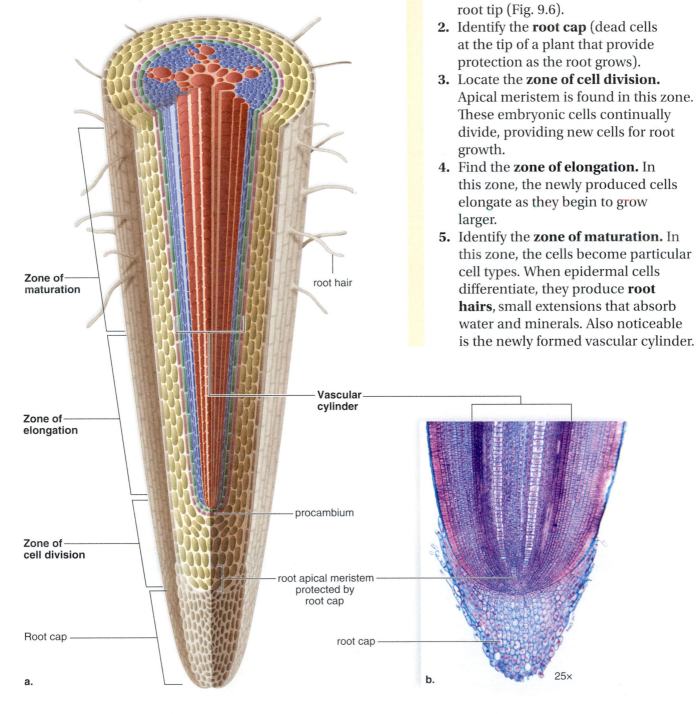

Zone of maturation

Zone of elongation

Zone of cell division

Root cap

root hair

Vascular cylinder

procambium

root apical meristem protected by root cap

root cap

a.

b.

25×

1. Examine a model and/or a slide of a root tip (Fig. 9.6).
2. Identify the **root cap** (dead cells at the tip of a plant that provide protection as the root grows).
3. Locate the **zone of cell division.** Apical meristem is found in this zone. These embryonic cells continually divide, providing new cells for root growth.
4. Find the **zone of elongation.** In this zone, the newly produced cells elongate as they begin to grow larger.
5. Identify the **zone of maturation.** In this zone, the cells become particular cell types. When epidermal cells differentiate, they produce **root hairs**, small extensions that absorb water and minerals. Also noticeable is the newly formed vascular cylinder.

Anatomy of Eudicot and Monocot Roots

Eudicot and monocot roots differ in the arrangement of their vascular tissue.

Observation: Cross-Section Anatomy of Eudicot and Monocot Roots

Eudicot Root

1. Obtain a prepared cross-section slide of a buttercup *(Ranunculus)* root. Use both low power and high power to identify the **epidermis** (the outermost layer of small cells that gives rise to root hairs). The epidermis protects inner tissues and absorbs water and minerals.
2. Locate the **cortex,** which consists of several layers of thin-walled cells (Fig. 9.7*a, b*). In Figure 9.7*b,* note the many stained starch grains in the cortex cells. The cortex is ground tissue that functions in food storage.
3. Find the **endodermis,** a single layer of cells whose walls are thickened by a layer of waxy material known as the **Casparian strip.** (It is as though these cells are glued together with a waxy glue.) Because of the Casparian strip, the only access to the xylem is through the living endodermal cells. The endodermis regulates what materials that enter a plant through the root? _____

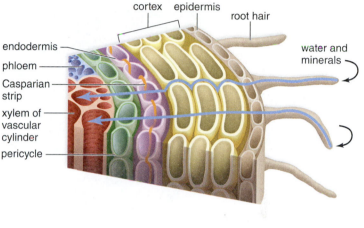

_____

Use this illustration to trace the path of water and minerals from the root hairs to xylem. _____

4. Identify the **pericycle,** a layer one or two cells thick just inside the endodermis. Branch roots originate from this tissue.
5. Locate the **xylem** in the vascular cylinder of the root. Xylem has several "arms" that extend like the spokes of a wheel. This tissue conducts water and minerals from the roots to the stem.
6. Find the **phloem,** located between the arms of the xylem. Phloem conducts organic nutrients from the leaves to the roots and other parts of the plant.

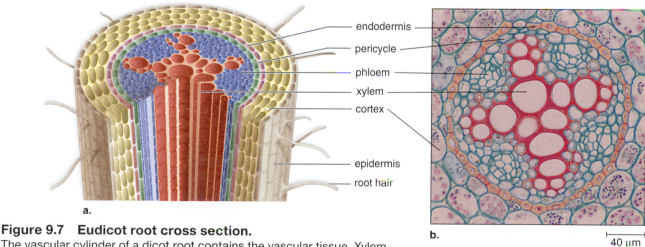

Figure 9.7   Eudicot root cross section.
The vascular cylinder of a dicot root contains the vascular tissue. Xylem is typically star-shaped, and phloem lies between the points of the star.
a. Drawing. **b.** Micrograph.

Monocot Root

1. Obtain a prepared cross-section slide of a corn *(Zea mays)* root (Fig. 9.8a, b). Use both low power and high power to identify the six tissues mentioned for the eudicot root.
2. In addition, identify the **pith,** a centrally located ground tissue that functions in food storage.

Figure 9.8 Monocot root cross section.
a. Micrograph of a monocot root cross section. **b.** An enlarged portion.

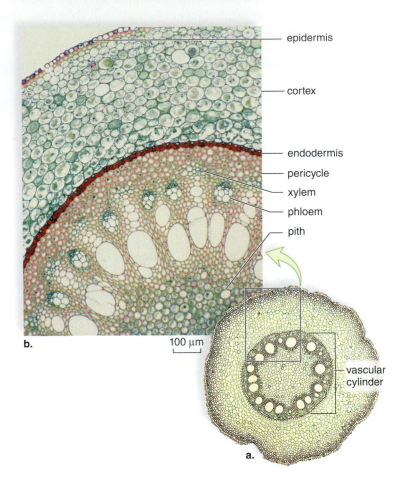

epidermis

cortex

endodermis

pericycle

xylem

phloem

pith

vascular cylinder

b.

100 µm

a.

Comparison

Contrast the arrangement of vascular tissue (xylem and phloem) in the vascular cylinder of monocot roots and eudicot roots by writing "monocot" or "eudicot" on the appropriate line.

_____ Xylem has the appearance of a wheel. Phloem is between the spokes of the wheel.

_____ Ring of xylem (inside) and phloem (outside) surrounds pith.

Root Diversity

Roots are quite diverse, and we will take this opportunity to become acquainted with only a few select types.

Observation: Root Diversity

Taproots and fibrous roots. Most plants have either a taproot or a fibrous root. Note in Figure 9.9*a* that carrots have a **taproot**. The main root is many times larger than the branch roots. Grasses such as in Figure 9.9*b* have a **fibrous root**: All the roots are approximately the same size.

Examine the taproots on display, and name one or two in which the taproot is enlarged for storage.

_____

The dandelion on display has a tap root. Describe. _____

_____

Taproots in particular function in food storage.

Adventitious roots. Some plants have **adventitious** roots. Roots that develop from nonroot tissues, such as nodes of stems, are called adventitious roots. Examples include the prop roots of corn (Fig. 9.9*c*) and the aerial roots of ivy that attach this plant to structures such as stone walls.

Which plants on display have adventitious roots? _____

Other types of roots. Mangroves and other swamp-dwelling trees have roots called pneumatophores that rise above the water line. Pneumatophores have numerous lenticels, which are openings that allow gas exchange to occur.

What root modifications not noted here are on display in your laboratory? _____

_____

Figure 9.9 Root diversity.
a. Carrots have a taproot. **b.** Grass has a fibrous root system. **c.** A corn plant has prop roots, and **d.** black mangroves have pneumatophores to increase their intake of oxygen.

a. Taproot system

b. Fibrous root system

c. Prop roots, a type of adventitious root

d. Pneumatophores of black mangrove trees

9.4 Stems

Stems are usually found aboveground where they provide support for leaves and flowers. Vascular tissue extends from the roots, through the stem and its branches to the leaves. Therefore, what function do botanists assign to stems in addition to support for branches and leaves? _____

_____ Explain why a branch cannot live if severed from the rest of the plant.

_____

Stems that do not contain wood are called **herbaceous,** or nonwoody, stems. Usually, monocots remain herbaceous throughout their lives. Some eudicots, such as those that live a single season, are also herbaceous. Other eudicots, namely trees, become woody as they mature.

Anatomy of Herbaceous Stems

Herbaceous stems undergo primary growth. **Primary growth** results in an increase in length due to the activity of the apical meristem located in the terminal bud (see Fig. 9.3*a*) of the shoot system.

Observation: Anatomy of Eudicot and Monocot Herbaceous Stems

Eudicot Herbaceous Stem

1. Examine a prepared slide of a eudicot herbaceous stem (Fig. 9.10), and identify the **epidermis** (the outer protective layer). *Label the epidermis in Figure 9.10a.*
2. Locate the **cortex,** which may photosynthesize or store nutrients.
3. Find a **vascular bundle,** which transports water and organic nutrients. The vascular bundles in a eudicot herbaceous stem occur in a ring pattern. *Label the vascular bundle in Figure 9.10a.* Which vascular tissue (xylem or phloem) is closer to the surface?_____

4. Label the central **pith,** which stores organic nutrients. Both cortex and pith are composed of which tissue type listed in Table 9.1? _____

Figure 9.10   Eudicot herbaceous stem.
The vascular bundles are in a definite ring in this photomicrograph of a eudicot herbaceous stem. Complete the labeling as directed by the Observation.

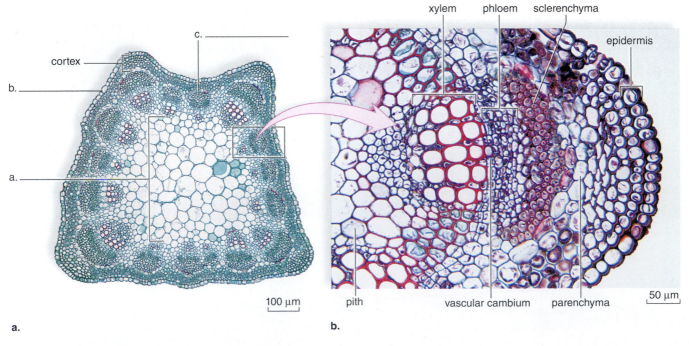

a.

b.

1. Examine a prepared slide of a monocot herbaceous stem (Fig. 9.11). Locate the epidermal, ground, and vascular tissues in the stem.

2. The vascular bundles in a monocot herbaceous stem are said to be scattered. Explain. _____

_____

Figure 9.11   Monocot stem.
The vascular bundles, one of which is enlarged, are scattered in this photomicrograph of a monocot herbaceous stem.

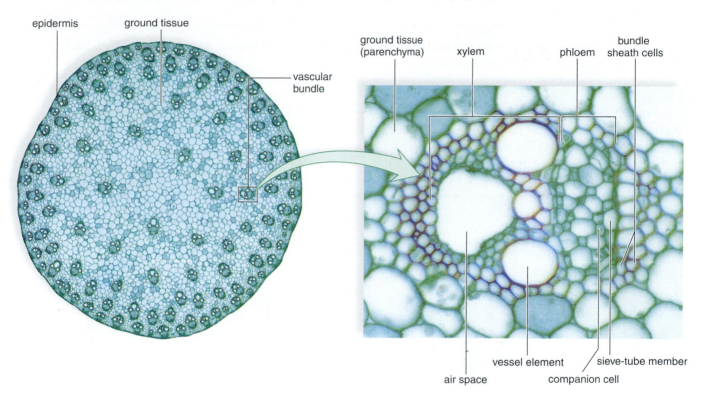

Comparison

1. Compare the arrangement of ground tissue in eudicot and monocot stems. _____

_____

2. Compare the arrangement of vascular bundles in the stems of eudicots and monocots. _____

_____

Stem Diversity

Stems can be quite diverse as well. We take this opportunity to become acquainted with those that allow a plant to accomplish vegetative reproduction and/or function in food storage. Several of these and other types of stems may be on display in the laboratory.

Stolons. The strawberry plant in Figure 9.12*a* has a horizontal aboveground stem called a runner or **stolon**. The stolon produces adventitious roots and new shoots at nodes. *Label an adventitious root and a new shoot in Figure 9.12a.*

List any other plants on display that spread and produce new shoots by sending out stolons. _____

_____

Rhizomes. An iris has a belowground horizontal stem called a **rhizome** which functions as a fleshy food storage organ (Fig. 9.12*b*). New plants can grow from a single piece of the rhizome.

List any other plants on display that have the same belowground horizontal stems (rhizomes) as

the iris. _____

Tubers. A white potato has a belowground rhizome that gives off food storage **tubers** (Fig. 9.12*c*). Each eye is a node that can produce new plants.

List any other types of plants on display whose belowground stems have tubers. _____

_____

Corms. A gladiolus has a belowground vertical stem called a **corm**, which functions in food storage and has thin, papery leaves (Fig. 9.12*d*).

List any other types of plants on display that have a vertical stem called a corm. _____

_____

Figure 9.12 Stem diversity.
a. Stolon of a strawberry plant. **b.** Rhizome of an iris. **c.** Tuber of a potato. **d.** Corm of a gladiolus.

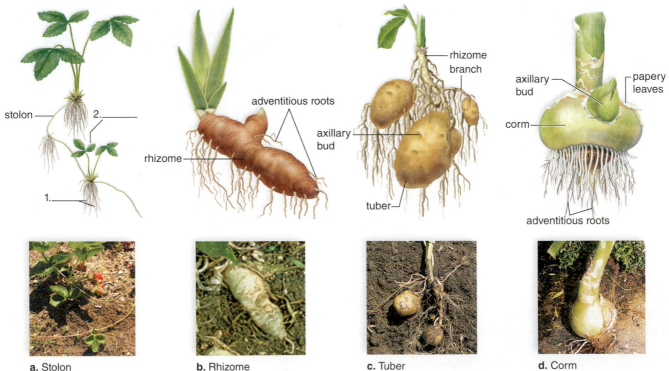

a. Stolon   **b.** Rhizome   **c.** Tuber   **d.** Corm

Anatomy of Woody Stems

Woody stems undergo both primary growth (increase in length) and secondary growth (increase in girth). When *primary growth* occurs, the apical meristem within a terminal bud is active. When *secondary growth* occurs, the vascular cambium is active. **Vascular cambium** is meristem tissue, which produces new xylem and phloem called **secondary xylem** and **phloem** each year. The buildup of secondary xylem year after year is called **wood.** Complete Table 9.3 to distinguish between primary growth and secondary growth of a stem.

Table 9.3 Primary Growth Versus Secondary Growth		
	Primary Growth	**Secondary Growth**
Active meristem		
Result		

Observation: Anatomy of a Winter Twig

1. A winter twig typically shows several past years' primary growth. Examine several examples of winter twigs (Fig. 9.13), and identify the **terminal bud** located at the tip of the twig. This is where new primary growth will originate. During the next growing season, the terminal bud produces new tissues including vascular bundles and leaves.
2. Locate a **terminal bud scar.** Marks left on stem from terminal bud scales (modified leaves protecting the bud). The distance between two adjacent terminal bud scars equals one year's primary growth.
3. Find a **leaf scar.** Mark where a leaf was attached to the stem.
4. Note the **bundle scars.** Complete this sentence: Marks left in the leaf scar where the vascular tissue

   _____ .

5. Identify a **node.** This is the region where you find leaf scars and bundle scars. The region between

   nodes is called an _____ .

6. Locate an **axillary bud.** This is where new branch growth can occur.

7. Note the numerous lenticels, breaks in the outer surface where gas exchange can occur.

Figure 9.13   External structure of a winter twig.
Counting the terminal bud scars tells the age of a particular branch.

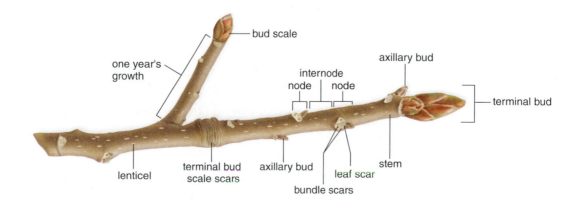

Laboratory 9   Organization of Flowering Plants   **117**

1. Examine a prepared slide of a cross section of a woody stem (Fig. 9.14), and identify the **bark** (the dark outer area), which contains **cork,** a protective outer layer; **cortex,** which stores nutrients; and **phloem,** which transports organic nutrients.

Figure 9.14   Woody eudicot stem cross section.
Because xylem builds up year after year, it is possible to count the annual rings to determine the age of a tree. This tree is three years old.

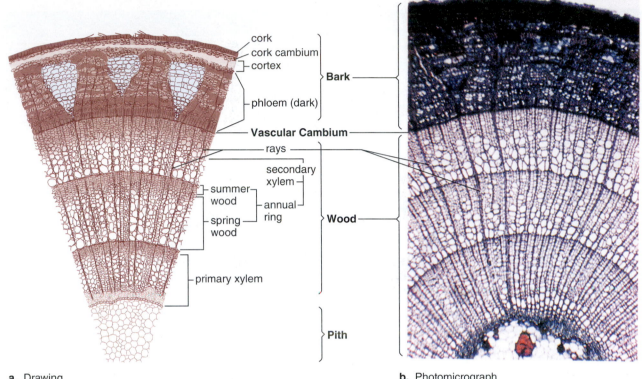

a. Drawing
b. Photomicrograph

2. Locate the **vascular cambium** at the inner edge of the bark, between the bark and the wood. Vascular cambium is meristem tissue whose activity accounts for secondary growth, which causes increased girth of a tree. Secondary phloem (which disappears) and secondary xylem (which builds up) are produced by vascular cambium each growing season.

3. Find the **wood,** which contains annual rings. An **annual ring** is the amount of xylem added to the plant during one growing season. Rings appear to be present because spring wood has large xylem vessels and looks light in color, while summer wood has much smaller vessels and appears much darker. How old is the stem you are observing? _____ Are all the rings the same width? _____

4. Identify the **pith,** a ground tissue at the center of a woody stem that stores organic nutrients and may disappear.

5. Locate **rays,** groups of small, almost cuboid cells that extend out from the pith laterally.

9.5 Leaves

A **leaf** is the organ that produces food for the plant by carrying on photosynthesis. Leaves are generally broad and quite thin. An expansive surface facilitates the capture of solar energy and gas exchange. Water and nutrients are transported to the cells of a leaf by leaf veins, extensions of the vascular bundles from the stem.

Anatomy of Leaves

Observation: Anatomy of Leaves

1. Examine a model of a leaf. With the help of Figure 9.15, identify the waxy **cuticle,** the outermost layer that protects the leaf and prevents water loss.
2. Locate the **upper epidermis** and **lower epidermis,** single layers of cells at the upper and lower surfaces. Trichomes are hairs that grow from the upper epidermis and help protect the leaf from insects and water loss.
3. Find the leaf veins in your model. The bundle sheath is the outer boundary of a vein; its cells surround and protect the vascular tissue. If this is a model of a monocot, all the leaf veins will be

   _____. If this is a model of a eudicot, some leaf veins will be circular and some

   will be oval. Why? _____
4. Identify the **palisade mesophyll,** located near the upper epidermis. These cells contain chloroplasts and carry on most of the plant's photosynthesis. Locate the **spongy mesophyll,** located near the lower epidermis. These cells have air spaces that facilitate the exchange of gases across the plasma membrane. *Label the layers of mesophyll in Figure 9.15.* Collectively, the mesophyll represents which

   of the three types of tissue found in all parts of a plant (see Table 9.1)? _____
5. *Label the two layers of epidermis in Figure 9.15.* Find a **stoma** (pl., stomata), an opening through which gas exchange occurs. Stomata are more numerous in the lower epidermis. Each stoma has two guard cells that regulate opening and closing of the opening.

Figure 9.15   Leaf anatomy.

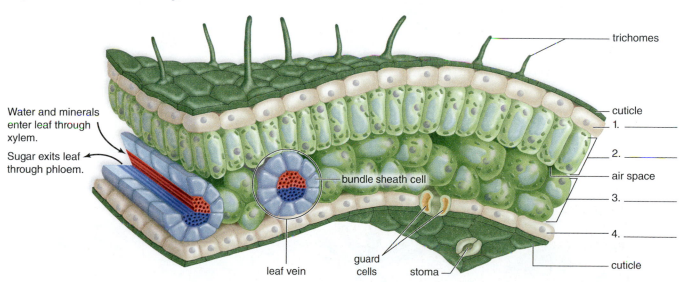

Water and minerals enter leaf through xylem.

Sugar exits leaf through phloem.

leaf vein

guard cells

stoma

bundle sheath cell

trichomes

cuticle

1. _____

2. _____

air space

3. _____

4. _____

cuticle

Leaf Diversity

A eudicot leaf consists of a flat blade and a stalk, called the petiole. An axillary bud appears at the point where the petiole attaches a leaf to the stem. In other words, an axillary bud is a tip-off that you are looking at a single leaf.

Observation: Leaf Diversity

Several types of eudicot leaves will be on display in the laboratory. Examine them using these directions.

- A leaf may be **simple**, in which case it consists of a single blade; or a leaf may be **compound**, meaning its single blade is divided into leaflets. *In Figure 9.16a write "simple" or "compound" next to the word leaf in 1–3.* (Among the leaves on display, find one that is simple and one that is compound.)
- A compound leaf can be **palmately** compound, meaning the leaflets are spread out from one point. Which leaf in Figure 9.16a is palmately compound? *Add "palmately" in front of "compound" where appropriate.* (See if you can find a palmately compound leaf among those on display.)
- A compound leaf can be **pinnately** compound, meaning the leaflets are attached at intervals along the petiole. Which leaf in Figure 9.16a is pinnately compound? *Add "pinnately" in front of "compound" where appropriate.* (See if you can find a pinnately compound leaf among those on display.)
- As shown in Figure 9.16b, leaves can be in various positions on a stem. On which stem in Figure 9.16b, 4–6, are the leaves **opposite** one another? *Write the word "opposite" where appropriate.* On which stem do the leaves **alternate** along the stem? *Write the word "alternate" where appropriate.* On which stem do the leaves **whorl** about a node? *Add the word "whorl" where appropriate in Figure 9.16b.* (See if you can find different arrangements of leaves among the stems on display.)

Figure 9.16   Classification of leaves.

1. _____ leaf, magnolia

a. Simple versus compound leaves

2. _____
leaf, buckeye

3. _____
leaf, black walnut

4. _____ leaves, beech

b. Arrangement of leaves on stem

5. _____ leaves, bedstraw

6. _____ leaves, maple

9.6 Xylem Transport

Xylem (Fig. 9.17), which transports water from the roots to the leaves, contains two types of conducting cells: tracheids and vessel elements. Both types of conducting cells are hollow and nonliving; the vessel elements are larger, they lack transverse end walls, and they are arranged to form a continous pipeline for water and mineral transport.

Water Column

The water column in xylem is continuous because water molecules are cohesive (they cling together) and because water molecules adhere to the sides of xylem cells. Therefore, water evaporation from leaf surfaces creates a negative pressure that pulls water upward.

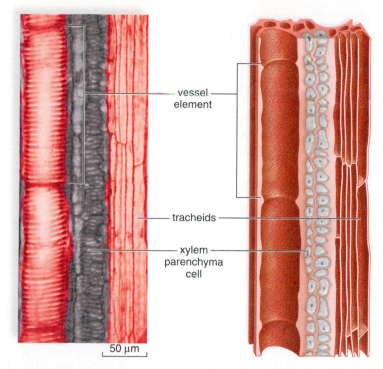

Figure 9.17 Xylem structure.
Xylem contains two types of conducting cells: tracheids and vessel elements. Tracheids have pitted walls, but vessel elements are larger and form a continuous pipeline from the roots to the leaves.

Experimental Procedure: The Water Column

1. Place a small amount of red-colored water in two beakers. Label one beaker "wet" and the other beaker "dry."
2. Transfer a stalk of celery (which was cut and then immediately placed in a container of water) into the "wet" beaker so that the large end is in the colored water.
3. Transfer a stalk of celery of approximately the same length and width (that was kept in the air after being cut) into the "dry" beaker so that the large end is in the colored water.
4. With scissors, cut off the top end of each stalk, leaving about 10 cm total length.
5. Time how long it takes for the red-colored water to reach the top of each stalk, and record these data in Table 9.4.
6. In which celery stalk was the water column broken? _____

   Use this information to write a conclusion in Table 9.4.
7. Make a cross-sectional wet mount of the stalk in the "wet" beaker. Observe this slide under the microscope. What type of tissue has been stained by the dye? _____

Table 9.4 Celery Stalk Experiment

Stalk	Speed of Dye (Minutes)	Conclusion
Cut end placed in water prior to experiment		
Cut end kept in air prior to experiment		

Transpirational Pull at Leaves

Evaporation of water from leaves is called **transpiration.** As transpiration occurs, the continuous water column is pulled upward—first within the leaf, then from the stem, and finally from the roots.

Virtual Lab **Plant Transpiration**   A virtual laboratory called Plant Transpiration is available on the *Inquiry into Life* website **www.mhhe.com/maderinquiry14**. After opening, you will follow the directions provided in the virtual lab (see Procedure in the write-up next to the experimental items). To save time, reduce the number of different plants selected but be sure to test any selected plant in the four available ways. Before you begin the experiment,

Hypothesize which of these—a heater, a fan, or a lamp—will have the greatest effect on the rate of transpiration.

Hypothesis: _____
_____

Results

Record your data in this table.
Conclusion

Amount of Water Transpired in 1 Hour (mL)				
	Normal Conditions (21°C)	With Fan (21°C)	With Heater (27°C)	With Lamp (21°C)
Arrowhead				
English Ivy				
Geranium				
Zebra Plant				

Did your results support or fail to support your hypothesis? _____
_____

Speculate on why the heater and the fan produced approximately the same results. _____
_____

Speculate on why the rate of transpiration with the lamp approximated the rate under normal conditions. _____
_____

Click on the Journal icon at the bottom of the screen and answer the questions that appear at the arrow heads from 1–7.

1. _____

2. _____

3. _____

4. _____

5. _____

6. _____

7. _____

_____   1. The leaves are attached to what portion of a stem?

_____   2. What type of venation do monocot leaves have?

_____   3. State the function of the structures called stomata that are present in leaf epidermis.

_____   4. What are the cells between the upper and lower epidermis of the leaf called?

_____   5. On a slide, what structures cause stomata to be open or closed?

_____   6. What is the pattern for vascular bundle distribution in a monocot stem?

_____   7. In woody stems, the bark is divided from the wood by what tissue?

_____   8. Identify the slide if annual rings are present.

_____   9. Pith is likely to be present in the (root, stem or leaf) _____ of what

   type plant (monocot or eudicot) _____.

_____   10. What zone follows the zone of cell division in a root tip?

_____   11. Lateral roots develop from which cell layer?

_____   12. Epidermis is modified in a root by the addition of _____.

_____   13. Identify the slide if the center tissue has several "arms" that extend like the spokes of a wheel.

_____   14. What tissue type transports materials in a plant?

_____   15. What tissue type allows the shoot tip and root tip to produce new cells?

Thought Questions

16. How could you identify a plant as a herbaceous eudicot by looking at root and stem slides?

17. Contrast how water arrives at the inside of a leaf with how carbon dioxide arrives at the inside of a leaf.

18. What advantage does the waxy cuticle provide to the leaf?

19. List two similarities and two differences between monocots and eudicots.

10

Reproduction in Flowering Plants

Introduction

The evolution of the flower and the involvement of animals in the plant life cycle helps account for the great success of flowering plants, also called **angiosperms.** The flower is the center of sexual reproduction for angiosperms and it produces seeds that contain an embryo which gives rise to the next generation after dispersal. The flower also produces a fruit that covers the seeds.

Flowering plants are stationary but through the process of evolution, often motile animals, such as insects, carry pollen from one flower to the other. Pollen contains sperm that then fertilize an egg and this triggers the development of a seed. The relationship between the pollinator (the insect) and the flower (Fig. 10.1) is mutualistic because before taking off to the next flower, the insect acquires nectar, a nutrient substance. Animals also help flowering plants disperse their seeds. When animals feed on fruits they inadvertently take the seeds to new locations.

Figure 10.1   Pollinator on a flower.
In gathering nectar from the same type of flower one after the other, the insect inadvertently distributes pollen which contains sperm, necessary to complete the plant's life cycle.

10.1 The Flower

The structure of a flower allows a flowering plant to produce seeds whose dispersal accounts for the success of these plants.

Structure of a Flower

The flower contains the reproductive structures necessary to the life cycle of a flowering plant.

Observation: Structure of a flower

1. Examine a model and identify these parts of a flower (Fig. 10.2).

Figure 10.2   Flower structure.

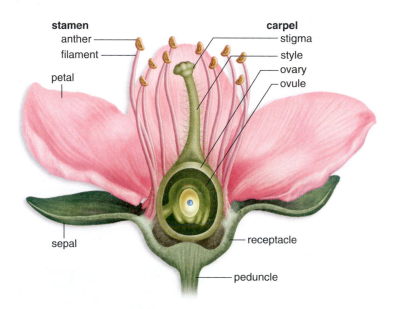

Peduncle:	The flower stalk.
Receptacle:	The portion of a flower that bears the other parts.
Sepals:	Collectively called the **calyx,** the **sepals** protect the flower bud before it opens. The sepals may drop off or may be colored like the petals. Usually, sepals are green and remain attached to the receptacle.
Petals:	Collectively called the **corolla,** the **petals** are quite diverse in size, shape, and color. Their color or arrangement can attract a pollinator.
Stamens:	Each **stamen** consists of two parts—the **anther,** and the **filament,** a slender stalk. Pollen grains develop in the anther.
Carpel:	A vaselike structure with three major regions from top to bottom: the **stigma,** an enlarged, sticky knob; the **style,** a slender stalk; and the **ovary,** an enlarged base that contains one or more ovules. **Ovules** develop into seeds. The ovule wall becomes the seed coat, which protects an embryonic plant and its stored food. An ovary becomes a fruit. Fruit is instrumental in the distribution of seeds. For example, after a bird eats a berry, it may likely fly to a new location where it defecates the seed.

Flower parts in threes and multiples of three

Table 10.1　Monocots and Eudicots

Monocots	Eudicots
One cotyledon	Two cotyledons
Flower parts in threes or multiples of three	Flower parts in fours or fives or multiples of four or five
Usually herbaceous	Woody or herbaceous
Usually parallel venation	Usually net venation
Scattered bundles in stem	Vascular bundles in a ring
Never woody	Can be woody

Flower parts in fours or fives and their multiples

2. As we studied in the previous lab, some flowering plants are monocots (monocotyledons) and some are eudicots (eudicotyledons) as described in Table 10.1. Note the difference between monocot and eudicot flowers. Is the flower model you have been examining a monocot or a eudicot? _____

Observation: Live Flower

1. Each student will bring a live flower to lab. It should be possible to see the sepals, petals, stamens, and carpel(s) of this flower as in Figure 10.3. Sketch the overall appearance of the whole flower here and label as much as possible using the terms in Figure 10.2.

Figure 10.3　Lily (*Lilium longiflorum*).

2. Describe the color and scent of your flower, if any.

_____

3. Fill in Table 10.2 as you complete your observation of the flower.

Table 10.2　Live Flower

Parts of Flower	Number	Description
Sepals		
Petals		
Stamens		
Carpel(s)		
Is this flower a monocot or a eudicot? _____		

4. Determine where the petals and sepals are attached and remove them. What color are the sepals? _____ Do they resemble the petals in any way? _____ If yes, how? _____

5. Count the number of stamens and record the number in Table 10.2. Is this the same as the petal number or a multiple of the petal number? _____ Remove the stamens and examine one to locate the anther and filament. Place a stamen on a slide and use a scalpel to remove and open the anther to disperse pollen on the slide. Remove all but a few pollen grains from the slide. Add a coverslip and examine microscopically, using high power. Can you find two different nuclei? _____ At this point, a pollen grain contains two cells. One of these cells will divide to produce the sperm and the other will produce the pollen tube through which the sperm will travel to the embryo sac in an ovule of ovary.

6. Before removing the carpel, identify the stigma, style, and ovary. Remove a carpel and place it on a slide. Use a scalpel to slit it longitudinally. Do you see any ovules? _____ Describe. _____

   Use the scalpel to make a cross section of the ovary. Does this ovary contain chambers, each with ovules? _____

7. If available, examine the fruit produced by this type of flower. The exterior wall came from what part of the carpel? _____ The seeds came from what structure? _____

10.2 The Flowering Plant Life Cycle

Land plants, including flowering plants, follow the **alternation of generation life cycle** (Fig. 10.4). Just as your arms are shaped differently from your legs, land plants have two generations that look different and have different functions:

1. The **sporophyte** (diploid generation) produces spores in a structure called a sporangium by meiosis. **Spores** are reproductive structures that result in haploid individuals. In plants, a spore develops into a haploid generation.
2. The **gametophyte** (haploid generation) produces gametes (eggs and sperm) by mitosis. The gametes then unite to form a diploid zygote.

In humans, meiosis is involved in the production of _____. What does meiosis produce in plants? _____

_____

Alternation of Generation in Flowering Plants

What does the sporophyte and gametophyte of a flowering plant look like? The sporophyte is the dominant generation in flowering plants. It is the generation we can easily see because we recognize

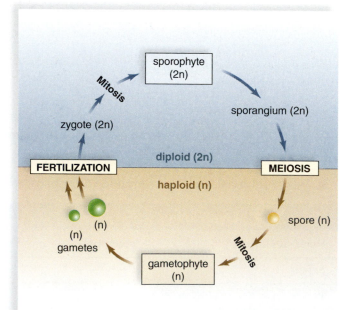

Figure 10.4   Alternation of generations in nonseed plants.

it as the plant. An oak tree, a rose bush, a daffodil plant—each is the sporophyte generation. A flowering plant produces spores in sporangia but they are microscopic and reside in the body of the dominant generation. Further the life cycle is modified. Flowering plants have two types of sporangia (Fig. 10.5). A megasporangium inside the ovule produces a megaspore and microsporangia inside pollen sacs produce microspores. **Microspores** develop into pollen grains, the **male gametophytes,** which contain two sperm each. The **megaspore** within an ovule develops into an 8-cell **female gametophyte** (embryo sac) which produces an egg.

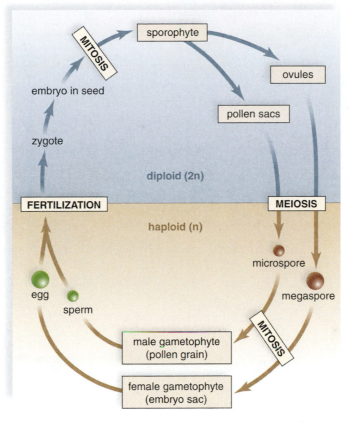

Figure 10.5 Alternation of generations in flowering plants, which are seed plants.

Flowering plants practice **double fertilization.** When a pollen grain of the correct species lands on a stigma, it develops a **pollen tube** that carries the sperm to the embryo sac in the ovule, where one sperm fertilizes the egg and the other participates in development of the **endosperm,** which is food for the developing embryo. An ovule develops into a seed.

Explain why a pollen grain is called the *male* gametophyte ? _____

_____

Explain why an embryo sac is called the *female* gametophyte? _____

_____

A seed contains which generation? _____

Figure 10.6 Flowering plant life cycle.

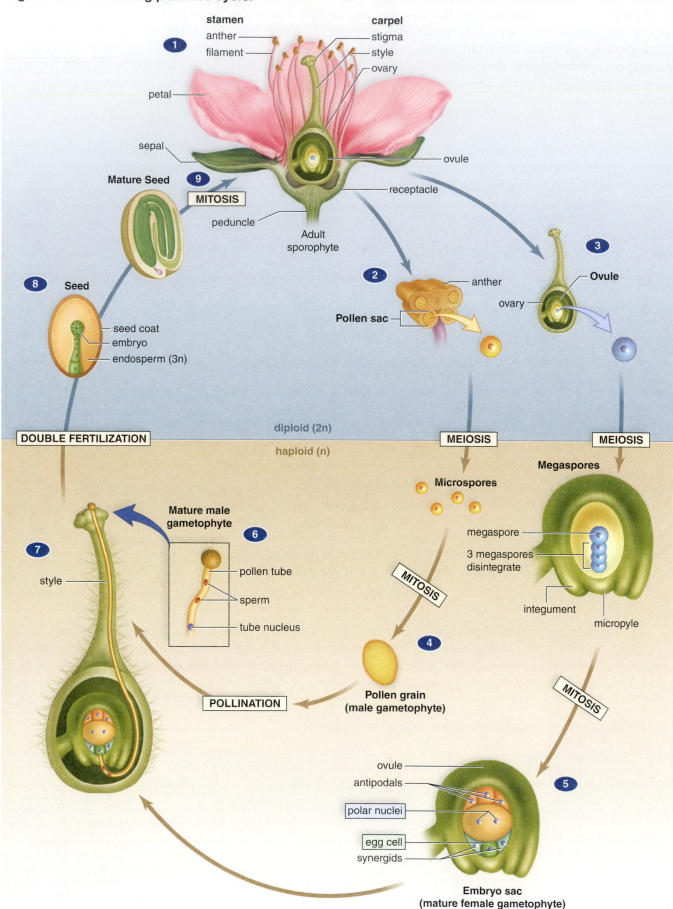

Use Figure 10.6 as a guide to describe the life cycle of flowering plants.

1. The parts of the flower involved in reproduction are the _____ and the _____.

2. The anther at the top of the stamen has two _____ sacs, which produce _____ by meiosis.

3. Within an ovule, a megaspore mother cell undergoes meiosis to produce four _____, three of which die.

4. A microspore undergoes mitosis and becomes a _____, the male gametophyte.

5. One megaspore undergoes mitosis and develops into a(n) _____, which contains, in addition to other cells, two _____ nuclei and the _____ cell. (These terms are shaded.)

6. **Pollination** is the transport of pollen by wind or by animals such as insects (see Fig. 10.1) from the pollen sacs to the stigma of carpels. After a pollen grain lands on the stigma of a carpel, it develops a pollen tube that passes down the style and takes _____ sperm to the embryo sac.

7. During double fertilization, one sperm from the pollen tube fertilizes the egg within the embryo sac, and the other joins with the two _____ nuclei.

8. The fertilized egg becomes an _____, and the joining of polar nuclei and sperm becomes the triploid (3n) **endosperm.** The ovule wall becomes the seed coat. A **seed** contains a sporophyte embryo, stored food, and a seed coat. In angiosperms, seeds are enclosed by _____ (not shown). A fruit assists the dispersal of seeds (e.g., when animals eat fleshy fruits, they may ingest the seeds and disperse by defecation sometime later).

9. After dispersal of the seed, it germinates and begins to grow, eventually becoming an adult sporophyte that bears flowers.

10.3 Pollination and Development of the Embryo

Pollination, the movement of pollen from the pollen sacs to the stigmas of carpels, is an important event in flowering plants. In some plants, wind accomplishes pollination and these plants produce copious amounts of pollen (Fig. 10.7). Why would you expect wind to carry out pollination in a tree and not in a garden plant? _____

_____

_____

_____

Figure 10.7 Wind pollination of a grass, with SEM of pollen grains.

Plants and Their Pollinators

Through the course of evolution, many plants—even trees—have come to depend on animals (e.g., beetles, bats, insects) called **pollinators** to carry out pollination. The relationship is mutualistic because both the plant and the pollinator benefit from the relationship. **Nectar** is a sugary substance produced by the plant for the pollinator. The pollinator goes from plant to plant to gather food (nectar and even pollen) and in the process carries pollen from flower to flower.

It might seem as if a pollinator such as a bee goes to all types of flowers, but it doesn't. Bees visit only certain flowers—the ones that provide them with nectar, a sugary liquid that serves as their food. Bee-pollinated flowers have a shape and appearance preferred by bees, even a particular type of bee.

How could it happen that particular bees and particular flowers are suited to one another? _____

_____

_____

How is this advantageous to both the plant and the bee? _____

_____

_____

Observation: Plants and Their Pollinators

1. The description you gave can help you decide what type of pollinator would be attracted to your flower. See also Fig. 10.8.

   - Bees and moths can smell. But bees like a delicate, sweet smell, while moths prefer a strong smell that can allow them to find flowers in the dark. Moths are nocturnal and feed at night. If the flower

     has a smell, which of these two pollinators might pollinate your flower? _____

   - Bees and butterflies generally need a landing platform. Bees can land on a small petal, but butterflies typically walk around on a cluster of flowers.

   - Moths and hummingbirds do not need a landing platform because they hover (flap their wings to stay in one place). Moths and hummingbirds have a long tongue to reach nectar at the bottom of the floral tube. Based on this information, which type of pollinator mentioned so far might pollinate

     your flower? _____

   - Moth-pollinated flowers are typically white; bees can't see red but can see yellow; butterflies like brightly colored flowers; and hummingbirds prefer the color red. Which of these pollinators might

     prefer your flower? _____

2. Your instructor may have suggested that you bring other flowers to lab. Examine several available flowers and tell how each of the features listed below help decide the pollinator. These features will help you decide the pollinator for each type flower. Then you will be ready to complete Table 10.3.

   How does each of these flower features help decide the pollinator?

   Landing platform _____

   Color _____

   Smell _____

   Shape of flower _____

a.

b.

c.

d.

Figure 10.8 Pollinators.
a. The bee pollinated flower has a landing platform, a bright color, and a sweet smell. **b.** The butterfly-pollinated flower is often a brightly colored composite containing many individual flowers. The broad exposure provides room for the butterfly to land. **c.** A moth-pollinated flower is usually light in color; a moth depends on scent to find the flower it prefers in the dark. **d.** Hummingbird-pollinated flowers are curved back, allowing the bird to insert its beak to reach the rich supply of nectar.

Table 10.3 Plants and Their Pollinators		
Description of Flower	**Possible Pollinator**	**Explanation**
1.		
2.		
3.		
4.		
5.		
6.		

Development of Eudicot Embryo

Stages in the development of a eudicot embryo are shown in Figure 10.9. During development, the **suspensor** anchors the embryo and transfers nutrients to it from the mature plant. The **cotyledons** store nutrients that the embryo uses as nourishment. An embryo consists of the **epicotyl,** which becomes the leaves; the **hypocotyl,** which becomes the stem; and the **radicle,** which becomes the roots.

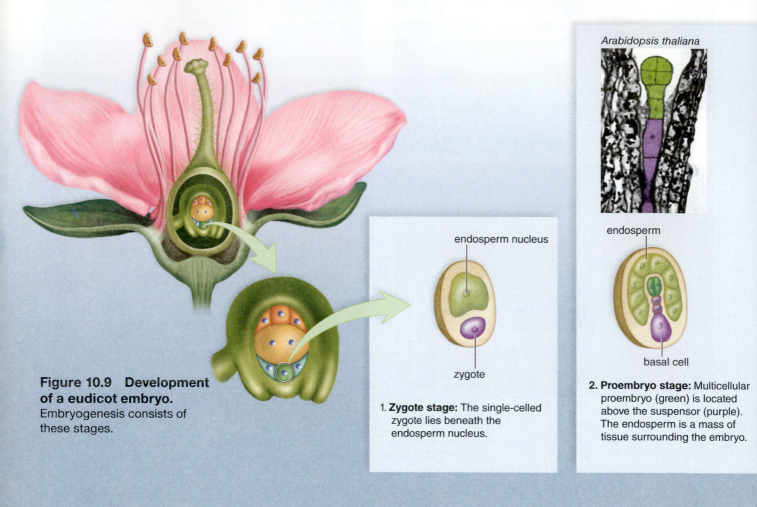

Figure 10.9   Development of a eudicot embryo. Embryogenesis consists of these stages.

1. **Zygote stage:** The single-celled zygote lies beneath the endosperm nucleus.

2. **Proembryo stage:** Multicellular proembryo (green) is located above the suspensor (purple). The endosperm is a mass of tissue surrounding the embryo.

Observation: Development of Eudicot Embryo

Study prepared slides and identify the stages described in Figure 10.9. List the stages you were able to identify. _____

_____

_____

_____

_____

A. thaliana

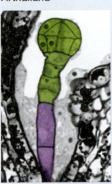

endosperm

3. Globular stage: As cell division continues, the proembryo (green) becomes globe-shaped. The stalklike suspensor (purple) anchors the embryo.

A. thaliana

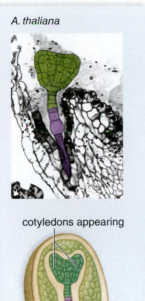

cotyledons appearing

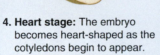

4. Heart stage: The embryo becomes heart-shaped as the cotyledons begin to appear.

Capsella

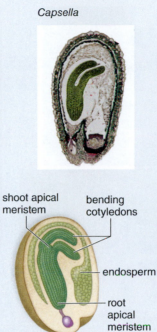

shoot apical meristem   bending cotyledons

endosperm

root apical meristem

5. Torpedo stage: The embryo becomes torpedo-shaped as the cotyledons enlarge. The endosperm lessens, and tissues become differentiated.

Capsella

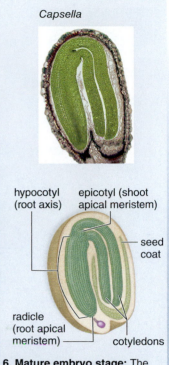

hypocotyl (root axis)   epicotyl (shoot apical meristem)

seed coat

radicle (root apical meristem)

cotyledons

6. Mature embryo stage: The embryo consists of the epicotyl (represented here by the shoot apex), the hypocotyl, and the radicle (which contains the root apex).

10.4 Fruits

Because fruits are defined as the mature ovary, what people normally call a vegetable can be a fruit to a botanist. For example, in addition to a blackberry, an almond, a pea pod, and a tomato are fruits (Fig. 10.10).

As a fruit develops from an ovary, the ovary wall thickens to become the **pericarp**. The pericarp can have as many as three layers: exocarp, mesocarp, and endocarp. The **exocarp** forms the outermost skin of a fruit. The **mesocarp** between the exocarp and endocarp can be fleshy. The receptacle of a flower

a. Almond, *Prunus*

seed

b. Tomato, *Lycopersicon*

one chamber of ovary

c. Pea, *Pisum*

developing fruit

seed

d. Blackberry, *Rubus*

many carpels

Figure 10.10 Fruit diversity.
a. The almond fruit is fleshy, with a single seed enclosed by a hard covering. **b.** The tomato is derived from a compound ovary. **c.** Dry fruit of the pea plant develops from a simple ovary. **d.** The blackberry fruit is an aggregate fruit derived from a flower that had many ovaries. Each ovary produces one of the "berries" in the aggregate fruit.

often contributes to the flesh of a fruit also. The **endocarp** serves as the boundary around the seed(s). The endocarp may be fleshy (as in tomatoes), hard (as in peach pits), or papery (as in apples). Botanists classify fruits in the manner described in the key. This is a simplified key but it does contain some of the most common types of fruits.

Dichotomous Key to Major Types of Fruit

I. Fleshy fruits

    A. Simple fruits (i.e., from a single ovary)

        1. Flesh mostly of ovary tissue, particularly the mesocarp

            a) Endocarp is hard and stony; ovary superior and single-seeded (cherry, olive, coconut): **drupe**

            b) Endocarp is fleshy or slimy; ovary usually many-seeded (tomato, grape, green pepper): **berry**

        2. Flesh mostly of receptacle tissue (apple, pear, quince): **pome**

    B. Complex fruits (i.e., from more than one ovary)

        1. Fruit from many carpels on a single flower (strawberry, raspberry, blackberry): **aggregate fruit**

        2. Fruit from carpels of many flowers fused together (pineapple, mulberry): **multiple fruit**

II. Dry fruits

    A. Fruits that split open at maturity (usually more than one seed)

        1. Split occurs along two seams in the ovary. Seeds borne on one of the halves of the split ovary (pea and bean pods, peanuts): **legume**

        2. Seeds released through pores or multiple seams (poppies, irises, lilies): **capsule**

    B. Fruits that do not split open at maturity (usually one seed)

        1. Pericarp hard and thick, with a cup at its base (acorn, chestnut, hickory): **nut**

        2. Pericarp thin and winged (maple, ash, elm): **samara**

        3. Pericarp thin and not winged (sunflower, buttercup): **achene** (cereal grains): **caryopsis**

Source: Vodopich and Moore: *Biology Laboratory Manual* 8/e, p. 344.

To use this key understand that at each step you have at least two choices and you pick one of these in order to proceed further. Therefore, first determine if the fruit is fleshy or dry. If fleshy, proceed further with I and if dry, proceed further with II. Next choose either A or B and so forth, until you arrive at the particular type of fruit. The last column in Table 10.4 will be one of the boldface terms from the key.

Observation: Fruits

1. Again examine the fruit of your flower. Use the key to determine the fruit type and list this fruit as an example in Table 10.4.

2. Examine an apple that has been sliced to give a longitudinal and a cross-section view of the interior. The flesh of an apple is from the receptacle and only the core of an apple is from the ovary. Locate the outer limit of the pericarp and of the endocarp. An ovary can be simple (have one chamber) or can be compound (have more than one chamber). What type of ovary does an apple have? _____

   How could an animal, such as a deer, help disperse the seeds of an apple? _____

   _____

   Use the apple as an example of a fruit in Table 10.4.

3. Examine the pod of a string bean or pea plant. How many seeds (beans or peas) are in the pod? _____ Would it help disperse the seeds of a pea plant if an animal were to eat the peas? _____ Why or why not? _____

Split the pea or bean and look for the embryo. Is this plant a monocot or eudicot plant? _____ How do you know? _____

Use a pea pod as an example of a fruit in Table 10.4.

4. Examine a sunflower fruit and remove the seed. The outer coat of a sunflower seed is actually _____. How can examining the seed tell you that the sunflower plant is a eudicot? _____

Except for the apple, all the fruits you have examined are dry fruits. What does this mean? _____

_____

Add sunflower fruit to Table 10.4.

5. Examine other available fruits and complete Table 10.4.

Table 10.4   Identification of Simple Fruits

Common Name	Fleshy or Dry?	Eaten as a Vegetable, Fruit, Other?	Type of Fruit (from Dichotomous Key)
1			
2			
3			
4			
5			
6			
7			
8			
9			
10			

10.5 Seeds and Seed Germination

The seeds of flowering plants develop from ovules. As noted previously in Figures 10.6 and 10.9, a seed contains an embryonic plant, stored food, and a seed coat. Monocot seeds have one **cotyledon** (seed leaf); eudicots have two cotyledons.

Observation: Eudicot and Monocot Seeds

Bean Seed

1. Obtain a presoaked bean seed (eudicot). Carefully dissect it, using Figure 10.11 to help you identify the following:

   a. **Seed coat:** The outer covering. Remove the seed coat with your fingernail.

   b. **Cotyledons:** Food storage organs. The endosperm was absorbed by the cotyledons during development. What is the function of cotyledons? _____

   _____

   c. **Epicotyl:** The small portion of the embryo located above the attachment of the cotyledons. The first true leaves (**plumules**) develop from the epicotyl.

   d. **Hypocotyl:** The small portion of embryo located below the attachment of the cotyledons. The lower end develops into the embryonic root, or **radicle.**

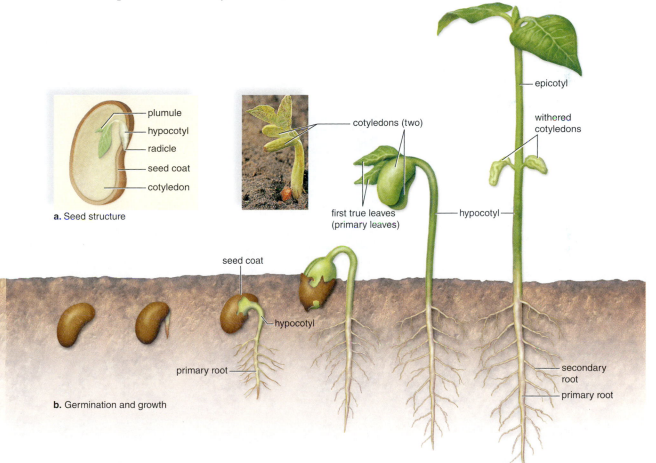

Figure 10.11 Eudicot seed structure and germination.
a. A eudicot seed has two cotyledons as exemplified by bean seed. The cotyledons (stored food) are attached to the embryo (plumule, hypocotyl, and epicotyl with radicle). Following **(b)** germination, a bean seedling grows to become a mature plant.

2. As observed in Figure 10.11, which organ emerges first from a dicot seed—the plumule or the radicle (i.e., root)? _____

Of what advantage is this to the plant? _____

_____

3. The hypocotyl is the first part to emerge from the soil. What is the advantage of the hypocotyl pulling the plumule up out of the ground instead of pushing it up through the ground? _____

_____

4. Do cotyledons stay beneath the ground in eudicots? _____

Corn Kernel

1. Obtain a presoaked corn kernel (monocot). Lay the seed flat, and with a razor, carefully slice it in half. A corn kernel is a fruit, and the seed coat is tightly attached to the pericarp (Fig. 10.12).
2. Identify the cotyledon, plumule, and radicle. In addition, identify the
   a. **Endosperm:** Stored food for the embryo; the nutrients pass into the cotyledon as the seedling grows.
   b. **Coleoptile:** A sheath that covers the emerging leaves. The coleorhiza is a protective sheath for the radicle.
3. As observed in Figure 10.12, does the cotyledon of a corn seed, our example of a monocot, stay

   beneath the ground? _____

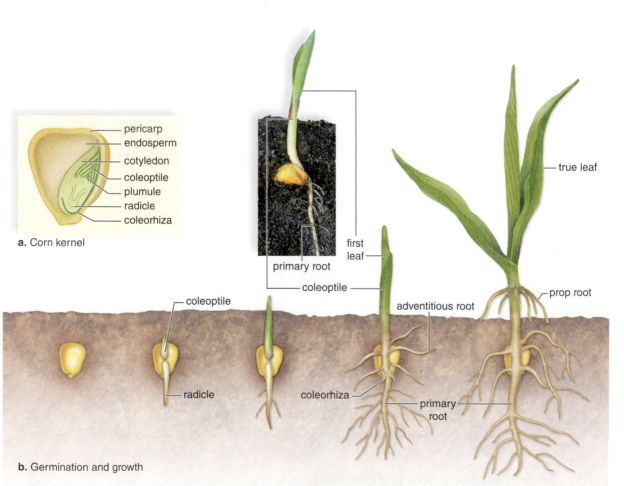

a. Corn kernel

b. Germination and growth

Figure 10.12 Monocot seed structure and germination.
a. A monocot seed has only one cotyledon as exemplified by a corn kernel. A corn kernel is a fruit—the seed is covered by a pericarp. Following **(b)** germination, a corn seedling grows to become a mature plant.

Seed Germination

Mature seeds contain an embryo that does not resume growth until after germination, which requires the proper environmental conditions. Mature seeds are dry, and for germination to begin, the dry tissues must take up water in a process called **imbibition.** After water has been imbibed, enzymes break down the food source into small molecules that can provide energy or be used as building blocks until the seedling is ready to photosynthesize.

Experimental Procedure: Seed Germination

Your instructor has placed sunflower seeds in five containers and watered them. The seedlings are in increasing stages of growth.

1. Order the seedlings in a series according to increasing stages of growth. What criteria did you use to order the seedlings? _____

   _____

2. Is this plant a monocot or a eudicot? _____ What criteria did you use to decide? _____

   _____

3. Can you see the cotyledons? _____ Explain why the cotyledons of a eudicot seedling shrivel as the seedling grows. _____

4. Use the space below to draw two contrasting stages of growth and add these labels to your drawings: hypocotyl, cotyledons, epicotyl, leaves, stem, terminal bud, node. For terminal bud and node, see Figure 9.2.

Laboratory Review 10

_____ 1. What structure transports sperm to the ovule in flowering plants?

_____ 2. Angiosperms are either monocots or _____.

_____ 3. What type of cell division produces microspores and megaspores in flowering plants?

_____ 4. Which angiosperm group has flower parts in threes or multiples of three?

_____ 5. Name the part of a flower that has a filament topped by the anther.

_____ 6. What kinds of spores are produced by flowering plants?

_____ 7. What structure disperses the offspring in flowering plants?

_____ 8. How many sperm are in the pollen grain of a flowering plant?

_____ 9. What is the 3n nutritive tissue in the seed of angiosperms called?

_____ 10. In flowering plants, the ovule becomes the seed, and the ovary becomes what structure?

_____ 11. Name the three pericarp layers in a fruit. Which layer is the fleshy one we eat?

_____ 12. Blackberries, raspberries, and strawberries are all examples of what kind of fruit?

_____ 13. Pineapples and mulberries are both examples of what kind of fruit?

_____ 14. Apples and tomatoes are both examples of what kind of fruit?

_____ 15. Peas and milkweeds are dry fruits that _____.

_____ 16. What process results in pollinators that are specific to a particular type of flower?

Thought Questions

17. What is the difference between pollination and fertilization in angiosperms?

18. Pollen sacs are analogous to what structures in human males? Explain.

19. Contrast the male gametophyte with the female gametophyte (e.g.,term, location, number of cells, gamete produced) of flowering plants.

20. In flowering plants the female gametophyte is dependent on the sporophyte. Explain.

11

Animal Organization

Introduction

Humans, as well as all other living things, are made up of **cells.** Groups of cells that have the same structural characteristics and perform the same functions are called **tissues.** Figure 11.1 shows the four categories of tissues in the human body. An **organ** is composed of different types of tissues, and various organs form **organ systems.** Humans thus have the following levels of biological organization: cells $\longrightarrow$ tissues $\longrightarrow$ organs $\longrightarrow$ organ systems.

The photomicrographs of tissues in this laboratory were obtained by viewing prepared slides with a light microscope. Preparation required the following sequential steps:

1. **Fixation:** The tissue is immersed in a preservative solution to maintain the tissue's existing structure.
2. **Embedding:** Water is removed with alcohol, and the tissue is impregnated with paraffin wax.
3. **Sectioning:** The tissue is cut into extremely thin slices by an instrument called a microtome. When the section runs the length of the tissue, it is called a longitudinal section (l.s.); when the section runs across the tissue, it is called a cross section (c.s.).
4. **Staining:** The tissue is immersed in dyes that stain different structures. The most common dyes are hematoxylin and eosin stains (H & E). They give a differential blue and red color to the basic and acidic structures within the tissue. Other dyes are available for staining specific structures. Because tissues can be stained various colors, the color of the tissues on your slides may not match those in the manual. Colors chosen to represent tissues in diagrams can vary also.

Figure 11.1 The major tissues in the human body.

The many kinds of tissues in the human body are grouped into four types: epithelial tissue, muscular tissue, nervous tissue, and connective tissue.

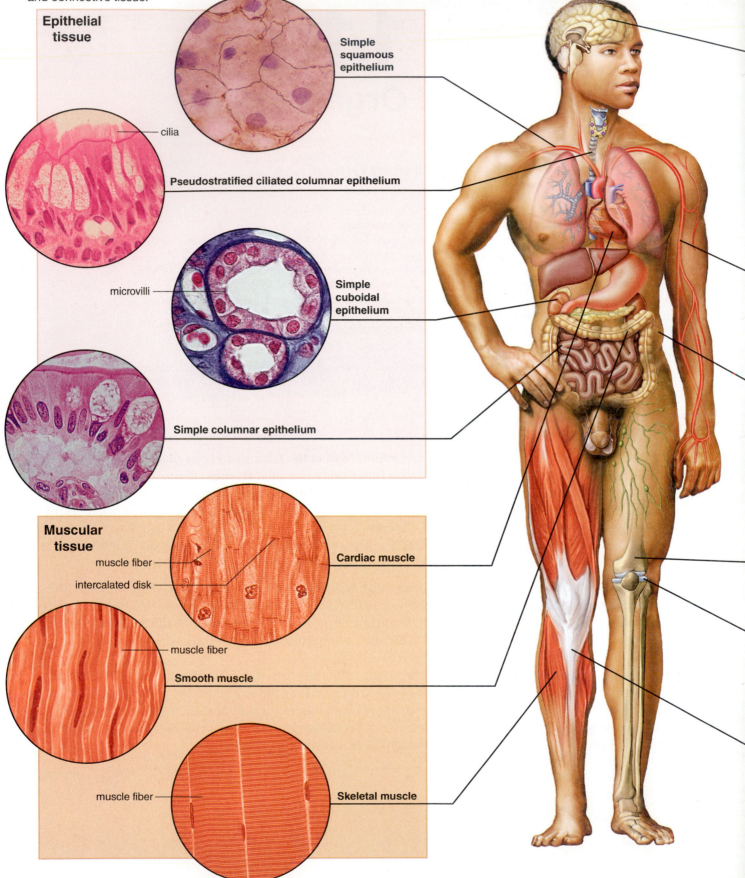

Epithelial tissue

Simple squamous epithelium

cilia

Pseudostratified ciliated columnar epithelium

microvilli

Simple cuboidal epithelium

Simple columnar epithelium

Muscular tissue

muscle fiber

intercalated disk

Cardiac muscle

muscle fiber

Smooth muscle

muscle fiber

Skeletal muscle

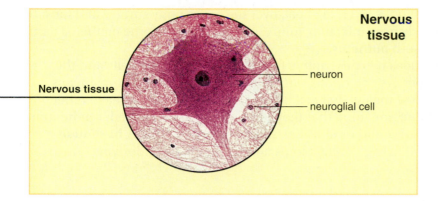

Nervous tissue

Nervous tissue

neuron

neuroglial cell

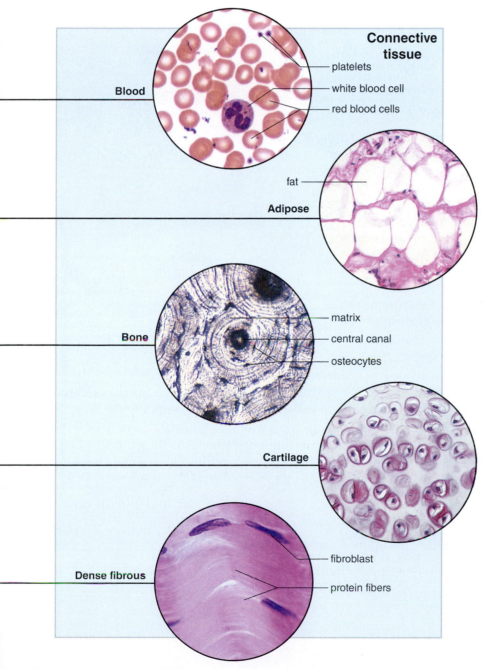

Connective tissue

Blood

platelets

white blood cell

red blood cells

fat

Adipose

Bone

matrix

central canal

osteocytes

Cartilage

Dense fibrous

fibroblast

protein fibers

11.1 Tissue Level of Organization

Epithelial tissue (epithelium) forms a continuous layer, or sheet, over the entire body surface and most of the body's inner cavities. It protects the body from infection, injury, and drying out. Some epithelial tissues produce and release secretions. Others absorb nutrients.

The name of an epithelial tissue includes two descriptive terms: the shape of the cells and the number of layers. The three possible shapes are *squamous, cuboidal,* and *columnar.* **Simple** means that there is only one layer of cells; **stratified** means that cell layers are placed on top of each other. Some epithelial tissues are **pseudostratified,** meaning that they only appear to be layered. Epithelium may also have cellular extensions called **microvilli** or hairlike extensions called **cilia.** A **basement membrane** consisting of glycoproteins and collagen fibers joins an epithelium to underlying connective tissue.

Observation: Simple and Stratified Squamous Epithelium

Simple Squamous Epithelium

Simple squamous epithelium is a single layer of thin, flat, many-sided cells, each with a central nucleus. It lines internal cavities, the heart, and all the blood vessels. It also lines parts of the urinary, respiratory, and male reproductive tracts.

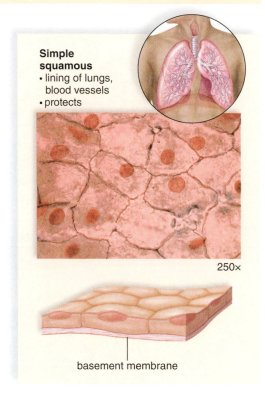

Simple squamous
• lining of lungs, blood vessels
• protects

250×

basement membrane

1. Study a model or diagram of simple squamous epithelium. What does squamous mean? _____

2. Examine a prepared slide of squamous epithelium. Under low power, note the close packing of the flat cells. What shapes are the cells?

   _____

3. Under high power, examine an individual cell, and identify the plasma membrane, cytoplasm, and nucleus.

4. Add a sketch of this tissue to Table 11.1 (p. 148).

Stratified Squamous Epithelium

As would be expected from its name, stratified squamous epithelium consists of many layers of cells. The innermost layer produces cells that are first cuboidal or columnar in shape, but as the cells push toward the surface, they become flattened.

The outer region of the skin, called the epidermis, is stratified squamous epithelium. As the cells move toward the surface, they flatten, begin to accumulate a protein called **keratin,** and eventually die. Keratin makes the outer layer of epidermis tough, protective, and able to repel water.

The linings of the mouth, throat, anal canal, and vagina are stratified epithelium. The outermost layer of cells surrounding the cavity is simple squamous epithelium. In these organs, this layer of cells remains soft, moist, and alive.

1. Either now or when you are studying skin in Section 11.2, examine a slide of skin and find the portion of the slide that is stratified squamous epithelium.

2. Approximately how many layers of cells make up this portion of skin? _____

3. Which layers of cells best represent squamous epithelium? _____

4. Add a sketch of this tissue to Table 11.1 (p. 148).

Observation: Simple Cuboidal Epithelium

Simple cuboidal epithelium is a single layer of cube-shaped cells, each with a central nucleus. It is found in tubules of the kidney and in the ducts of many glands, where it has a protective function. It also occurs in the secretory portions of some glands—that is, where the tissue produces and releases secretions.

1. Study a model or diagram of simple cuboidal epithelium.
2. Examine a prepared slide of simple cuboidal epithelium. Move the slide until you locate cube-shaped cells that line a lumen (cavity). Are these cells ciliated? _____
3. Add a sketch of this tissue to Table 11.1 (p. 148).

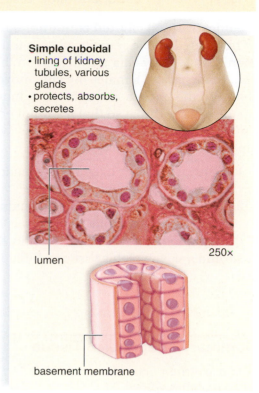

Simple cuboidal
- lining of kidney tubules, various glands
- protects, absorbs, secretes

250×

lumen

basement membrane

Observation: Simple Columnar Epithelium

Simple columnar epithelium is a single layer of tall, cylindrical cells, each with a nucleus near the base. This tissue, which lines the digestive tract from the stomach to the anus, protects, secretes, and allows absorption of nutrients.

1. Study a model or diagram of simple columnar epithelium.
2. Examine a prepared slide of simple columnar epithelium. Find tall and narrow cells that line a lumen. Under high power, focus on an individual cell. Identify the plasma membrane, the cytoplasm, and the nucleus. Epithelial tissues are attached to underlying tissues by a basement membrane composed of extracellular material containing protein fibers.
3. The tissue you are observing contains mucus-secreting cells. Search among the columnar cells until you find a **goblet cell,** so named because of its goblet-shaped, clear interior. This region contains mucus, which may be stained a light blue. In the living animal, the mucus is discharged into the gut cavity and protects the lining from digestive enzymes.
4. Add a sketch of this tissue to Table 11.1 (p. 148).

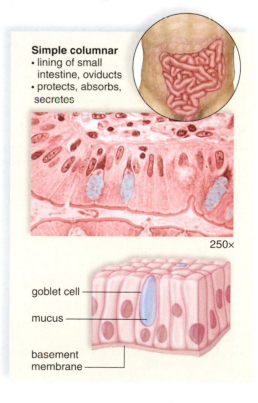

Simple columnar
- lining of small intestine, oviducts
- protects, absorbs, secretes

250×

goblet cell

mucus

basement membrane

Pseudostratified ciliated columnar epithelium appears to be layered, while actually all cells touch the basement membrane. Many cilia are located on the free end of each cell. In males and females the cilia move sex cells along tubes. In the trachea, the cilia move mucus and debris up toward the throat so that it cannot enter the lungs. Smoking destroys these cilia, but they will grow back if smoking is discontinued.

1. Study a model or diagram of pseudostratified ciliated columnar epithelium.
2. Examine a prepared slide of pseudostratified ciliated columnar epithelium. Concentrate on the part of the slide that resembles the model. Identify the cilia.
3. Add a sketch of this tissue to Table 11.1.

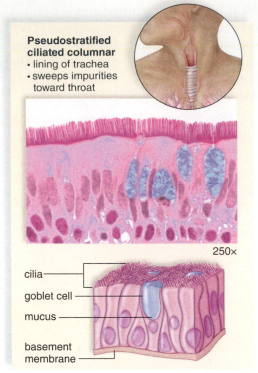

Pseudostratified ciliated columnar
- lining of trachea
- sweeps impurities toward throat

250×

cilia
goblet cell
mucus
basement membrane

Summary of Epithelial Tissue

Add a sketch in first column of Table 11.1 under each type of epithelial tissue.

Table 11.1   Epithelial Tissue

Sketch	Structure	Function	Location
Simple squamous	Flat, pancake-shaped	Filtration, diffusion, osmosis	Walls of capillaries, lining of blood vessels, air sacs of lungs, lining of internal cavities
Stratified squamous	Innermost layers are cuboidal or columnar; outermost layers are flattened.	Protection, repel water	Skin, linings of mouth, throat, anal canal, vagina
Simple cuboidal	Cube-shaped	Protection, secretion, absorption	Surface of ovaries, linings of ducts and glands, lining of kidney tubules
Simple columnar	Columnlike—tall, cylindrical nucleus at base	Protection, secretion, absorption	Lining of uterus, tubes of digestive tract
Pseudostratified ciliated columnar	Looks layered but is not; ciliated	Protection, secretion, movement of mucus and sex cells	Linings of respiratory passages

Connective Tissue

Connective tissue joins different parts of the body together. There are four general classes of connective tissue: connective tissue proper, bone, cartilage, and blood. All types of connective tissue consist of cells surrounded by a matrix that usually contains fibers. Elastic fibers are composed of a protein called elastin. Collagenous fibers contain the protein collagen.

Observation: Connective Tissue

There are several different types of connective tissue. We will study loose fibrous connective tissue, dense fibrous connective tissue, adipose tissue, bone, cartilage, and blood. **Loose fibrous connective tissue** (sometimes called areolar tissue) supports epithelium and also many internal organs, such as muscles, blood vessels, and nerves. Its presence allows organs to expand. **Dense fibrous connective tissue** (sometimes called white fibrous tissue) contains many collagenous fibers packed together, as in tendons, which connect muscles to bones, and in ligaments, which connect bones to other bones at joints.

1. Examine a slide of loose fibrous connective tissue, and compare it to the figure below *(left)*. What is the function of loose fibrous connective tissue? _____

2. Examine a slide of dense fibrous connective tissue, and compare it to the figure below *(right)*. What two kinds of structures in the body contain dense fibrous connective tissue? _____
   _____

3. Add sketches of these tissues to Table 11.2 (p. 152).

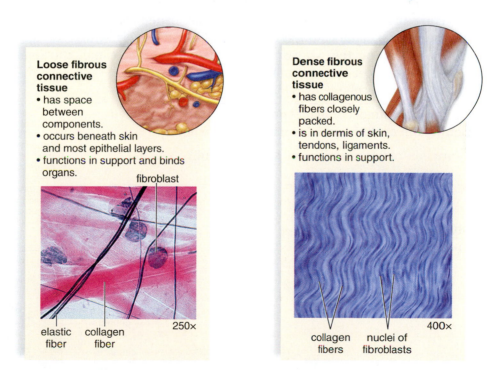

Loose fibrous connective tissue
- has space between components.
- occurs beneath skin and most epithelial layers.
- functions in support and binds organs.

fibroblast

elastic fiber collagen fiber 250×

Dense fibrous connective tissue
- has collagenous fibers closely packed.
- is in dermis of skin, tendons, ligaments.
- functions in support.

collagen fibers nuclei of fibroblasts 400×

Observation: Adipose Tissue

In **adipose tissue,** the cells have a large, central, fat-filled vacuole that causes the nucleus and cytoplasm to be at the perimeter of the cell. Adipose tissue occurs beneath the skin, where it insulates the body, and around internal organs, such as the kidneys and heart. It cushions and helps protect these organs.

1. Examine a prepared slide of adipose tissue. Why is the nucleus pushed to one side? _____

_____

2. State a location for adipose tissue in the body.

_____

_____

   What are two functions of adipose tissue at this location?

_____

_____

3. Add a sketch of this tissue to Table 11.2 (p. 152).

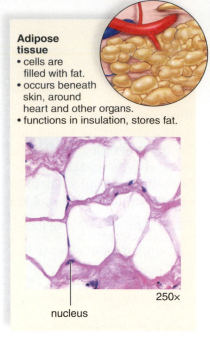

Adipose tissue
• cells are filled with fat.
• occurs beneath skin, around heart and other organs.
• functions in insulation, stores fat.

250×

nucleus

Observation: Compact Bone

Compact bone is found in the bones that make up the skeleton. It consists of **osteons** (Haversian system) with a **central canal,** and concentric rings of spaces called **lacunae,** connected by tiny crevices called **canaliculi.** The central canal contains a nerve and blood vessels, which service bone. The lacunae contain bone cells called **osteocytes,** whose processes extend into the canaliculi. Separating the lacunae is a matrix that is hard because it contains minerals, notably calcium salts. The matrix also contains collagenous fibers.

1. Study a model or diagram of compact bone. Then look at a prepared slide and identify the central canal, lacunae, and canaliculi.

2. What is the function of the central canal and canaliculi?

_____

_____

3. Add a sketch of this tissue to Table 11.2 (p. 152).

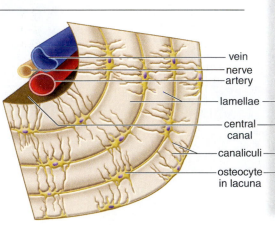

vein
nerve
artery
lamellae
central canal
canaliculi
osteocyte in lacuna

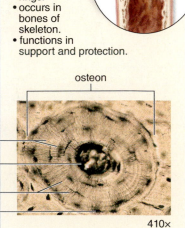

Compact bone
• has cells in concentric rings.
• occurs in bones of skeleton.
• functions in support and protection.

osteon

410×

Observation: Hyaline Cartilage

In **hyaline cartilage,** cells called **chondrocytes** are found in twos or threes in lacunae. The lacunae are separated by a flexible matrix containing weak collagenous fibers.

1. Study the diagram and photomicrograph of hyaline cartilage in the figure at the right. Then study a prepared slide of hyaline cartilage, and identify the matrix, lacunae, and chondrocytes.

2. Compare compact bone and hyaline cartilage. Which of these types of connective tissue is more organized? _____

   Why? _____

3. Which of these two types of connective tissue lends more support to body parts? Why? _____

   _____

4. Add a sketch of this tissue to Table 11.2 (p. 152).

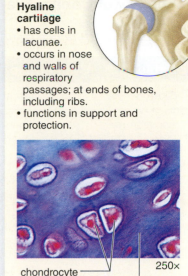

Hyaline cartilage
- has cells in lacunae.
- occurs in nose and walls of respiratory passages; at ends of bones, including ribs.
- functions in support and protection.

chondrocyte within lacunae   matrix   250×

Observation: Blood

Blood is a connective tissue in which the matrix is an intercellular fluid called **plasma. Red blood cells** (erythrocytes) have a biconcave appearance and lack a nucleus. These cells carry oxygen combined with the respiratory pigment hemoglobin. **White blood cells** (leukocytes) have a nucleus and are typically larger than the more numerous red blood cells. These cells fight infection. Platelets (thrombocytes) are tiny cell fragments which are involved in blood clotting.

1. Study a prepared slide of human blood. With the help of Figure 11.2, identify the red blood cells and the white blood cells, which appear faint because of the stain.
2. Try to identify a neutrophil, which has a multilobed nucleus, and a lymphocyte, which is the smallest of the white blood cells, with a spherical or slightly indented nucleus.
3. Add a sketch of this tissue to Table 11.2 (p. 152).

Figure 11.2 Blood cells.
Red blood cells are more numerous than white blood cells. White blood cells can be separated into five distinct types.

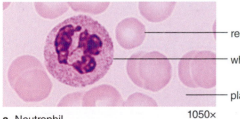

red blood cell
white blood cell
plasma

a. Neutrophil   1050×

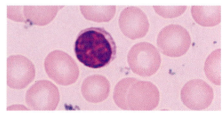

b. Lymphocyte   1050×

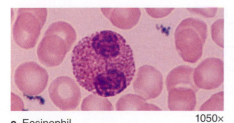

c. Eosinophil   1050×

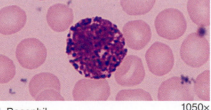

d. Basophil   1050×

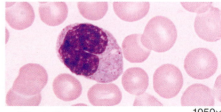

e. Monocyte   1050×

Summary of Connective Tissue

1. *Add a sketch in first column of Table 11.2 under each type of connective tissue.*

Table 11.2 Connective Tissue			
Sketch	**Structure**	**Function**	**Location**
Loose fibrous	Fibers are widely separated.	Binds organs together	Between the muscles; beneath the skin; beneath most epithelial layers
Dense fibrous	Fibers are closely packed.	Binds organs together, binds muscle to bone, binds bone to bone	Tendons, ligaments
Adipose tissue	Large cell with fat-filled vacuole; nucleus pushed to one side	Insulation, fat storage, cushioning, and protection	Beneath the skin; around the kidney and heart; in the breast
Compact bone	Concentric circles	Support, protection	Bones of skeleton
Hyaline cartilage	Cells in lacunae	Support, protection	Nose, ends of bones, rings in walls of respiratory passages; between ribs and sternum
Blood	Red and white cells floating in plasma	RBCs carry oxygen and hemoglobin for respiration; WBCs fight infection.	Blood vessels

2. Working with others in a group, decide how the structure of each connective tissue suits its function.

Loose fibrous connective tissue _____

Dense fibrous connective tissue _____

Adipose tissue _____

Compact bone _____

Hyaline cartilage _____

Blood _____

Muscular Tissue

Muscular (contractile) tissue is composed of cells called muscle fibers. Muscular tissue has the ability to contract, and contraction usually results in movement. The body contains skeletal, cardiac, and smooth muscle.

Observation: Skeletal Muscle

Skeletal muscle occurs in the muscles attached to the bones of the skeleton. The contraction of skeletal muscle is said to be **voluntary** because it is under conscious control. Skeletal muscle is striated; it contains light and dark bands. The striations are caused by the arrangement of contractile filaments (actin and myosin filaments) in muscle fibers. Each fiber contains many nuclei, all peripherally located.

1. Study a model or diagram of skeletal muscle, and note that striations are present. You should see several muscle fibers, each marked with striations.
2. Examine a prepared slide of skeletal muscle. Using high power, locate the striations. Bringing the slide in and out of focus may also help.
3. Add a sketch of this tissue to Table 11.3 (p. 154).

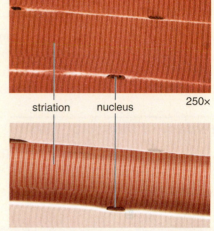

Skeletal muscle
• has striated cells with multiple nuclei.
• occurs in muscles attached to skeleton.
• functions in voluntary movement of body.

striation   nucleus   250×

Observation: Cardiac Muscle

Cardiac muscle is found only in the heart. It is called **involuntary** because its contraction does not require conscious effort. Cardiac muscle is striated in the same way as skeletal muscle. However, the fibers are branched and bound together at **intercalated disks,** where their folded plasma membranes touch. This arrangement aids communication between fibers.

1. Study a model or diagram of cardiac muscle, and note that striations are present.
2. Examine a prepared slide of cardiac muscle. Using high power, find an intercalated disk. What is the function of cardiac muscle?

_____

_____

3. Add a sketch of this tissue to Table 11.3 (p. 154).

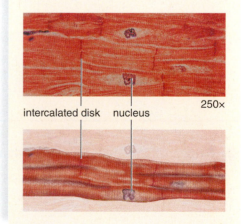

Cardiac muscle
• has branching, striated cells, each with a single nucleus.
• occurs in the wall of the heart.
• functions in the pumping of blood.
• is involuntary.

intercalated disk   nucleus   250×

Smooth muscle is sometimes called **visceral muscle** because it makes up the walls of the internal organs, such as the intestines and the blood vessels. Smooth muscle is involuntary because its contraction does not require conscious effort.

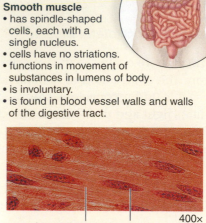

Smooth muscle
- has spindle-shaped cells, each with a single nucleus.
- cells have no striations.
- functions in movement of substances in lumens of body.
- is involuntary.
- is found in blood vessel walls and walls of the digestive tract.

1. Study a model or diagram of smooth muscle, and note the shape of the cells and the centrally placed nucleus. Smooth muscle has spindle-shaped cells. What does *spindle-shaped* mean? _____

_____

_____

2. Examine a prepared slide of smooth muscle. Distinguishing the boundaries between the different cells may require you to take the slide in and out of focus.

3. Add a sketch of this tissue to Table 11.3.

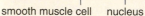

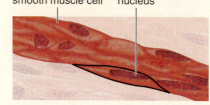

smooth muscle cell   nucleus   400×

Summary of Muscular Tissue

1. *Add a sketch in first column of Table 11.3 under each type of muscular tissue.*

Table 11.3 Muscular Tissue			
Sketch	**Striations (Yes or No)**	**Branching (Yes or No)**	**Conscious Control (Yes or No)**
Skeletal			
Cardiac			
Smooth			

2. How does it benefit an animal that skeletal muscle is voluntary, while cardiac and smooth muscle are involuntary? _____

_____

Nervous Tissue

Nervous tissue is found in the brain, spinal cord, and nerves. Nervous tissue receives and integrates incoming stimuli before conducting nerve impulses, which control the glands and muscles of the body. Nervous tissue is composed of two types of cells: **neurons** that transmit messages and **neuroglia** that support and nourish the neurons. Motor neurons, which take messages from the spinal cord to the muscles, are often used to exemplify typical neurons. Motor neurons have several **dendrites,** processes that take signals to a **cell body,** where the nucleus is located, and an **axon** that takes nerve impulses away from the cell body.

Observation: Nervous Tissue

1. Study a model or diagram of a neuron, and identify the dendrites, cell body, nucleus, and axon (Fig. 11.3*a*). Long axons are called nerve fibers.

2. *In Figure 11.3b, label the dendrites, cell body, nucleus and axon. Also label neuroglia.*

3. Explain the appearance and function of the parts of a motor neuron:

   a. Dendrites _____

   _____

   b. Cell body _____

   _____

   c. Axon _____

   _____

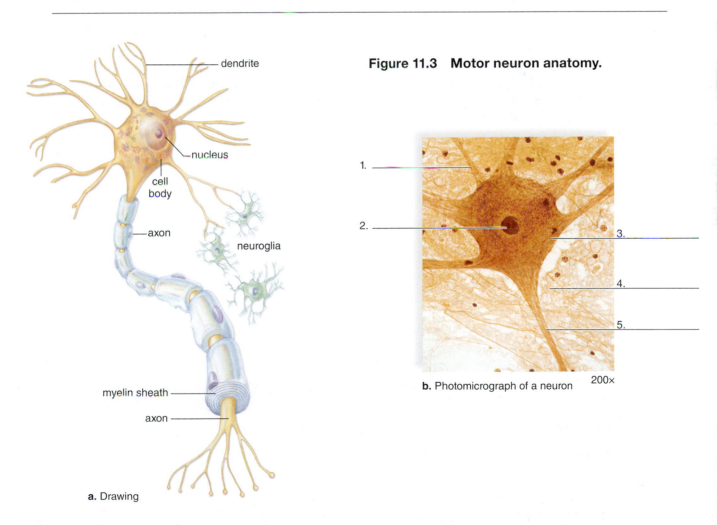

Figure 11.3   Motor neuron anatomy.

a. Drawing

b. Photomicrograph of a neuron   200×

11.2 Organ Level of Organization

Organs are structures composed of two or more types of tissue that work together to perform particular functions. You may tend to think that a particular organ contains only one type of tissue. For example, muscular tissue is usually associated with muscles and nervous tissue with the brain. However, muscles and the brain also contain other types of tissue—for example, loose connective tissue and blood. Here we will study the compositions of two organs—the intestine and the skin.

Intestine

The **intestine,** a part of the digestive system, processes food and absorbs nutrient molecules.

Observation: Intestinal Wall

Study a slide of a cross section of intestinal wall. With the help of Figure 11.4, identify the following layers:

1. **Mucosa** (mucous membrane layer): This layer, which lines the central lumen (cavity), is made up of columnar epithelium overlying a layer of connective tissue. The epithelium is glandular—that is, it secretes mucus from goblet cells and digestive enzymes from the rest of the epithelium. The membrane is arranged in deep folds (fingerlike projections) called **villi,** which increase the small intestine's absorptive surface.
2. **Submucosa** (submucosal layer): This connective tissue layer contains nerve fibers, blood vessels, and lymphatic vessels. The products of digestion are absorbed into these blood and lymphatic vessels.
3. **Muscularis** (smooth muscle layer): Circular muscular tissue and then longitudinal muscular tissue are found in this layer. Rhythmic contraction of these muscles causes **peristalsis,** a wavelike motion that moves food along the intestine.
4. **Serosa** (serous membrane layer): In this layer, a thin sheet of connective tissue underlies a thin, outermost sheet of squamous epithelium. This membrane is part of the **peritoneum,** which lines the entire abdominal cavity.

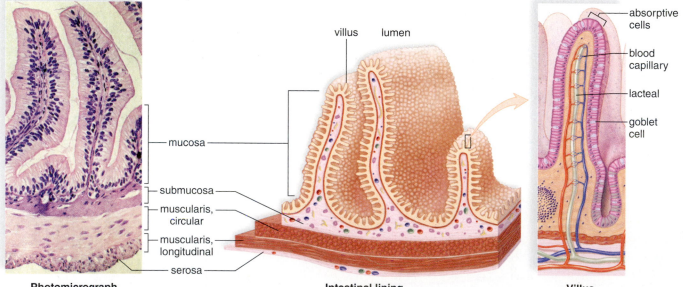

Photomicrograph   Intestinal lining   Villus

Figure 11.4 Tissues of the intestinal wall.

Skin

The skin covers the entire exterior of the human body. Skin functions include protection, water retention, sensory reception, body temperature regulation, and vitamin D synthesis.

Observation: Skin

Study a model or diagram and also a prepared slide of the skin. With the help of Figure 11.5, identify the two skin regions and the subcutaneous layer.

1. **Epidermis:** This region is composed of stratified squamous epithelial cells. The outer cells of the epidermis are nonliving and create a waterproof covering that prevents excessive water loss. These cells are always being replaced because an inner layer of the epidermis is composed of living cells that constantly produce new cells.

2. **Dermis:** This region is a connective tissue containing blood vessels, nerves, sense organs, and the expanded portions of oil (sebaceous) and sweat glands and hair follicles.

   List the structures you can identify on your slide:

   _____

   _____

3. **Subcutaneous layer:** This is a layer of loose connective tissue and adipose tissue that lies beneath the skin proper and serves to insulate and protect inner body parts. This layer is not part of the skin.

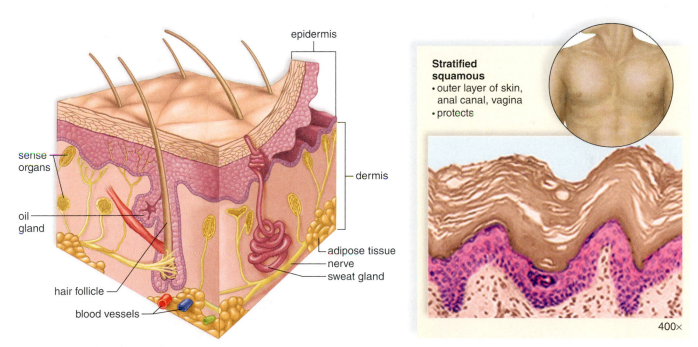

Figure 11.5 Human skin.
Human skin contains two regions, the epidermis and the dermis.

_____ 1. What is the name for a group of cells that has the same structural characteristics and performs the same functions?

_____ 2. Which type of epithelium has flattened cells?

_____ 3. Name a body location for pseudostratified ciliated columnar epithelium.

_____ 4. What is the function of goblet cells?

_____ 5. What type of tissue occurs in the epidermis of the skin?

_____ 6. Name a body location for hyaline cartilage.

_____ 7. Name the tissue in which the cells contain a large, central vacuole filled with fat.

_____ 8. What type of muscular tissue is involuntary and striated?

_____ 9. Name a body location for smooth muscle.

_____ 10. What types of muscular tissue are striated?

_____ 11. Name a body location for nervous tissue.

_____ 12. Where is the nucleus located in a nerve cell?

_____ 13. What type of tissue accounts for the movement of food along the digestive tract?

_____ 14. Which skin layer contains blood vessels?

_____ 15. What portion of a nerve cell transmits information away from the cell body?

Thought Questions

16. **a.** List the four major types of human body tissues and list the distinguishing characteristics of each.

   b. How does the human body benefit from having four types of tissue rather than a single type?

17. List the five types of epithelial tissue and the distinguishing characteristics of each.

18. Your lab instructor gives you a slide containing prepared muscle tissue. How would you identify the type of muscle tissue located on the slide?

19. How does the organization of an organ differ from that of a tissue? Give an example.

20. Why might an injury to a bone have a faster recovery/healing time when compared to a muscle or nerve cell?

12

Basic Mammalian Anatomy I

Learning Outcomes

Introduction

In this laboratory, you will dissect a fetal pig. Alternately your instructor may choose to have you observe a pig that has already been dissected. Both pigs and humans are mammals; therefore, you will be studying mammalian anatomy. The period of pregnancy, or gestation, in pigs is approximately 17 weeks (compared with an average of 40 weeks in humans). The piglets used in class will usually be within 1 to 2 weeks of birth.

The pigs may have a slash in the right neck region, indicating the site of blood drainage. A red latex solution may have been injected into the **arterial system,** and a blue latex solution may have been injected into the **venous system** of the pigs. If so, when a vessel appears red, it is an artery, and when a vessel appears blue, it is a vein.

As a result of this laboratory, you should gain an appreciation of which organs work together. For example, the liver and the pancreas aid the digestion of fat in the small intestine.

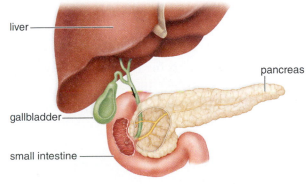

12.1 External Anatomy

Mammals are characterized by the presence of mammary glands and hair. Mammals also occur in two distinct sexes, males and females, often distinguishable by their external **genitals,** the reproductive organs.

Both pigs and humans are placental mammals, which means that development occurs within the uterus of the mother. An **umbilical cord** stretches externally between the fetal animal and the **placenta,** where carbon dioxide and organic wastes are exchanged for oxygen and organic nutrients.

Pigs and humans are tetrapods—that is, they have four limbs. Pigs walk on all four of their limbs; in fact, they walk on their toes, and their toenails have evolved into hooves. In contrast, humans walk only on the feet of their legs.

Observation: External Anatomy

Body Regions and Limbs

1. Place a pig in a dissecting pan, and observe the following body regions: the rather large head; the short, thick neck; the cylindrical trunk with two pairs of appendages (forelimbs and hindlimbs); and the short tail (Fig. 12.1a). The tail is an extension of the vertebral column.

⚠️ **Latex gloves:** Wear protective latex gloves when handling preserved animal organs. Use protective eyewear and exercise caution when using sharp instruments during this experiment. Wash hands thoroughly upon completion of this experiment.

2. Examine the four limbs, and feel for the joints of the digits, wrist, elbow, shoulder, hip, knee, and ankle.
3. Determine which parts of the forelimb correspond to your arm, elbow, forearm, wrist, and hand.
4. Do the same for the hindlimb, comparing it with your thigh and leg.
5. The pig walks on its toenails, which would be like a ballet dancer on "tiptoe." Notice how your heel touches the ground when you walk. Where is the heel of the pig? _____

Umbilical Cord

1. Locate the umbilical cord arising from the ventral (toward the belly) portion of the abdomen.
2. Note the cut ends of the umbilical blood vessels. If they are not easily seen, cut the umbilical cord near the end and observe this new surface.
3. What is the function of the umbilical cord? _____

_____

Nipples and Hair

1. Locate the small **nipples,** the external openings of the **mammary glands.** The nipples are *not* an indication of sex, since both males and females possess them. How many nipples does a pig have? _____ When is it advantageous for a pig to have so many nipples?

_____

2. Can you find hair on the pig? _____ Where? _____

_____

Directional Terms for Dissecting Fetal Pig

Anterior: toward the head end   Ventral: toward the belly
Posterior: toward the hind end   Dorsal: toward the back

Figure 12.1   External anatomy of the fetal pig.
a. Body regions and limbs. **b, c.** The sexes can be distinguished by the external genitals.

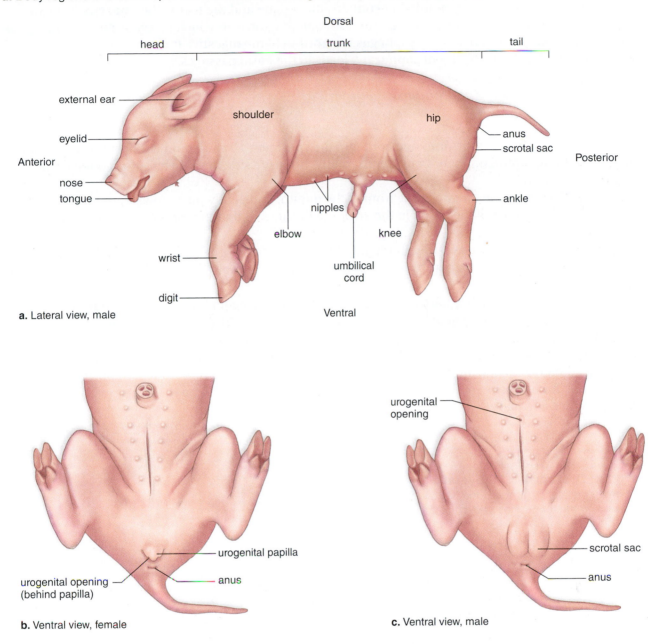

a. Lateral view, male

b. Ventral view, female

c. Ventral view, male

Anus and External Genitals

1. Locate the **anus** under the tail. Name the organ system that ends in the opening called the anus. _____

2. In females, locate the **urogenital opening,** just anterior to the anus, and a small, fleshy **urogenital papilla** projecting from the urogenital opening (Fig. 12.1*b*).

3. In males, locate the urogenital opening just posterior to the umbilical cord. The duct leading to it runs forward from between the legs in a long, thick tube, the **penis,** which can be felt under the skin. In males, the urinary system and the genital system are always joined (Fig. 12.1*c*).

4. You are responsible for identifying pigs of both sexes. What sex is the pig you are examining? _____

   Be sure to look at a pig of the opposite sex that another group of students is dissecting.

12.2 Oral Cavity and Pharynx

The **oral cavity** is the space in the mouth that contains the tongue and the teeth. The **pharynx** is dorsal to the oral cavity and has three openings: The glottis is an opening through which air passes on its way to the **trachea** (the windpipe) and lungs. The esophagus is a portion of the digestive tract that leads through the neck and thorax to the stomach. The **nasopharynx** leads to the nasal passages.

Observation: Oral Cavity and Pharynx

Oral Cavity

1. Insert a sturdy pair of scissors into one corner of the specimen's mouth, and cut posteriorly (toward the hind end) for approximately 4 cm. Repeat on the opposite side until the mouth is open as in Figure 12.2 below.
2. Place your thumb on the tongue at the front of the mouth, and gently push downward on the lower jaw. This will tear some of the tissue in the angles of the jaws so that the mouth will remain partly open (Fig. 12.2).
3. Note small, underdeveloped teeth in both the upper and lower jaws. Care should be taken because teeth can be very sharp. Other embryonic, nonerupted teeth may also be found within the gums. The teeth are used to chew food.
4. Examine the tongue, which is partly attached to the lower jaw region but extends posteriorly and is attached to a bony structure at the back of the oral cavity (Fig. 12.2). The tongue manipulates food for swallowing.
5. Locate the hard and soft palates (Fig. 12.2). The **hard palate** is the ridged roof of the mouth that separates the oral cavity from the nasal passages. The **soft palate** is a smooth region posterior to the hard palate. An extension of the soft palate—the **uvula**—hangs down into the throat in humans. (A pig does not have a uvula.)

Figure 12.2   Oral cavity of the fetal pig.
The roof of the oral cavity contains the hard and soft palates, and the tongue lies above the floor of the oral cavity.

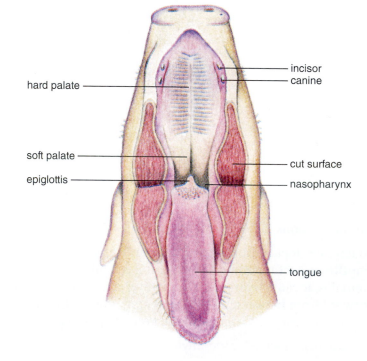

Pharynx

1. Push down on the jaws until they have opened far enough to reveal a slightly pointed flap of tissue that points dorsally (toward the back) (Fig. 12.2). This flap is the **epiglottis,** which covers the glottis. The **glottis** leads to the trachea (Fig. 12.3*a*).
2. Posterior and dorsal to the glottis, find the opening into the **esophagus,** a tube that takes food to the stomach. Note the proximity of the glottis and the opening to the esophagus. Each time the pig—or a human—swallows, the epiglottis instantly closes to keep food and fluids from going into the lungs via the trachea.
3. Insert a blunt probe into the glottis, and note that it enters the trachea. Remove the probe, insert it into the esophagus, and note the position of the esophagus beneath the trachea.
4. Make two lateral cuts at the edge of the hard palate.
5. Posterior to the soft palate, locate the openings to the nasal passages.
6. Explain why it is correct to say that the air and food passages cross in the pharynx.

Figure 12.3   Air and food passages in the fetal pig.
The air and food passages cross in the pharynx. **a.** Drawing. **b.** Dissection of specimen.

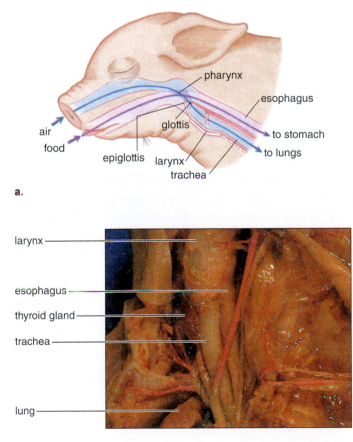

12.3 Thoracic and Abdominal Incisions

First, prepare your pig according to the following directions, and then make thoracic and abdominal incisions so that you will be able to study the internal anatomy of your pig.

Preparation of Pig for Thoracic and Abdominal Incisions

1. Place the fetal pig on its back in the dissecting pan.
2. Tie a cord around one forelimb, and then bring the cord around underneath the pan to fasten back the other forelimb.
3. Spread the hindlimbs in the same way.
4. With scissors always pointing up (never down), make the following incisions to expose the thoracic and abdominal cavities. The incisions are numbered on Figure 12.4 to correspond with the following steps.

Thoracic Incisions

1. Cut anteriorly up from the **diaphragm,** a structure that separates the thoracic cavity from the abdominal cavity, until you reach the clump of hair below the chin.
2. Make two lateral cuts, one on each side of the midline incision anterior to the forelimbs, taking extra care not to damage the blood vessels around the heart.
3. Make two lateral cuts, one on each side of the midline just posterior to the forelimbs and anterior to the diaphragm, following the ends of the ribs. Pull back the flaps created by these cuts to expose the

thoracic cavity. List the organs you find in the thoracic cavity. See Figures 12.5 and 12.6. _____

_____

Abdominal Incisions

4. With scissors pointing up, cut posteriorly from the diaphragm to the umbilical cord.
5. Make a flap containing the umbilical cord by cutting a semicircle around the cord and by cutting posteriorly to the left and right of the cord.
6. Make two cuts, one on each side of the midline incision posterior to the diaphragm. Examine the diaphragm, attached to the chest wall by radially arranged muscles. The central region of the diaphragm, called the **central tendon,** is a membranous area.
7. Make two more cuts, one on each side of the flap containing the umbilical cord and just anterior to the hindlimbs. Pull back the side flaps created by these cuts to expose the **abdominal cavity.**
8. Lifting the flap with the umbilical cord requires cutting the **umbilical vein.** Before cutting the umbilical vein, tie a thread on each side of the vein. Cut the vein but keep the threads in place for future reference.
9. As soon as you have opened the abdominal cavity, rinse out the pig. If you have a problem with excess fluid, obtain a disposable plastic pipette to suction off the liquid.
10. Name the two cavities separated by the diaphragm. _____
11. List the organs located in the abdominal cavity. See Figures 12.5 and 12.6. _____

_____

_____

Figure 12.4 Ventral view of the fetal pig indicating incisions.

These incisions are to be made preparatory to dissecting the internal organs. They are numbered here in the order they should be done.

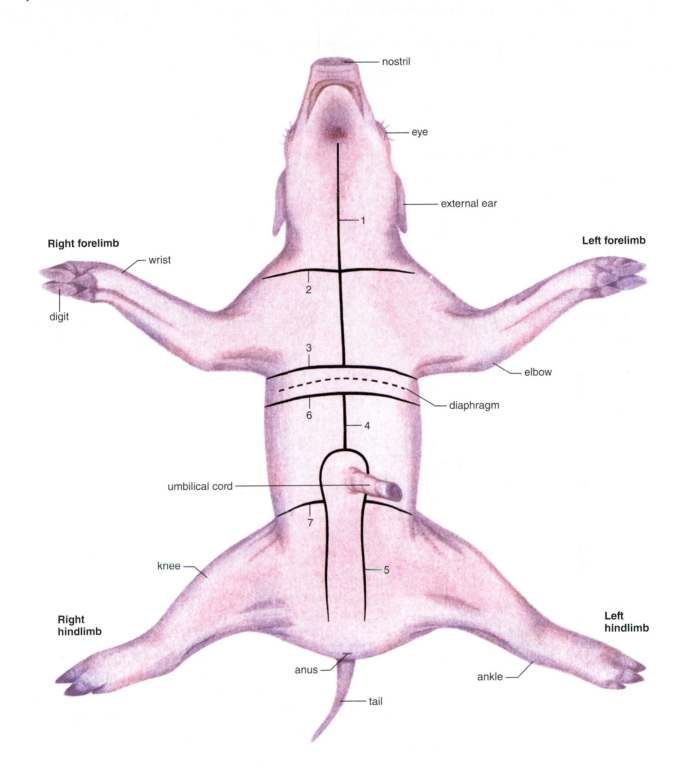

nostril

eye

external ear

Right forelimb

wrist

1

Left forelimb

digit

2

3

elbow

diaphragm

6

4

umbilical cord

7

Right hindlimb

knee

5

Left hindlimb

anus

ankle

tail

12.4 Neck Region

Several organs in the neck region are of interest. Use Figures 12.3b and 12.5 as a guide to locate these organs, but *keep all the flaps* in order to close the thoracic and abdominal cavities at the end of the laboratory session.

The **thymus gland** is a part of the lymphatic system. Certain white blood cells called T (for thymus) lymphocytes mature in the thymus gland and help us fight disease. The **larynx,** or voice box, sits atop the **trachea,** or windpipe. The esophagus is a portion of the digestive tract that leads to the stomach. The **thyroid gland,** a part of the endocrine system, secretes hormones that travel in the blood and act upon other body cells. These hormones (e.g., thyroxine) regulate the rate at which metabolism occurs in cells.

Observation: Neck Region

Thymus Gland

1. Move the skin apart in the neck region just below the hairs mentioned earlier. If necessary, cut the body wall laterally to make flaps.
2. If necessary, *cut through and clear away muscle* to expose the thymus gland, a diffuse gland that lies among the muscles. Later you will notice that the thymus flanks the thyroid and overlies the heart. The thymus is particularly large in fetal pigs, since their immune systems are still developing.

Larynx, Trachea, and Esophagus

1. Probe down into the deeper layers of the neck. Medially (toward the center), beneath several strips of muscle, find the hard-walled larynx and the trachea, which are parts of the respiratory passage. Dorsal to the trachea, find the esophagus.
2. Open the mouth and insert a probe into the glottis and esophagus from the pharynx to better understand the orientation of these two organs.

Thyroid Gland

Locate the thyroid gland just posterior to the larynx, lying ventral to (on top of) the trachea.

12.5 Thoracic Cavity

As previously mentioned, the body cavity of mammals, including human beings, is divided by the diaphragm into the thoracic cavity and the abdominal cavity. The heart and lungs are in the thoracic cavity (Figs. 12.5 and 12.6). The **heart** is a pump for the cardiovascular system, and the **lungs** are organs of the respiratory system where gas exchange occurs.

Observation: Thoracic Cavity

Heart and Lungs

1. In order to fold back the chest wall flaps, tear the thin membranes that divide the thoracic cavity into three compartments. The three compartments are the **left pleural cavity** containing the left lung, the **right pleural cavity** containing the right lung, and the **pericardial cavity** containing the heart.
2. Examine the lungs. Locate the four lobes of the right lung and the three lobes of the left lung. The trachea, dorsal to the heart, divides into the **bronchi,** which enter the lungs.
3. Trace the path of air from the nasal passages to the lungs.

_____

_____

Figure 12.5 Internal anatomy of the fetal pig.

The major organs are featured in this drawing. In the fetal pig, a vessel colored red is an artery, and a vessel colored blue is a vein. (The color does not indicate whether this vessel carries O_2-rich or O_2-poor blood.) Contrary to this drawing, do not cut the flaps, because they can be closed to protect the thoracic and abdominal cavities.

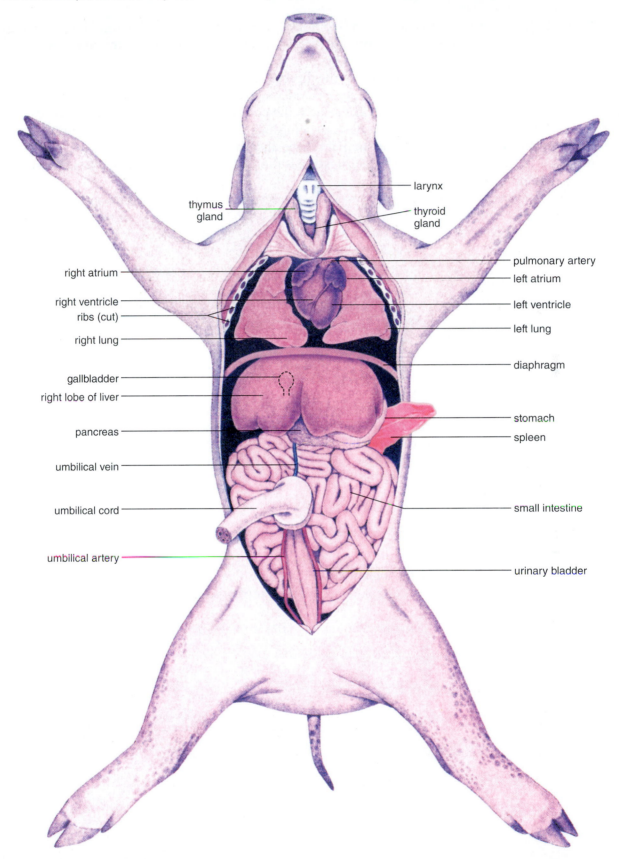

larynx

thymus gland

thyroid gland

pulmonary artery

right atrium

left atrium

right ventricle

left ventricle

ribs (cut)

left lung

right lung

diaphragm

gallbladder

right lobe of liver

stomach

pancreas

spleen

umbilical vein

umbilical cord

small intestine

umbilical artery

urinary bladder

12.6 Abdominal Cavity

The abdominal wall and organs are lined by a membrane called **peritoneum,** consisting of epithelium supported by connective tissue. Double-layered sheets of peritoneum, called **mesenteries,** project from the body wall and support the organs.

The **liver,** the largest organ in the abdomen (Fig. 12.6), performs numerous vital functions, including (1) disposing of worn-out red blood cells, (2) producing bile, (3) storing glycogen, (4) maintaining the blood glucose level, and (5) producing blood proteins.

The abdominal cavity also contains organs of the digestive tract, such as the stomach, small intestine, and large intestine. The **stomach** (see Fig. 12.5) stores food and has numerous gastric glands that secrete gastric juice, which digests protein. The **small intestine** is the part of the digestive tract that receives secretions from the pancreas and gallbladder. Besides being an area for the digestion of all components of food—carbohydrate, protein, and fat—the small intestine absorbs the products of digestion: glucose, amino acids, glycerol, and fatty acids. The **large intestine** is the part of the digestive tract that absorbs water and prepares feces for defecation at the anus.

The **gallbladder** stores and releases bile, which aids the digestion of fat. The **pancreas** (see Fig. 12.5) is both an exocrine and an endocrine gland. As an exocrine gland, it produces and secretes pancreatic juice, which digests all the components of food in the small intestine. Both bile and pancreatic juice enter the duodenum, the first, straight part of the small intestine, by way of ducts. As an endocrine gland, the pancreas secretes the hormones insulin and glucagon into the bloodstream. Insulin and glucagon regulate blood glucose levels.

The **spleen** (see Fig. 12.5) is a lymphoid organ in the lymphatic system that contains both white and red blood cells. It purifies blood and disposes of worn-out red blood cells.

Observation: Abdominal Cavity

Liver

1. If your particular pig is partially filled with dark, brownish material, take your animal to the sink and rinse it out. This material is clotted blood. Consult your instructor before removing any red or blue latex masses, since they may enclose organs you will need to study.
2. Locate the liver, a large, brown organ. Its anterior surface is smoothly convex and fits snugly into the concavity of the diaphragm.
3. Name several functions of the liver. _____

_____

Stomach and Spleen

1. Push aside and identify the stomach, a large sac dorsal to the liver on the left side.
2. Locate the point near the midline of the body where the **esophagus** penetrates the diaphragm and joins the stomach.
3. Find the spleen, a long, flat, reddish organ attached to the stomach by mesentery.
4. The stomach is a part of the _____ system.

   What is its function? _____

5. The spleen is a part of the _____ system.

   What is its function? _____

Figure 12.6 Internal anatomy of the fetal pig.

Most of the major organs are shown in this photograph. The stomach has been removed. The spleen, gallbladder, and pancreas are not visible. *Do not* remove any organs or flaps from your pig.

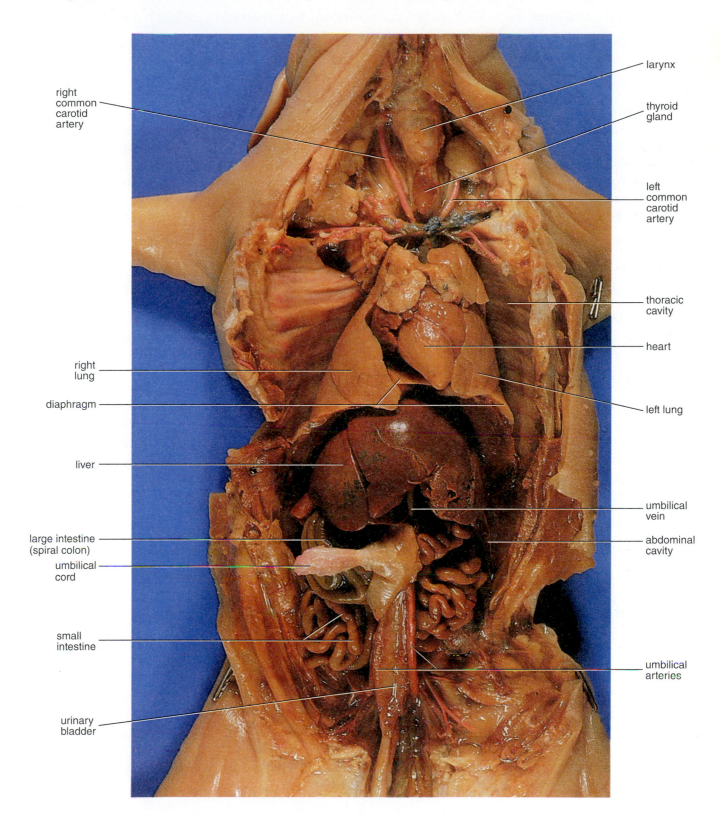

right
common
carotid
artery

right
lung

diaphragm

liver

large intestine
(spiral colon)

umbilical
cord

small
intestine

urinary
bladder

larynx

thyroid
gland

left
common
carotid
artery

thoracic
cavity

heart

left lung

umbilical
vein

abdominal
cavity

umbilical
arteries

Small Intestine

1. Look posteriorly where the stomach makes a curve to the right and narrows to join the anterior end of the small intestine called the **duodenum.**
2. From the duodenum, the small intestine runs posteriorly for a short distance and is then thrown into an irregular mass of bends and coils held together by a common mesentery.
3. The small intestine is a part of the _____ system.

   What is its function? _____

Gallbladder and Pancreas

1. Locate the **bile duct,** which runs in the mesentery stretching between the liver and the duodenum. Find the gallbladder, embedded in the liver on the underside of the right lobe. It is a small, greenish sac.
2. Lift the stomach and locate the pancreas, the light-colored, diffuse gland lying in the mesentery between the stomach and the small intestine. The pancreas has a duct that empties into the duodenum of the small intestine.
3. What is the function of the gallbladder? _____
4. What is the function of the pancreas? _____

   _____

Large Intestine

1. Locate the distal (far) end of the small intestine, which joins the large intestine posteriorly, in the left side of the abdominal cavity (right side in humans). At this junction, note the **cecum,** a blind pouch.
2. Compare the large intestine of a pig to Figure 12.7. The organ does not have the same appearance in humans.
3. Follow the main portion of the large intestine, known as the **colon,** as it runs from the point of juncture with the small intestine into a tight coil (spiral colon), then out of the coil anteriorly, then posteriorly again along the midline of the dorsal wall of the abdominal cavity. In the pelvic region, the **rectum** is the last portion of the large intestine. The rectum leads to the **anus.**
4. The large intestine is a part of the _____ system.
5. What is the function of the large intestine? _____
6. Trace the path of food from the mouth to the anus. _____

   _____

Storage of Pigs

1. Before leaving the laboratory, place your pig in the plastic bag provided.
2. Expel excess air from the bag, and tie it shut.
3. Write your name and section on the tag provided, and attach it to the bag. Your instructor will indicate where the bags are to be stored until the next laboratory period.
4. Clean the dissecting tray and tools, and return them to their proper location.
5. Wipe off your goggles.
6. Wash your hands.

12.7 Human Anatomy

Humans and pigs are both mammals, and their organs are similar. A human torso model shows the exact location of the organs in human beings (Fig. 12.7). Learn to associate these organs with their particular system. Four systems are color-coded in Figure 12.7.

Figure 12.7 Human internal organs.
The dotted lines indicate the full shape of an organ that is partially covered by another organ.

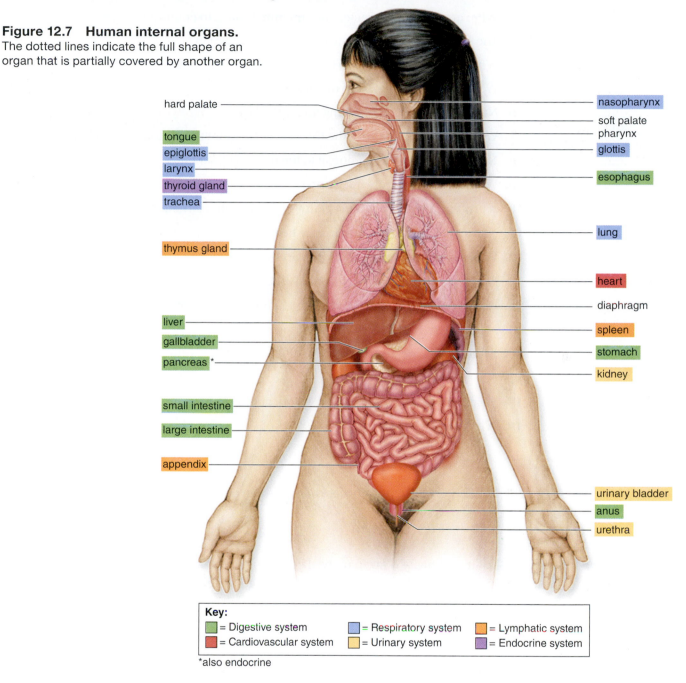

Key:
- ■ = Digestive system
- ■ = Cardiovascular system
- ■ = Respiratory system
- ■ = Urinary system
- ■ = Lymphatic system
- ■ = Endocrine system

*also endocrine

Observation: Human Torso

1. Examine a human torso model, and using Figure 12.7 as a guide, locate the same organs just dissected in the fetal pig.

2. Name any observed major differences between pig internal anatomy and human internal anatomy.

_____ 1. In the fetal pig, what sex has a urogenital opening beneath the papilla just superior to the anus?

_____ 2. What are the two characteristics mammals have in common?

_____ 3. The esophagus connects the pharynx to what organ?

_____ 4. What is the hard portion of the roof of the mouth called?

_____ 5. What is the opening to the trachea called?

_____ 6. Name the largest organ in the abdominal cavity.

_____ 7. What structure separates the thoracic cavity from the abdominal cavity?

_____ 8. Name the structure just dorsal to the thyroid gland.

_____ 9. What structure covers the glottis?

_____ 10. A probe placed through the glottis will enter what structure?

_____ 11. Name the cavity that contains the heart.

_____ 12. What organs are in the pleural cavity?

_____ 13. Name the specific part of the small intestine that attaches to the stomach.

_____ 14. Identify the gland that is located by lifting the stomach.

_____ 15. Name a lymphoid organ in the abdominal cavity.

_____ 16. Is the spleen located on the right or left side of the abdominal cavity?

_____ 17. The pancreas is a part of the _____ system.

_____ 18. Where do air and food passages cross one another?

_____ 19. What organ releases bile?

Thought Questions

20. What difficulty would probably arise if a person were born without an epiglottis?

21. A large portion of the abdominal cavity is taken up by digestive organs. Which organs are these?

22. The small intestine exists as a series of folds and coils. What might be the advantage of such a configuration?

23. Difficulties maintaining blood glucose level, bile production, and the production of blood proteins might be associated with problems in what organ?

13

Chemical Aspects of Digestion

Introduction

In Laboratory 12, you examined the organs of digestion in the fetal pig. Now we wish to further our knowledge of the digestive process by associating certain digestive enyzmes with particular organs, as shown in Figure 13.1. This laboratory will also give us an opportunity to study the action of enzymes, much as William Beaumont did when he removed food samples through a hole in the stomach wall of his patient, Alexis St. Martin. Every few hours, Beaumont would see how well the food had been digested.

In Laboratory 6 we learned that enzymes are very specific and usually participate in only one type of reaction. The active site of an enzyme has a shape that accommodates its substrate, and if an environmental factor such as a boiling temperature or a wrong pH alters this shape, the enzyme loses its ability to function well, if at all. We will have an opportunity to make these observations with controlled experiments. The box on the next page reviews what is meant by a controlled experiment.

> **Planning Ahead** Be advised that protein digestion (page 175) requires 1½ hours and fat digestion (page 177) requires 1 hour. Also a boiling water bath is required for starch digestion (page 179).

> **Virtual Lab Nutrition** This virtual laboratory available on the *Inquiry into Life* website, **www.mhhe.com/mader Inquiry14**, provides an opportunity to create a diet of varied foods. Be sure to click on the forward arrow for choices from all food groups; keep within the allotted calories for your build; and fulfill but do not exceed the percent daily requirements for nutrients.

Figure 13.1 Organs of the digestive tract (right) and accessory organs (left).

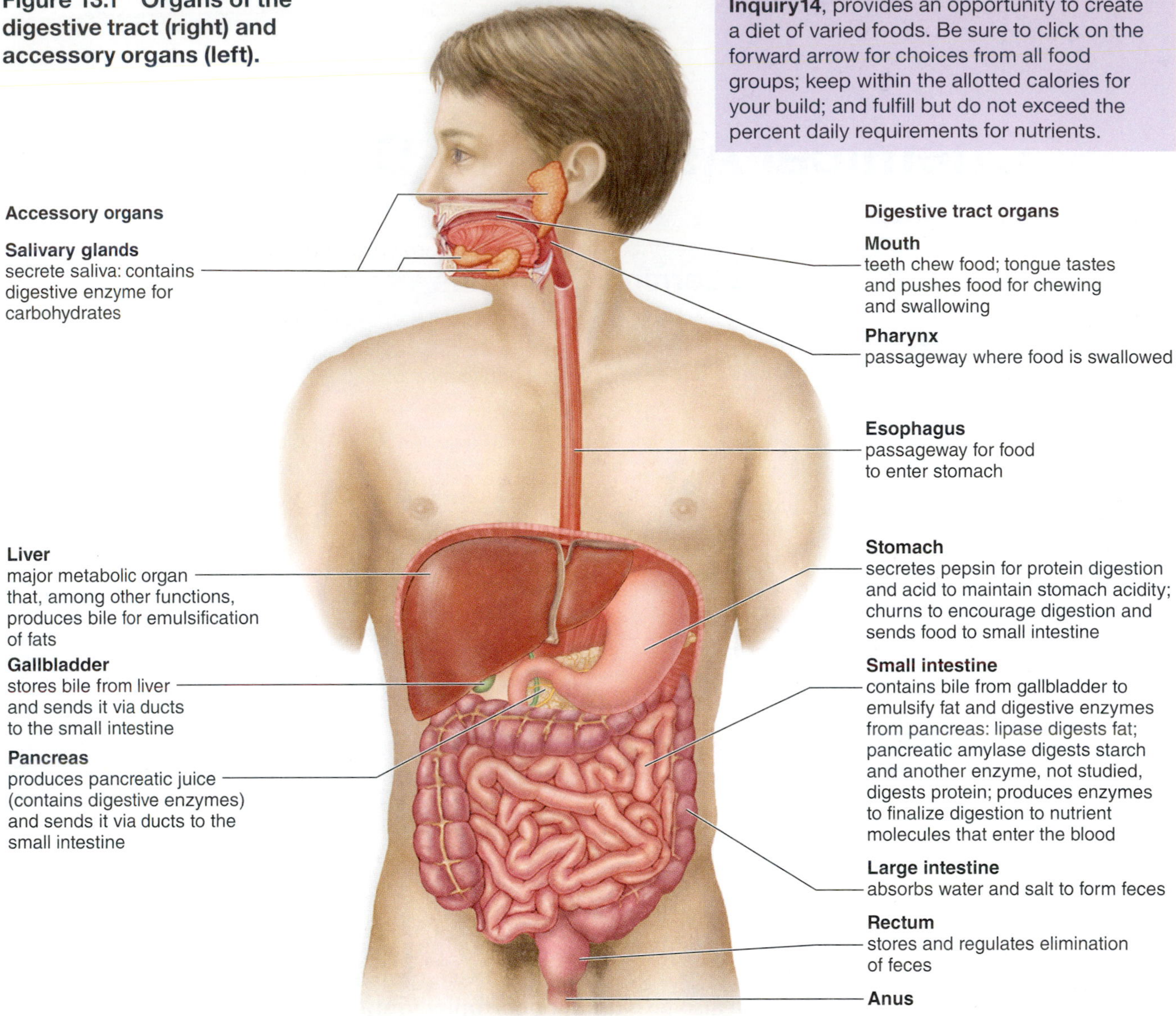

Accessory organs
Salivary glands
secrete saliva: contains digestive enzyme for carbohydrates
Liver
major metabolic organ that, among other functions, produces bile for emulsification of fats
Gallbladder
stores bile from liver and sends it via ducts to the small intestine
Pancreas
produces pancreatic juice (contains digestive enzymes) and sends it via ducts to the small intestine
Digestive tract organs
Mouth
teeth chew food; tongue tastes and pushes food for chewing and swallowing
Pharynx
passageway where food is swallowed
Esophagus
passageway for food to enter stomach
Stomach
secretes pepsin for protein digestion and acid to maintain stomach acidity; churns to encourage digestion and sends food to small intestine
Small intestine
contains bile from gallbladder to emulsify fat and digestive enzymes from pancreas: lipase digests fat; pancreatic amylase digests starch and another enzyme, not studied, digests protein; produces enzymes to finalize digestion to nutrient molecules that enter the blood
Large intestine
absorbs water and salt to form feces
Rectum
stores and regulates elimination of feces
Anus

What Is a Control?

The experiments in today's laboratory have both a positive control and a negative control, *which should be saved for comparison purposes until the experiment is complete.* The **positive control** goes through all the steps of the experiment and does contain the substance being tested. Therefore, positive results are expected. The **negative control** goes through all the steps of the experiment, except it does not contain the substance being tested. Therefore, negative results are expected.

For example, if a test tube contains glucose (the substance being tested) and Benedict's reagent (blue) is added, a red color develops upon heating. This test tube is the positive control; it tests positive for glucose. If a test tube does not contain glucose and Benedict's reagent is added, Benedict's is expected to remain blue. This test tube is the negative control; it tests negative for glucose.

What benefit is a positive control? Positive controls give you a standard by which to tell if the substance being tested is present (or acting properly) in an unknown sample. Negative controls ensure that the experiment is giving reliable results; after all, if a negative control should happen to give a positive result, then the entire experiment may be faulty and unreliable.

13.1 Protein Digestion by Pepsin

Certain foods, such as meat and egg whites, are rich in protein. Egg whites contain albumin, which is the protein used in this Experimental Procedure. Protein is digested by **pepsin** in the stomach (Fig. 13.2), a process described by the following reaction:

$$\text{protein} + \text{water} \xrightarrow{\text{pepsin (enzyme)}} \text{peptides}$$

The stomach has a very low pH. Does this indicate that pepsin works effectively in an acidic or a basic environment? _____ This is the pH that allows the enzyme to maintain its normal shape so that it will combine with the substrate. A warm temperature causes molecules to move about more rapidly and increases the encounters between enzyme and substrate. Therefore you would hypothesize that the yield from this enzymatic reaction will be higher if the pH is _____ and the temperature is _____ (body temperature 37˚C).

Test for Protein Digestion

Biuret reagent is used to test for protein digestion. If digestion has not occurred, biuret reagent turns purple, indicating that protein is present. If digestion has occurred, biuret reagent turns pinkish-purple, indicating that peptides are present.

> ⚠ **Biuret reagent** is highly corrosive. Exercise care in using this chemical. If any should spill on your skin, wash the area with mild soap and water. Follow your instructor's directions for its disposal.

Experimental Procedure: Protein Digestion

With a wax pencil, number four test tubes 1 through 4, and mark at the 2 cm, 4 cm, 6 cm, and 8 cm levels. Fill all tubes to the 2 cm mark with albumin solution. Albumin is a protein.

Tube 1 Fill to the 4 cm mark with pepsin solution, and to the 6 cm mark with 0.2% HCl. HCl simulates the acidic conditions of the stomach. Swirl to mix, and incubate at 37˚C. After 1½ hours, fill to the 8 cm mark with biuret reagent. Record the temperature and your results in Table 13.1.

Tube 2 Fill to the 4 cm mark with pepsin solution, and to the 6 cm mark with 0.2% HCl. Swirl to mix, and keep at room temperature. After 1½ hours, fill to the 8 cm mark with biuret reagent. Record the temperature and your results in Table 13.1.

Tube 3 Fill to the 4 cm mark with pepsin solution, and to the 6 cm mark with water. Swirl to mix, and incubate at 37˚C. After 1½ hours, fill to the 8 cm mark with biuret reagent. Record the temperature and your results in Table 13.1.

Tube 4 Fill to the 6 cm mark with water. Swirl to mix, and incubate at 37˚C. After 1½ hours, fill to the 8 cm mark with biuret reagent. Record the temperature and your results in Table 13.1.

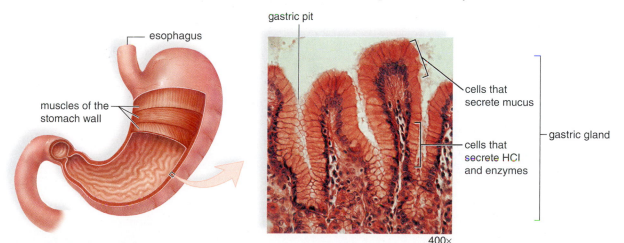

Figure 13.2 Digestion of protein.
Pepsin, produced by the gastric glands of the stomach, helps digest protein.

Table 13.1 Protein Digestion by Pepsin

Tube	Contents	Temperature	Results	Explanation
1	Albumin Pepsin HCl Biuret reagent			
2	Albumin Pepsin HCl Biuret reagent			
3	Albumin Pepsin Water Biuret reagent			
4	Albumin Water Biuret reagent			

Conclusions: Protein Digestion

- Explain your results in Table 13.1 by giving an explanation why digestion did or did not occur. To be complete, consider all the requirements for an enzymatic reaction as listed in Table 13.4. Now show here that Tube 1 met all the requirements for digestion:

  Pepsin is the correct _____.

  Albumin is the correct _____.

  37°C is the optimum _____.

  HCL provides the optimum _____.

  1½ hours provides _____ for the reaction to occur.

- Review "What Is a Control?" on page 174. Which tube was the negative control? _____

  Explain. _____

  _____

  _____

- If this control tube had given a positive result for protein digestion, what could you conclude about this experiment? _____

  _____

  _____

13.2 Fat Digestion by Pancreatic Lipase

Lipids include fats (e.g., butterfat) and oils (e.g., sunflower, corn, olive, and canola). Lipids are digested by **pancreatic lipase** in the small intestine (Fig. 13.3).

Figure 13.3   Emulsification and digestion of fat.
Bile from the liver (stored in the gallbladder) enters the small intestine, where lipase in pancreatic juice from the pancreas digests fat.

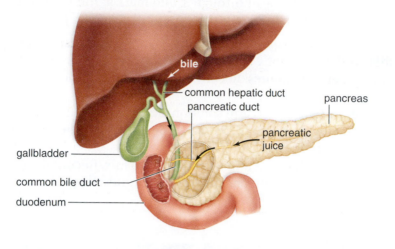

The following two reactions describe fat digestion:

1. $$\text{fat} \xrightarrow{\quad \text{bile (emulsifier)} \quad} \text{fat droplets}$$

2. $$\text{fat droplets} + \text{water} \xrightarrow{\quad \text{lipase (enzyme)} \quad} \text{glycerol} + \text{fatty acids}$$

With regard to the first step, consider that fat is not soluble in water; yet, lipase makes use of water when it digests fat. Therefore, bile is needed to emulsify fat—cause it to break up into fat droplets that disperse in water. The reason for dispersal is that bile contains molecules with two ends. One end is soluble in fat, and the other end is soluble in water. Bile can emulsify fat because of this.

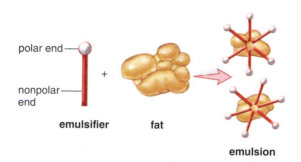

With regard to the second step, would the pH of the solution be lower before or after the enzymatic reaction? (*Hint:* Remember that an acid decreases pH and a base increases pH.) _____

Test for Fat Digestion

In the test for fat digestion, you will be using a pH indicator, which changes color as the solution in the test tube goes from basic conditions to acidic conditions. Phenol red is a pH indicator that is red in basic solutions and yellow in acidic solutions.

Experimental Procedure: Fat Digestion

With a wax pencil, number three clean test tubes 1 through 3, and mark at the 1 cm, 3 cm, and 5 cm levels. Fill all the tubes to the 1 cm mark with vegetable oil, and to the 3 cm mark with phenol red.

Tube 1   Fill to the 5 cm mark with pancreatin solution (pancreatic lipase). Add a pinch of bile salts, the emulsifier, and invert gently to mix. Record the initial color in Table 13.2. Incubate at 37°C, and check every 20 minutes. Record any color change and the time taken for the change.

Tube 2   Fill to the 5 cm mark with pancreatin solution, and invert gently to mix. Record the initial color in Table 13.2. Incubate at 37°C, and check every 20 minutes. Record any color change and the time taken for the change.

Tube 3   Fill to the 5 cm mark with water, and invert gently to mix. Record the initial color in Table 13.2. Incubate at 37°C, and check every 20 minutes. Record any color change and the time taken for the change.

Table 13.2   Fat Digestion by Pancreatic Lipase

Tube	Contents	Color		Time Taken	Explanation
		Initial	Final		
1	Vegetable oil Phenol red Pancreatin Bile salts				
2	Vegetable oil Phenol red Pancreatin				
3	Vegetable oil Phenol red Water				

Conclusions: Fat Digestion

- Explain your results in Table 13.2 by giving an explanation why digestion did or did not occur.

- What role did bile salts play in this experiment? _____

- What role did phenol red play in this experiment? _____

- Review "What Is a Control?" on page 174. Which test tube in this experiment could be considered a negative control? _____

13.3   Starch Digestion by Pancreatic Amylase

Starch is present in bakery products and in potatoes, rice, and corn. Starch is digested by **pancreatic amylase** in the small intestine, a process described by the following reaction:

$$\text{starch + water} \xrightarrow{\text{amylase (enzyme)}} \text{maltose}$$

1. If digestion *does not* occur, which will be present—starch or maltose? _____

2. If digestion *does* occur, which will be present—starch or maltose? _____

Tests for Starch Digestion

You will be using two tests for starch digestion:

1. If digestion has not taken place, the iodine test for starch will be positive (+). If digestion has occurred, the iodine test for starch will be negative (–).

2. If digestion has taken place, the Benedict's test for sugar (maltose) will be positive (+). If digestion has not taken place, the Benedict's test for sugar will be negative (–). To test for sugar, add an equal amount of Benedict's reagent to each test tube. Place the tube in a boiling water bath for 2 to 5 minutes, and note any color changes (see Table 3.5 on page 32). Boiling the test tube is necessary for the Benedict's reagent to react.

> ⚠️ **Benedict's reagent** is highly corrosive. Use protective eyewear when performing this experiment. Exercise care in using this chemical. If any should spill on your skin, wash the area with mild soap and water. Follow your instructor's directions for disposal of this chemical.

Experimental Procedure: Starch Digestion

Preparation

1. With a wax pencil, number six clean test tubes 1 through 6 above the level of boiling water bath, and mark at the 1 cm and 2 cm levels.
2. Fill tubes 1 through 4 to the 1 cm mark with pancreatic-amylase solution. Fill tubes 5 and 6 to the 1 cm mark with water. See Testing for tubes 1 and 2.
3. Shake the starch suspension well each time before dispensing. After tubes 3 through 6 have received the starch suspension, at least 30 minutes will lapse before testing occurs.
4. Fill tubes 3 and 4 to the 2 cm mark with starch suspension, and allow them to stand at room temperature for 30 minutes.
5. Fill tubes 5 and 6 to the 2 cm mark with starch suspension. Allow the tubes to stand for 30 minutes.

Testing

Tube 1   Fill to the 2 cm mark with starch suspension, and test for starch *immediately*, using the iodine test described previously. As an example, all but explanation has been completed for you for tube 1 in Table 13.3.

Tube 2   Fill to the 2 cm mark with starch suspension, and test for sugar *immediately*, using Benedict's reagent, described earlier, which requires boiling. Complete all but explanation in Table 13.3.

Why do you expect tube 1 to have a positive test for starch and tube 2 to have a negative test for sugar? _____

_____

Record your explanation for tubes 1 and 2 in Table 13.3.

Tubes 3 and 5   After 30 minutes, test for starch using the iodine test. Complete all but explanation in Table 13.3.

Tubes 4 and 6   After 30 minutes, test for sugar using the Benedict's test. Complete all but explanation in Table 13.3.

Why do you expect tube 3 to have a negative test for starch and tube 4 to have a positive test for sugar?

_____

_____ .

Why do you expect tube 5 to have a positive test for starch and tube 6 to have a negative test for sugar?

_____

_____ .

Record your explanations for tubes 3–6 in Table 13.3.

Table 13.3 Starch Digestion by Amylase					
Tube	Contents	Time*	Type of Test	Results	Explanation
1	Pancreatic amylase Starch	0	Iodine	+	
2	Pancreatic amylase Starch				
3	Pancreatic amylase Starch				
4	Pancreatic amylase Starch				
5	Water Starch				
6	Water Starch				

* Enter either 0 for immediately or T for after 30 minutes.

Conclusions: Starch Digestion

- This experiment demonstrated that for an enzymatic reaction to occur, an active _____ must be present, and _____ must pass to allow the reaction to occur.
- Which test tubes served as a negative control in this experiment? _____
  Explain. _____

  _____

Absorption of Sugars and Other Nutrients

Figure 13.4 shows that the folded lining of the small intestine has many fingerlike projections called villi. The small intestine not only digests food; it also absorbs the products of digestion, such as sugars from carbohydrate digestion, amino acids from protein digestion, and glycerol and fatty acids from fat digestion at the villi.

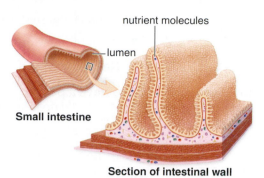

Figure 13.4 Anatomy of the small intestine.
Nutrients enter the bloodstream across the much-convoluted walls of the small intestine.

13.4 Requirements for Digestion

Explain in Table 13.4 how each of the requirements listed influences effective digestion.

Table 13.4 Requirements for Digestion	
Requirement	**Explanation**
Specific enzyme	
Specific substrate	
Warm temperature	
Specific pH	
Time	
Fat emulsifier	

To lose weight, some obese individuals undergo an operation in which (1) the stomach is reduced to the size of a golf ball, and (2) food bypasses the duodenum (first 2 feet) of the intestine. Answer these questions to explain how this operation would affect the requirements for digestion.

1. How is the amount of substrate reduced? _____

2. How is the amount of digestive enzymes reduced? _____

3. How is time reduced? _____

4. What makes the pH of the small intestine higher than before? _____

5. How is fat emulsification reduced? _____

6. How does surgery to reduce obesity sometimes result in malnutrition? _____

_____ 1. When iodine (IKI) solution turns blue-black, what substance is present?

_____ 2. What color is Benedict's reagent originally?

_____ 3. Name the enzyme in saliva.

_____ 4. As oil is digested, why does the first tube turn from red to yellow?

_____ 5. What temperature promotes enzymatic action?

_____ 6. Name the type of sample that goes through all the steps of an experiment but lacks the factor being tested.

_____ 7. What role do bile salts play in the digestion of fat?

_____ 8. What color does Biuret reagent turn when peptides are present?

_____ 9. Is the optimal pH for pepsin acidic or basic?

_____ 10. Why would you predict that pepsin would not digest starch?

_____ 11. In addition to pepsin and water, what is needed to digest protein?

_____ 12. Name the enzyme responsible for the hydrolysis of starch.

Thought Questions

13. Which of the following two combinations is most likely to result in digestion?
    a. Pepsin, protein, water, body temperature
    b. Pepsin, protein, hydrochloric acid (HCl), body temperature
    Explain.

14. An unknown sample is tested with both Biuret reagent and Benedict's reagent. Both tests result in a blue color. What has been learned?

15. What would you conclude if the tests for starch and maltose were both positive?

16. A test tube contains boiled pancreatic amylase and starch, and a student tests immediately for sugar. Should the student have waited for 30 minutes before testing?

14

Cardiovascular System

Introduction

Blood must circulate to serve the body. The **heart,** located in the thoracic cavity, pumps the blood, which moves away from the heart in **arteries** and **arterioles** and returns to the heart in **venules** and **veins.** The arteries receive blood from the heart under pressure and they have thicker walls than veins. Muscular contraction pushes on the thin walls of veins, which have valves to keep the blood moving toward the heart. Capillaries, which connect arterioles to venules, have thin walls that allow exchange of molecules with tissue fluid.

In today's laboratory, you will first learn the anatomy of the heart and how to trace the path of blood through the heart into two major circuits. In the **pulmonary circuit,** blood moves to and from the lungs and in the **systemic circuit,** the blood moves to and from the other organs of the body. You will learn to trace the path of blood in both circuits before dissecting major blood vessels in the fetal pig. Humans and pigs are both mammals and therefore your work today will give you an appreciation of your own heart and blood vessels and how the blood moves through these vessels to service your organs.

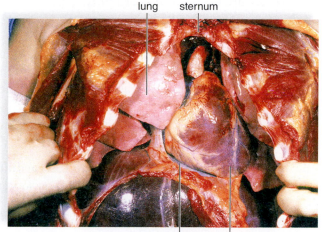

lung   sternum

pericardium   heart

14.1 The Heart

The heart has right and left sides divided by the **septum.** To tell the right from the left side, position Figure 14.1 so it corresponds to your body. There are four **chambers:** two upper, thin-walled atria and two lower, thick-walled ventricles. The heart has valves that keep the blood flowing in one direction; special muscles and tendons secure the valves to prevent backflow from the ventricles to the atria. The right side of the heart sends blood through the lungs, and the left side sends blood into the body. Therefore, the heart is called a double pump.

Anatomy of the Heart

A heart model will be used to study the anatomy of the heart (Fig. 14.1).

Observation: External Anatomy of the Heart

1. Identify the **right atrium** and its attached blood vessels, the superior and inferior **venae cavae.** The superior vena cava and the inferior vena cava return blood from the head and body, respectively, to

   the right atrium. Is the blood that enters the right atrium O_2-poor or O_2-rich? _____

   How do you know? _____

   _____

2. Identify the **right ventricle** and its attached blood vessel, the **pulmonary trunk.** The pulmonary trunk leaves the ventral side of the heart from the top of the right ventricle and then passes diagonally forward, before branching into the **right** and **left pulmonary arteries.** Is the blood in the pulmonary

   arteries O_2-poor or O_2-rich? _____ How do you know? _____

   _____

3. Identify the **left atrium** and its attached blood vessels, the left and right **pulmonary veins.** The pulmonary veins return blood from the lungs to the left atrium. Is the blood that enters the

   left atrium O_2-poor or O_2-rich? _____ How do you know? _____

   _____

4. Identify the **left ventricle** and its attached blood vessel, the **aorta,** which arises from the anterior end of the left ventricle, just dorsal to the origin of the pulmonary trunk. The aorta soon bends to the animal's left as the aortic arch. The aorta carries blood to the body proper. Is the blood in the aorta

   O_2-poor or O_2-rich? _____ How do you know? _____

   _____

5. Identify the **coronary arteries** and the **cardiac veins,** which service the needs of the heart wall. The coronary arteries branch off the aorta as soon as it leaves the heart and appear on the surface of the heart. The cardiac veins, also on the surface of the heart, join and then enter the right atrium through the coronary sinus on the dorsal side of the heart.

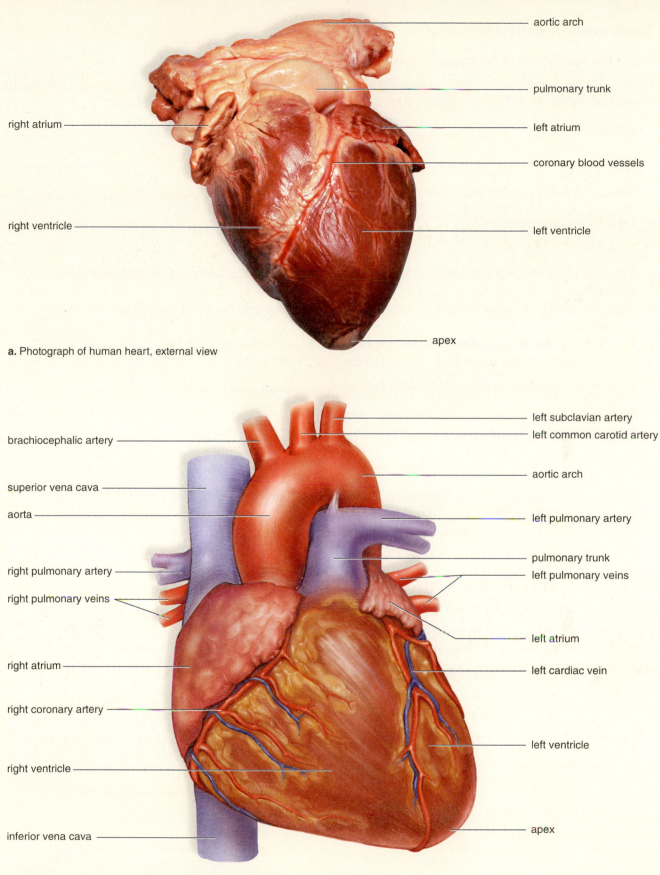

a. Photograph of human heart, external view

- aortic arch
- pulmonary trunk
- left atrium
- coronary blood vessels
- left ventricle
- apex
- right atrium
- right ventricle

- left subclavian artery
- left common carotid artery
- brachiocephalic artery
- aortic arch
- superior vena cava
- aorta
- left pulmonary artery
- pulmonary trunk
- right pulmonary artery
- left pulmonary veins
- right pulmonary veins
- left atrium
- right atrium
- left cardiac vein
- right coronary artery
- left ventricle
- right ventricle
- inferior vena cava
- apex

b. Drawing of human heart, external view

Figure 14.1 External view of heart.
Externally, notice the coronary arteries and cardiac veins that serve the heart.

Remove the ventral half of the human heart model (Fig. 14.2).

1. Identify the four chambers of the heart in longitudinal section: right atrium, right ventricle, left atrium, and left ventricle. Which chambers contain O_2-poor and which contain O_2-rich blood?

   O2-poor: _____

   O2-rich: _____

2. Which ventricle is more muscular? _____

   Why is this appropriate? _____

   _____

3. Find the **right atrioventricular** (tricuspid) valve, located between the right atrium and the right ventricle.
4. Find the **left atrioventricular** (bicuspid or mitral) valve, located between the left atrium and the left ventricle.
5. Find the **pulmonary semilunar** valve, located in the base of the pulmonary trunk.
6. Find the **aortic semilunar** valve, located in the base of the aorta.

   What is the function of the atrioventricular valves? _____

   _____

   What is the function of the semilunar valves? _____

   _____

7. Note the **chordae tendineae** ("heartstrings") that hold the atrioventricular valves in place while the heart contracts. These extend from the papillary muscles. The chordae tendineae prevent the atrioventricular valves from inverting into the atria when the ventricles contract.

Path of Blood Through the Heart

To demonstrate that O_2-poor blood is kept separate from O_2-rich blood, trace the path of blood from the right side of the heart to the aorta by filling in the following blanks. The blood passes through the lungs to go from the right to the left sides of the heart. Which side of the heart pumps O_2-poor blood? _____ Which side pumps O_2-rich blood? _____

Venae Cavae   *Lungs*

_____   _____

_____ valve   _____

_____   _____ valve

_____ valve   _____

_____   _____ valve

Lungs   Aorta

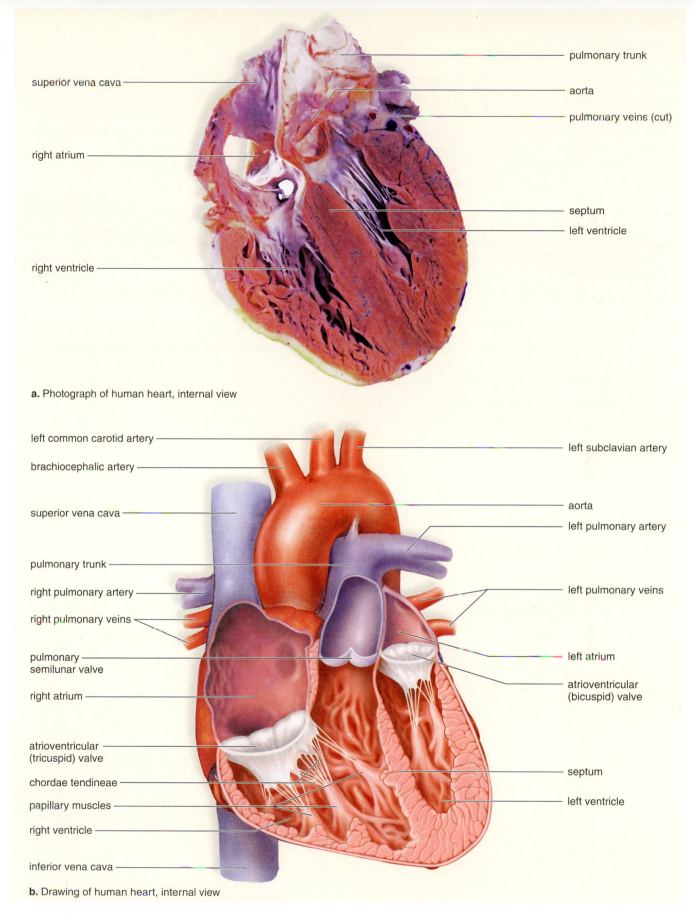

a. Photograph of human heart, internal view

Labels (photograph): pulmonary trunk, superior vena cava, aorta, pulmonary veins (cut), right atrium, septum, left ventricle, right ventricle

Labels (drawing): left common carotid artery, left subclavian artery, brachiocephalic artery, superior vena cava, aorta, left pulmonary artery, pulmonary trunk, right pulmonary artery, left pulmonary veins, right pulmonary veins, pulmonary semilunar valve, left atrium, right atrium, atrioventricular (bicuspid) valve, atrioventricular (tricuspid) valve, chordae tendineae, septum, papillary muscles, left ventricle, right ventricle, inferior vena cava

b. Drawing of human heart, internal view

Figure 14.2   Internal view of heart.
Internally, the heart has four chambers and there is a septum that separates the left side from the right side.

Path of Blood in the Body

In adult mammals, including humans, the heart is a double pump. The right ventricle pumps blood into the **pulmonary circuit**—that is, to the lungs (i.e., the capillaries in the lungs) and back to the heart (Fig. 14.3). While the blood is in the lungs, it gives up carbon dioxide and gains oxygen. The left ventricle pumps blood to the **systemic circuit**—that is, throughout the whole body except to the lungs. Blood in the systemic circuit gives up oxygen and gains carbon dioxide.

Figure 14.3 shows how to trace the path of blood in both the pulmonary and systemic circuits in adult humans. It also helps you to learn the names of some of the major blood vessels.

Pulmonary Circuit

1. Trace the path of blood in the pulmonary circuit from the heart to the lungs (i.e., the capillaries of the lungs) and then from the lungs to the heart. Follow the arrows in Figure 14.3, and use the label names provided there.

   right ventricle of heart ⟶ _____ ⟶ lungs ⟶ _____ ⟶

   _____ of heart

   Notice that in the adult there is no connection between the right and left sides of the heart except via the lungs.

Systemic Circuit

2. Trace the path of blood in the systemic circuit from the heart to the kidneys (i.e., the capillaries of the kidneys), and then from the kidneys to the heart.

   left ventricle of heart ⟶ _____ ⟶ _____ ⟶ kidneys ⟶

   _____ ⟶ _____ ⟶ _____ of heart

Names of Blood Vessels

3. Use Figure 14.3 to complete Table 14.1.

Table 14.1	Major Blood Vessels in the Systemic Circuit	
Body Part	**Artery**	**Vein**
Heart		
Head		
Arms		
Kidney		
Legs		
Intestines		

What vessel lies between the digestive tract and the liver? _____

A portal system lies between two sets of capillaries.

Figure 14.3 Diagram of the human cardiovascular system.

In the pulmonary circuit, the pulmonary arteries take O_2-poor blood to the lungs (i.e., the capillaries in the lungs), and the pulmonary veins return O_2-rich blood to the heart. In the systemic circuit, the aorta branches into the various arteries that go to all other parts of the body. After blood passes through arterioles, capillaries, and venules, it enters various veins and then the superior and inferior venae cavae, which return it to the heart.

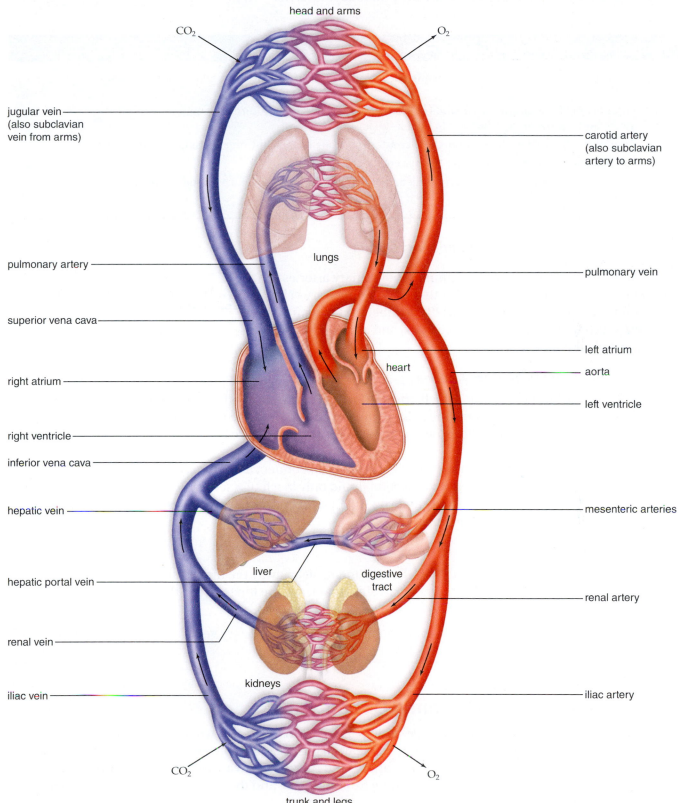

14.2 Vessels of the Pulmonary Circuit

In this section, we will use the fetal pig to examine the pulmonary arteries and veins that occur in mammals, including humans. The fetal pig will have pulmonary arteries and veins, even though they are not functional until after the pig is born. Fetal circulation has two features, numbered #1 and #2 in Table 14.2 that keep blood from entering the pulmonary arteries.

Table 14.2   Unique Features of Fetal Circulation
1. Oval opening (foramen ovale): an opening between the atria of the heart that allows blood to bypass the lungs
2. Arterial duct (ductus arteriosus): a short, stout vessel leading directly from the pulmonary trunk to the aorta; provides another way to bypass the lungs
3. Umbilical arteries: take blood from the iliac arteries to the placenta, the organ that nourishes the fetus
4. Umbilical vein: returns blood from the placenta to the liver
5. Venous duct (ductus venosus): a continuation of the umbilical vein that takes blood to the inferior vena cava

Observation: Vessels of the Pulmonary Circuit

In the pulmonary circuit of adult mammals, pulmonary arteries take blood away from the heart to the lungs, and pulmonary veins take blood from the lungs to the heart. Remember that in your pig, *all arteries* have been injected with red latex, and *all veins* have been injected with blue latex.

Wear protective latex gloves and eyewear when handling preserved animal organs. Exercise caution when using sharp instruments during this laboratory. Wash hands thoroughly upon completion of this laboratory.

Pulmonary Trunk and Pulmonary Arteries

1. Locate the **pulmonary trunk,** which arises from the ventral side of the heart. It may appear white because the thick wall prevents the color of the red latex from showing through.
2. If necessary, remove the pericardial sac from the heart to reveal the blood vessels entering the heart. Veins take blood to the heart.
3. Trace the pulmonary trunk, and notice that it seems to connect directly with the aorta, the major artery. This is the **arterial duct**, a connection found only in a fetus (see Table 14.2).
4. In addition to this duct, look closely and you will find the **pulmonary arteries,** which leave the pulmonary trunk and go to the lungs.

Pulmonary Veins

1. The pulmonary veins are hard to find. To view them, clean away the membrane dorsal to the heart, and carefully note the vessels (pulmonary veins) that leave the lungs. Trace these to the left atrium of the heart.

2. In the adult, which blood vessels—pulmonary arteries or pulmonary veins—carry O_2-rich blood?

14.3 Vessels of the Systemic Circuit

In adult mammals, the **systemic circuit** serves all parts of the body except where gas exchange occurs in of the lungs. Arteries take O_2-rich blood from the heart to the organs, and veins take O_2-poor blood from the organs to the heart. The aorta is the major artery, and the **venae cavae** are the major veins. It will be possible for you to identify the arteries branching from the aorta and the corresponding veins branching from the venae cavae. In the fetal pig, the venae cavae are called the *anterior* vena cava and the *posterior* vena cava because the normal position of the body is horizontal rather than vertical. Use the figures to trace blood vessels, but **do not remove any organs. You will need them in Laboratory 15.**

1. Follow the aorta as it extends through the thoracic cavity (Fig. 14.4). To do this, gently move the lungs and heart to the right side of the thoracic cavity. Notice how the thoracic or dorsal aorta is a large, whitish vessel extending through the thoracic cavity and then through the diaphragm to become the abdominal aorta. Also notice the esophagus, a smaller, flattened tube more toward the midline. The esophagus goes through the diaphragm to join with the stomach. Locate the anterior vena cava coming off the top of the heart and the posterior vena cava just to the right of the midline. The posterior vena cava also passes through the diaphragm.

2. Name three structures that pass through the diaphragm. 1. _____

   2. _____   3. _____

Blood Vessels of the Upper Body

The **coronary arteries** and **cardiac veins** lie on the surface of the heart. The **carotid arteries** and **jugular veins** serve the neck and head regions of the body.

Observation: Blood Vessels of the Upper Body

Coronary Arteries and Cardiac Veins

1. Locate the coronary arteries and cardiac veins, which are easily visible on the heart's surface (see Fig. 14.1).
2. The coronary arteries arise from the aorta just as it leaves the heart, and the cardiac veins go directly into the right atrium.

Carotid Arteries and Jugular Veins

1. Find the aorta as it leaves the left ventricle and curves downward. As it arches it gives off branches. Its first branch, the **brachiocephalic arterial trunk,** divides almost immediately into the **right subclavian** (to the pig's right shoulder) **artery** and the **right** and **left common carotid arteries** (Fig. 14.4 and 14.5).
2. Find the second branch of the aorta, which is the **left subclavian artery.**
3. Both the right and left subclavian arteries have branches called the right and left brachial arteries, which serve the upper limbs.
4. Find the **jugular veins** alongside the carotid arteries (Fig. 14.4 and 14.6). Trace the carotid arteries and jugular veins as far as possible toward the head. What part of the body is serviced by the carotid

   arteries and the jugular veins? _____

Subclavian Arteries and Veins

1. Locate the **subclavian arteries** (Fig. 14.4 and 14.5) and **subclavian veins** (Fig. 14.4 and 14.6), which serve the forelimbs and are easily identified.
2. Note that while the left subclavian artery branches from the aorta, the right subclavian artery branches from the brachiocephalic arterial trunk.
3. Note that the jugular veins and the subclavian veins join to form the brachiocephalic venous trunk which enters the anterior vena cava.

Figure 14.4 Ventral view of fetal pig arteries and veins.

Use this diagram to trace blood vessels, *but do not remove any organs.* You will need them in Laboratory 15.
(a. = artery; v. = vein.)

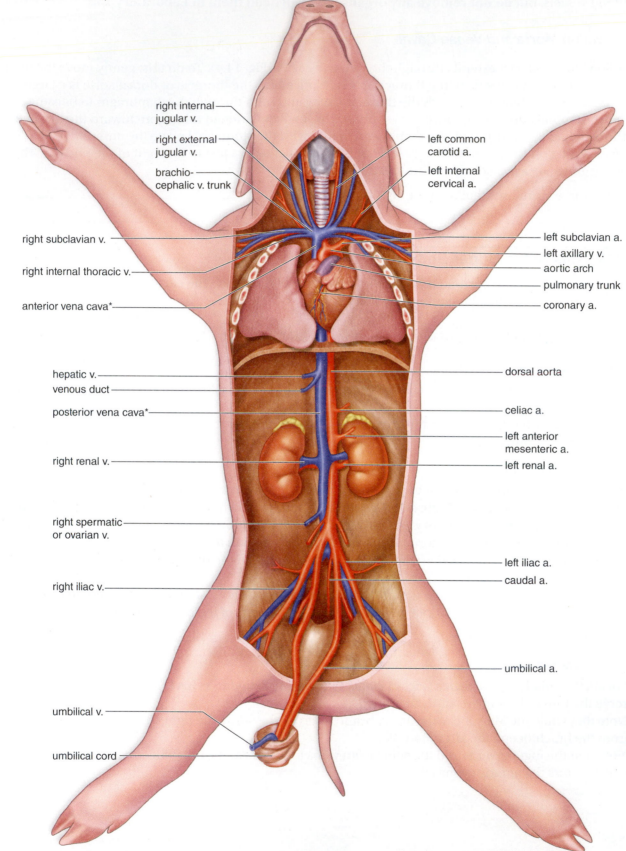

right internal jugular v.

right external jugular v.

brachio-cephalic v. trunk

right subclavian v.

right internal thoracic v.

anterior vena cava*

hepatic v.

venous duct

posterior vena cava*

right renal v.

right spermatic or ovarian v.

right iliac v.

umbilical v.

umbilical cord

left common carotid a.

left internal cervical a.

left subclavian a.

left axillary v.

aortic arch

pulmonary trunk

coronary a.

dorsal aorta

celiac a.

left anterior mesenteric a.

left renal a.

left iliac a.

caudal a.

umbilical a.

* Because the pig walks on all four limbs, the anterior vena cava in pigs is called the superior vena cava in humans, and the posterior vena cava in pigs is called the inferior vena cava in humans.

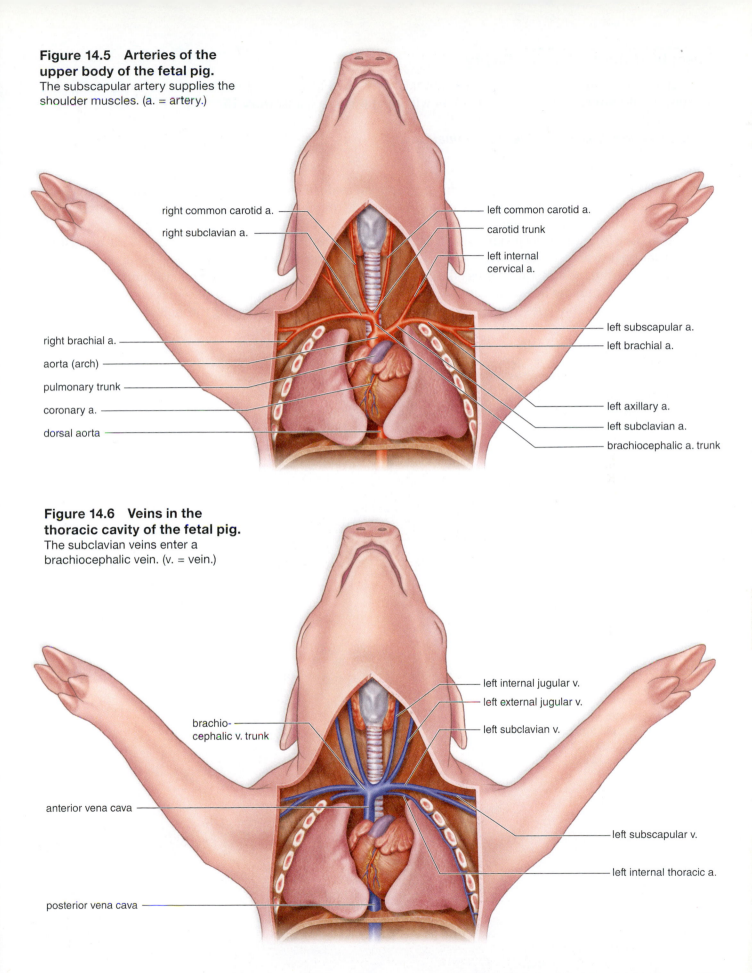

Figure 14.5 Arteries of the upper body of the fetal pig.
The subscapular artery supplies the shoulder muscles. (a. = artery.)

right common carotid a.
right subclavian a.

left common carotid a.
carotid trunk
left internal cervical a.

left subscapular a.
left brachial a.

right brachial a.
aorta (arch)
pulmonary trunk
coronary a.
dorsal aorta

left axillary a.
left subclavian a.
brachiocephalic a. trunk

Figure 14.6 Veins in the thoracic cavity of the fetal pig.
The subclavian veins enter a brachiocephalic vein. (v. = vein.)

brachio-cephalic v. trunk

left internal jugular v.
left external jugular v.
left subclavian v.

anterior vena cava

left subscapular v.

left internal thoracic a.

posterior vena cava

Vessels of the Abdominal Cavity

Several major blood vessels are in the abdominal cavity. Use Figures 14.7 and 14.8 to trace the blood vessels, but **do not remove any organs. You will need them in Laboratory 15.**

Observation: Vessels of the Abdominal Cavity

Celiac and Mesenteric Arteries

1. Carefully lift up the liver and stomach, and put them aside to your left. Dissect the dorsal mesentery to see the **celiac artery** as it leaves the aorta. Tributaries from this vessel eventually reach the stomach, duodenum, liver, and spleen.
2. Branching from the aorta just posterior to the celiac artery is a long, unpaired trunk, called the **anterior mesenteric artery,** which has tributaries to the pancreas and small intestine (see Fig. 14.7).
3. The celiac and mesenteric arteries take blood to the intestines. Thereafter, the hepatic portal vein takes blood from the intestinal capillaries to capillaries in the liver. A portal system is defined as a circulatory unit that goes from one capillary bed to another without passing through the heart (see Figure 14.3).

Renal Arteries and Veins

1. Locate the **renal arteries** as they branch from the aorta, and trace these arteries as they go into the kidneys.
2. Locate the renal veins as they leave the kidneys (Fig. 14.8), and trace these veins as they join the posterior vena cava.

Figure 14.7 Arteries in the abdominal cavity of the fetal pig. The celiac and mesenteric artery serve the digestive and associated organs. The external iliac arteries proceed into the hindlimbs from the caudal end of the dorsal aorta. They give rise to the umbilical arteries and continue as much smaller vessels. (aa. = arteries.) Use this diagram to trace blood vessels, *but do not remove any organs. You will need them in Laboratory 15.*

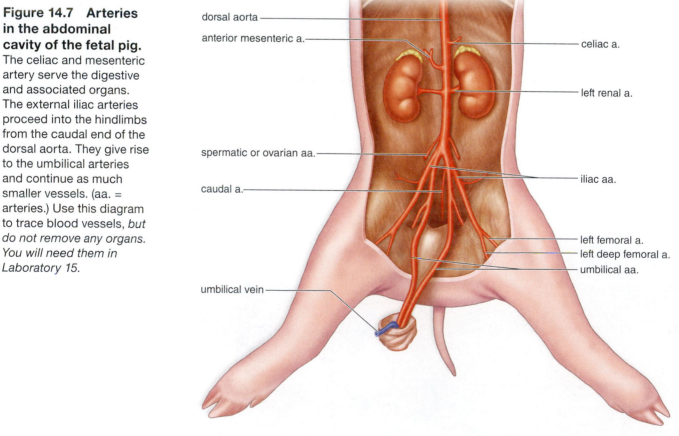

dorsal aorta
anterior mesenteric a.
celiac a.
left renal a.
spermatic or ovarian aa.
iliac aa.
caudal a.
left femoral a.
left deep femoral a.
umbilical aa.
umbilical vein

Iliac Arteries and Veins

1. At its posterior end, the aorta branches into the paired **iliac arteries.** Locate the iliac arteries at the posterior end of the aorta, and trace these arteries into the hindlimbs (see Fig. 14.7).
2. Find the **iliac veins** alongside the iliac arteries, and trace these veins as they join the posterior vena cava (see Fig. 14.8).

Umbilical Arteries and Veins

1. Locate the **umbilical arteries** on either side of the bladder. Trace these arteries as they branch from the iliac arteries and as they pass into the **umbilical cord** (see Fig. 14.7).
2. When you were exposing the abdominal cavity, you cut the **umbilical vein.** Trace the umbilical vein from the umbilical cord to the liver. The umbilical vein is joined to the posterior vena cava by the **venous duct** (ductus venosus), which passes through the posterior portion of the liver.

Posterior Vena Cava

1. Locate the **posterior vena cava,** which is easily seen as a large, blue vessel just ventral to the dorsal aorta (see Fig. 14.8).
2. Note that this vessel seems to disappear in the region of the liver. Here the posterior vena cava receives the hepatic veins coming from the liver. Scrape away some of the liver tissue in order to see these veins.
3. Locate the posterior vena cava as it passes through the diaphragm into the thoracic cavity and as it enters the right atrium.
4. Trace the posterior vena cava from the iliac veins to the right atrium of the heart (see Fig. 14.4).

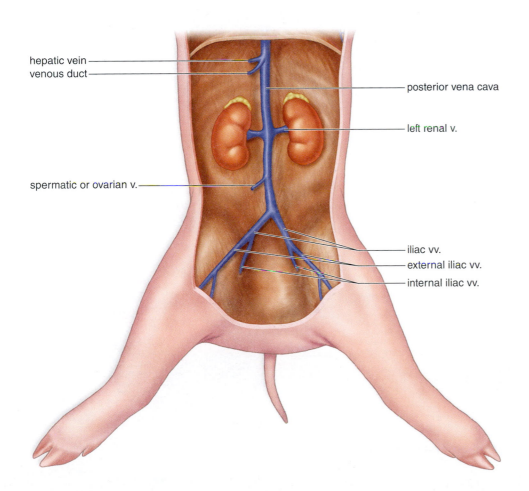

hepatic vein
venous duct
posterior vena cava
left renal v.
spermatic or ovarian v.
iliac vv.
external iliac vv.
internal iliac vv.

Figure 14.8   Veins in the abdominal cavity of the fetal pig.

The posterior vena cava divides into the common iliac veins. The common iliac veins branch into the external and internal iliac veins. (vv. = veins.) Use this diagram to trace blood vessels, but *do not remove any organs. You will need them in Laboratory 15.*

Storage of Pigs

1. Before leaving the laboratory, place your pig in the plastic bag provided.
2. Expel excess air from the bag, and tie it shut.
3. Write your name and section on the tag provided, and attach it to the bag. Your instructor will indicate where the bags are to be stored until the next laboratory period.
4. Clean the dissecting tray and tools, and return them to their proper location.
5. Wipe off your goggles.
6. Wash your hands.

14.4   Blood Vessel Comparison

Blood pressure keeps the blood in arteries moving away from the heart. Skeletal muscle contraction pushing in on veins keeps the blood in veins moving toward the heart. Do you predict that arteries or veins are generally more superficial in the body?

_____

Wall of an Artery Compared with Wall of a Vein

Both arteries and veins have three distinct layers, or **tunicas,** that form a wall around the lumen, the space through which blood flows. The three tunicas are called the inner layer, the middle layer, and the outer layer. Figure 14.9 shows that arteries have a thicker wall than veins because middle layer consisting of smooth muscle and elastic fibers is much thicker. The elastic fibers in the wall of an artery allows it to expand when the blood pours into it with each heart beat. The smooth muscle allows the arterial wall to constrict when needed to keep blood pressure normal.

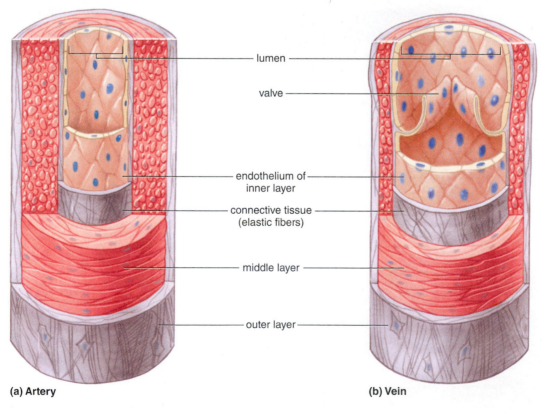

(a) Artery   (b) Vein

Figure 14.9   Blood vessel comparison.

1. Obtain a microscope slide that shows an artery and a vein in cross section.
2. View the slide, under both low and high power, and with the help of Figure 14.10, determine which is the artery and which is the vein.
3. Identify the **outer layer,** which contains many collagen and elastic fibers and often appears white in specimens.
4. Identify the **middle layer,** the thickest layer, which is composed of smooth muscle and elastic tissue.

   Does this layer appear thicker in arteries than in veins? _____

5. Identify the **inner layer,** a smooth lining of simple squamous epithelial cells called the endothelium. In veins, the endothelium forms valves that keep the blood moving toward the heart. Arteries do not have valves. Considering the relationship of arteries and veins to the heart, why do veins have valves, while

   arteries do not? _____

   _____

Figure 14.10   Photomicrograph of an artery and a vein.

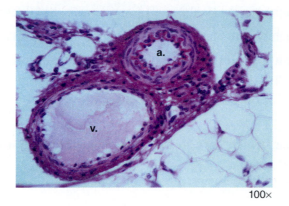

100×

Conclusions

- Which type of blood vessel (arteries or veins) has thicker walls? What makes the wall thicker? _____

  _____

  _____

- Which type of blood vessel has thinner walls? _____

  _____

- Which type of blood vessel is more apt to lose its elasticity, leading to a discoloration that can be

  externally observed? _____

  _____

  What is this condition called? _____

_____ 1. Which heart chamber pumps blood to the body proper?

_____ 2. Identify the vessel that conducts blood from the right ventricle.

_____ 3. Which type chambers are receiving chambers in the heart?

_____ 4. Do the pulmonary arteries carry O_2-rich or O_2-poor blood?

_____ 5. The coronary arteries and cardiac veins serve what organ?

_____ 6. Identify the blood vessels that conduct blood to the head.

_____ 7. Identify the artery that serves the kidney.

_____ 8. Identify the large artery that runs dorsally along the wall of the abdominal cavity.

_____ 9. Identify the arteries that take blood from the aorta to the hindlimbs.

_____ 10. What part of the human body is served by the subclavian vessels?

_____ 11. Identify the large abdominal vein that runs alongside the aorta and enters the right atrium.

_____ 12. What part of the body is not served by the systemic circuit?

_____ 13. Which type of blood vessel (artery or vein) has thicker walls?

_____ 14. Which type of blood vessel (artery or vein) has valves?

_____ 15. Identify the artery whose tributaries serve the stomach, duodenum, liver, and spleen.

Thought Questions

16. During a heart attack, cardiac muscle cells are deprived of their blood supply, yet the ventricles are still full of blood. Explain.

17. Trace the path of blood from the left ventricle to the kidneys and back to the right atrium.

18. Evaluate the following statement:

    All arteries carry oxygenated blood and all veins carry deoxygenated blood.

    Based on what you have learned in this laboratory, is this statement correct? Explain your answer.

19. Explain how the heart is a double pump.

15

Basic Mammalian Anatomy II

Learning Outcomes

15.1 Urinary System
- Locate and identify the organs of the urinary system.   200–1
- State a function for the organs of the urinary system.   200–1

15.2 Male Reproductive System
- Locate and identify the organs of the male reproductive system.   202–3
- State a function for the organs of the male reproductive system.   202–3
- Compare the pig reproductive system with that of the human male.   204

15.3 Female Reproductive System
- Locate and identify the organs of the female reproductive system.   205–6
- State a function for the organs of the female reproductive system.   205–6
- Compare the pig reproductive system with that of the human female.   207

15.4 Review of the Respiratory, Digestive, and Cardiovascular Systems
- Using preserved specimens, images, or charts, locate and identify the individual organs of the respiratory, digestive, and cardiovascular systems.   208–210
- Using preserved specimens, images, or charts, locate and identify the hepatic portal system. State a function for this system.   211

Introduction

The **urinary system** and the **reproductive system** are so closely associated in mammals that they are often considered together as the **urogenital system.** They are particularly associated in males, where certain structures function in both systems. In this laboratory, we will focus first on dissecting the urinary and reproductive systems in the fetal pig.

The kidneys of the urinary system produce urine, which is stored in the bladder before being released to the exterior. As the kidneys produce urine they also regulate the volume and the composition of the blood so that the water and salt balance and the acid-base balance of the blood stays within normal limits.

In mammalian reproductive systems, the testes (sing., testis) are the male gonads, and the ovaries (sing., ovary) are the female gonads. The testes produce sperm, and the ovaries produce oocytes that become eggs. We will compare the anatomy of the reproductive systems in pigs with those in humans.

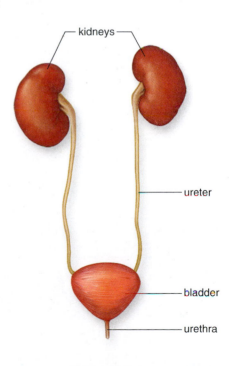

kidneys

ureter

bladder

urethra

Urinary system

Figure 15.1 Urinary system of the fetal pig.
In **(a)** females and **(b)** males, urine is made by the kidneys, transported to the bladder by the ureters, stored in the bladder, and then excreted from the body through the urethra.

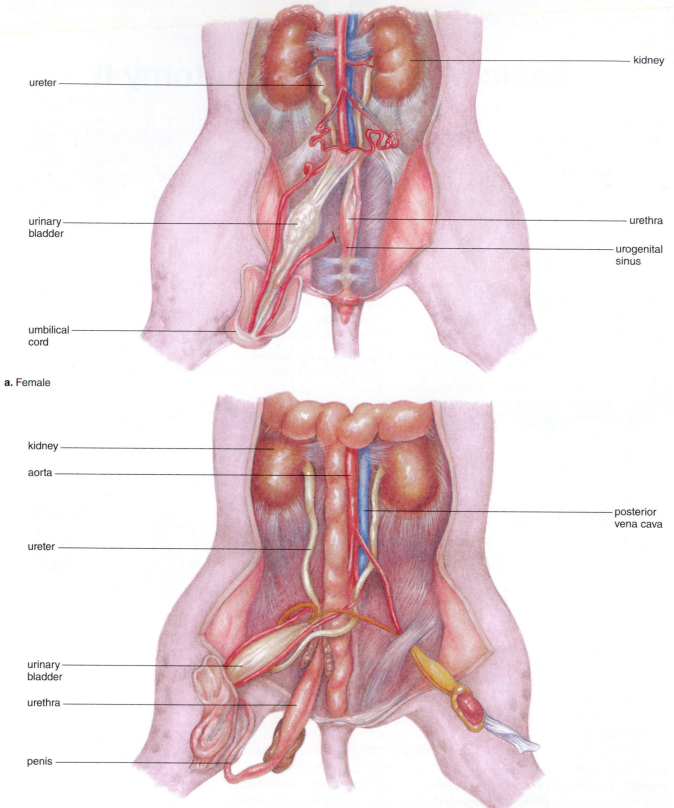

ureter

kidney

urinary bladder

urethra

urogenital sinus

umbilical cord

a. Female

kidney

aorta

posterior vena cava

ureter

urinary bladder

urethra

penis

b. Male

15.1 Urinary System

The urinary system consists of the **kidneys,** which produce urine; the **ureters,** which transport urine to the **urinary bladder,** where urine is stored; and the **urethra,** which transports urine to the outside. In males, the urethra also transports sperm during ejaculation.

During this dissection, compare the urinary system structures of both sexes of fetal pigs. Later in this laboratory period, exchange specimens with a neighboring team for a more thorough inspection.

Observation: Urinary System in Pigs

1. The large, paired kidneys (Fig. 15.1) are reddish organs covered by **peritoneum,** a membrane that anchors them to the dorsal wall of the abdominal cavity, sometimes called the **peritoneal cavity.** Clean the peritoneum away from one of the kidneys, and study it more closely.

> ⚠️ **Wear protective latex gloves and eyewear** when handling preserved animal organs. Exercise caution when using sharp instruments during this laboratory. Wash hands thoroughly upon completion of this laboratory.

2. Using a razor blade or scalpel, section one of the kidneys in place, cutting it lengthwise (Fig. 15.2). At the center of the medial portion of the kidney is an irregular, cavity-like reservoir, the **renal pelvis.** The outermost portion of the kidney (the **renal cortex**) shows many small striations perpendicular to the outer surface. This region and the more even-textured **renal medulla** region inward from it are composed of **nephrons** (excretory tubules).

3. Locate the **ureters,** which leave the kidneys and run posteriorly under the peritoneum (Fig. 15.1).

4. Clean the peritoneum away, and follow a ureter to the **urinary bladder,** which normally lies in the posterior ventral portion of the abdominal cavity. The urinary bladder is on the inner surface of the flap of tissue to which the umbilical cord was attached.

5. The **urethra,** which arises from the bladder posteriorly, runs parallel to the rectum. Follow the urethra until it passes from view into the ring formed by the pelvic girdle.

6. Trace the path of urine. _____

_____

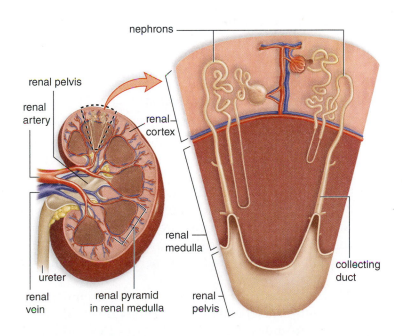

Figure 15.2 Anatomy of the kidney.
A kidney has a renal cortex, a renal medulla, a renal pelvis, and microscopic tubules called nephrons.

15.2 Male Reproductive System

The **male reproductive system** consists of the **testes** (sing., testis), which produce sperm, and the **epididymides** (sing., epididymis), which store sperm before they enter the **vasa deferentia** (sing., vas deferens). Just prior to ejaculation, sperm leave the vasa deferentia and enter the **urethra,** which eventually passes into the penis. The **penis** is the male organ of sexual intercourse. **Seminal vesicles,** the **prostate gland,** and the **bulbourethral glands** (Cowper's glands) add fluid to semen after sperm reach the urethra. Table 15.1 summarizes the male reproductive organs.

Table 15.1   Male Reproductive Organs and Functions

Organ	Function
Testis	Produces sperm and sex hormones
Epididymis	Stores sperm as they mature
Vas deferens	Conducts and stores sperm
Seminal vesicle	Contributes secretions to semen
Prostate gland	Contributes secretions to semen
Urethra	Conducts sperm
Bulbourethral glands	Contribute secretions to semen
Penis	Organ of copulation

The testes begin their development in the abdominal cavity, just anterior and dorsal to the kidneys. Before birth, however, they gradually descend into paired **scrotal sacs** within the scrotum, suspended anterior to the anus. Each scrotal sac is connected to the body cavity by an **inguinal canal,** the opening of which can be found in the pig. The passage of the testes from the body cavity into the scrotal sacs is called the descent of the testes. The testes in most of the male fetal pigs being dissected will probably be partially or fully descended.

Observation: Male Reproductive System in Pigs

Inguinal Canal, Testis, Epididymis, and Vas Deferens

1. Locate the opening of the left inguinal canal, which leads to the left scrotal sac (Fig. 15.3).
2. Expose the canal and sac by making an incision through the skin and muscle layers from a point over this opening back to the left scrotal sac.
3. Open the sac, and find the testis. Note the much-coiled tubule—the epididymis—that lies alongside the testis. This is continuous with the vas deferens, which passes back toward the abdominal cavity.
4. Trace a vas deferens as it loops over an umbilical artery and ureter and unites with the urethra dorsally at the posterior end of the urinary bladder.

Penis, Urethra, and Accessory Glands

1. Cut through the ventral skin surface just posterior to the umbilical cord. This will expose the rather undeveloped penis, which extends from this point posteriorly toward the anus. The central duct of the penis is the urethra.
2. Lay the penis to one side, and then cut down through the ventral midline, laying the legs wide apart in the process (Fig. 15.4). The cut will pass between muscles and through pelvic cartilage (bone has not developed yet). Do not cut any of the ducts or tracts in the region.
3. Now locate the urethra passing ventrally above the rectum. It is somewhat heavier in the male due to certain accessory glands:
   a. Bulbourethral glands (Cowper's glands), about 1 cm in diameter, lie laterally and well back toward the anal opening.
   b. The prostate gland, about 4 mm across and 3 mm thick, is located on the dorsal surface of the urethra, just posterior to the juncture of the urinary bladder with the urethra. It is often difficult to locate and is not shown in Figures 15.3 and 15.4.
   c. Small, paired seminal vesicles may be seen on either side of the prostate gland.

Figure 15.3 Male reproductive system of the fetal pig.

In males, the urinary system and the reproductive system are joined. The vasa deferentia (sing., vas deferens) enter the urethra, which also carries urine.

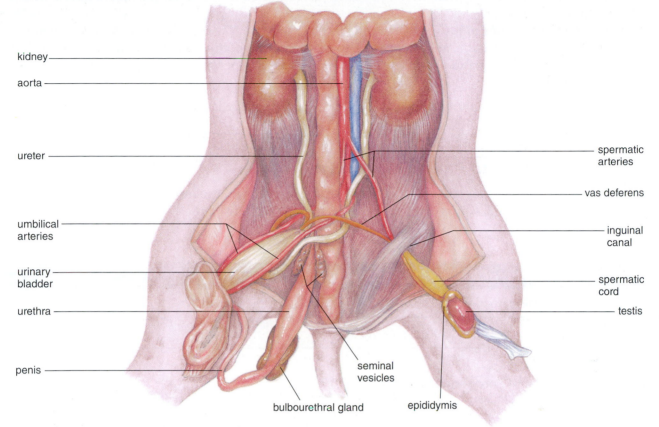

Figure 15.4 Photograph of the male reproductive system of the fetal pig.

Compare the diagram in Figure 15.3 to this photograph to help identify the structures of the male urogenital system.

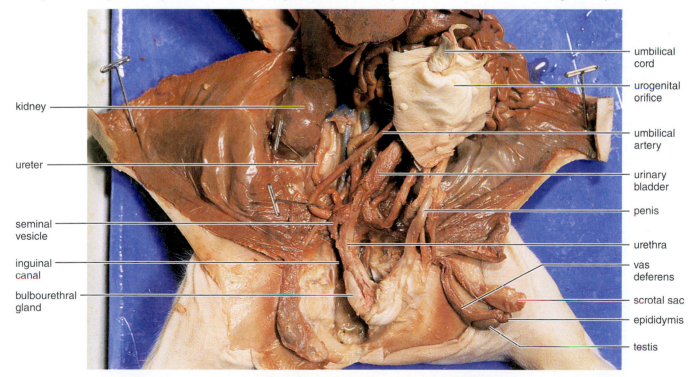

4. Trace the urethra as it leaves the urinary bladder. It proceeds posteriorly, but when it nears the posterior end of the body, it turns rather abruptly anterioventrally and runs forward just under the skin of the midventral body wall, where you have just dissected it. This latter portion of the urethra is, then, within the penis.

5. Now locate the entrance of the vasa deferentia into the urethra. If necessary, dissect these structures free from surrounding tissue, and expose the point of entrance of these ducts into the urethra near the location of the prostate gland. In males, the urethra transports sperm, as well as urinary wastes from the bladder.

6. Trace the path of sperm in the male. _____

_____

Comparison of Male Fetal Pig and Human Male

Use Figure 15.5 to help compare the male pig reproductive system with the human male reproductive system. Complete Table 15.2, which compares the location of the penis in these two mammals.

| Table 15.2 Location of Penis in Male Fetal Pig and Human Male | |
Fetal Pig	Human
Penis	

Figure 15.5   Human male urogenital system.
In the fetal pig, but not in the human male, the penis lies beneath the skin and exits at the urogenital opening.

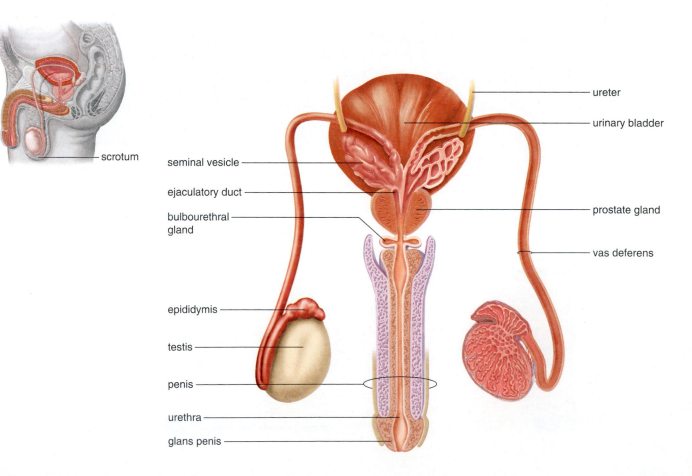

scrotum

seminal vesicle

ejaculatory duct

bulbourethral gland

epididymis

testis

penis

urethra

glans penis

ureter

urinary bladder

prostate gland

vas deferens

15.3 Female Reproductive System

The **female reproductive system** (Table 15.3) consists of the **ovaries,** which produce eggs, and the **oviducts,** which transport eggs to the **uterus,** where development occurs. In the fetal pig, the uterus does not form a single organ, as in humans, but is partially divided into external structures called **uterine horns,** which connect with the oviduct. The **vagina** is the birth canal and the female organ of sexual intercourse.

Table 15.3 Female Reproductive Organs and Functions

Organ	Function
Ovary	Produces egg and sex hormones
Oviduct (fallopian tube)	Conducts egg toward uterus
Uterus	Houses developing fetus
Vagina	Receives penis during copulation and serves as birth canal

Observation: Female Reproductive System in Pigs

Ovaries and Oviducts

1. Locate the paired ovaries, small bodies suspended from the peritoneal wall in mesenteries, posterior to the kidneys (Figs. 15.6 and 15.7).
2. Closely examine one ovary. Note the small, short, coiled oviduct, sometimes called the **Fallopian tube.** The oviduct does not attach directly to the ovary but ends in a funnel-shaped structure with fingerlike processes (fimbriae) that partially encloses the ovary.

Uterine Horns

1. Locate the **uterine horns.** (Do not confuse the uterine horns with the oviducts; the latter are much smaller and are found very close to the ovaries.)
2. Find the median body of the uterus, located at the joined posterior ends of the uterine horns.

Vagina

1. Separate the hindlimbs of your specimen, and cut down along the midventral line. The cut will pass through muscle and the cartilaginous pelvic girdle. With your fingers, spread the cut edges apart, and use blunt dissecting instruments to separate connective tissue.
2. Note three ducts passing from the body cavity to the animal's posterior surface. One of these is the urethra, which leaves the urinary bladder and passes into the **urogenital sinus.** The urethra is a part of the urinary system. The most dorsal of the three ducts is the **rectum,** which passes to its own opening, the **anus.** The rectum and anus are, of course, part of the digestive system, not the reproductive system.
3. Find the vagina, located dorsally to the urethra. The vagina is the organ of copulation and is the birth canal. Anteriorly, it connects to the uterus, and posteriorly it enters the urogenital sinus. This sinus is absent in adult humans and several other female mammals.

Figure 15.6   Female reproductive system of the fetal pig.
In the adult female, the urinary system and the reproductive system are separate. In the fetus, the vagina joins the urethra just before the urogenital sinus.

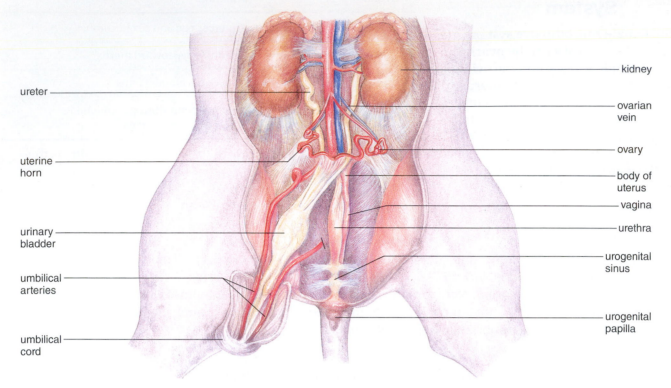

kidney

ureter

ovarian vein

ovary

uterine horn

body of uterus

vagina

urethra

urinary bladder

urogenital sinus

umbilical arteries

umbilical cord

urogenital papilla

Figure 15.7   Photograph of the female reproductive system of the fetal pig.
Compare the diagram in Figure 15.6 with this photograph to help identify the structures of the female urinary and reproductive systems.

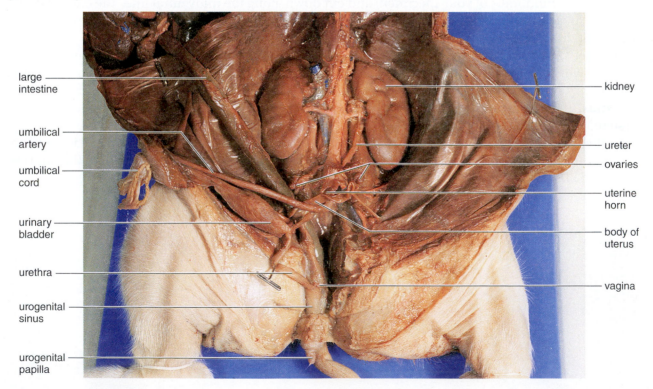

large intestine

kidney

umbilical artery

ureter

umbilical cord

ovaries

uterine horn

urinary bladder

body of uterus

urethra

vagina

urogenital sinus

urogenital papilla

Comparison of Female Fetal Pig with Human Female

Use Figure 15.8 to compare the female pig reproductive system with the human female reproductive system. Complete Table 15.4, which compares the appearance of the oviducts and the uterus, as well as the presence or absence of a urogenital sinus in these two mammals.

Figure 15.8   Human female reproductive system.

Especially compare the anatomy of the oviducts in humans with that of the uterine horns in a pig. In a pig, the fetuses develop in the uterine horns; in a human female, the fetus develops in the body of the uterus.

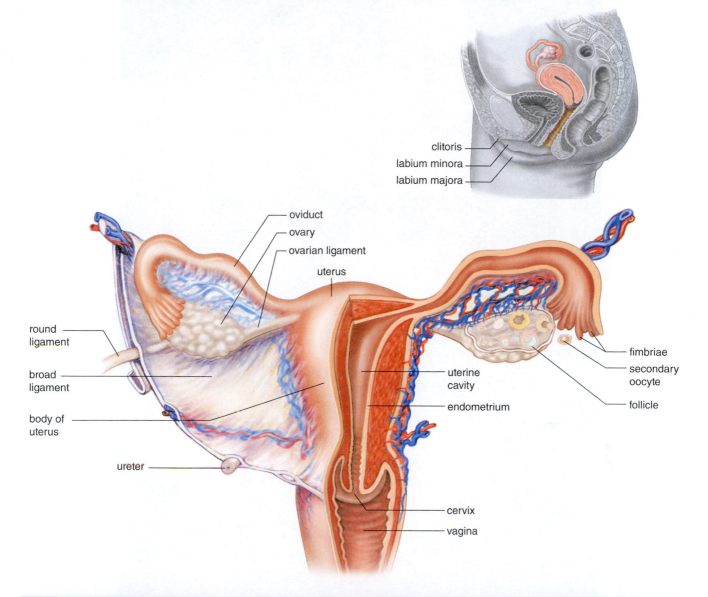

Table 15.4   Comparison of Female Fetal Pig with Human Female

	Fetal Pig	Human
Oviducts		
Uterus		
Urogenital sinus		

Figure 15.9 Internal anatomy of the fetal pig.

Most of the major organs are shown in this photograph. The stomach has been removed. The spleen, gallbladder, and pancreas are not visible.

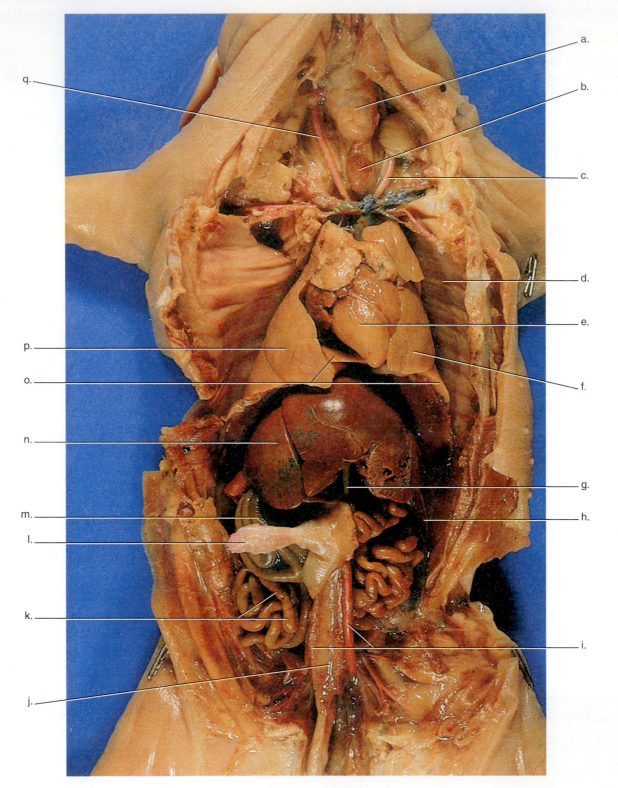

15.4 Review of the Respiratory, Digestive, and Cardiovascular Systems

In Laboratory 12, you dissected the respiratory, digestive, and cardiovascular systems of the fetal pig. Review your knowledge of these systems by reexamining your dissection of the fetal pig and *label Figure 15.9.* In this portion of today's lab, you will review each system and examine some organs in more detail. **Do not remove any organs** unless told to do so by your instructor.

Observation: Respiratory System in Pigs

1. Using these terms (bronchiole, bronchus, glottis, larynx, pharynx, trachea), trace the path of air from the nasal passages to the lungs. List the first three organs in the left column and the last three organs in the right column.

   nasal passages _____   _____

   _____   _____

   _____   _____

   _____   lungs

2. Make sure you have cut the corners of the mouth as directed in Lab 12, page 162. In the **pharynx,** you should be able to locate the **glottis,** an opening to the _____.

3. If necessary, make a midventral incision in the neck to expose the **larynx.**

4. Clear away the "straplike" muscles covering the **trachea.** Now you should be able to feel the cartilaginous rings that hold the trachea open. Locate the esophagus, which lies below the trachea.

5. If available, observe a slide on display showing a section through the trachea and esophagus. Notice in Figure 15.10 that the air and food pathways cross in the pharynx.

6. Open the pig's mouth, insert a blunt probe into the glottis, and carefully work the probe down through the larynx to the level of the **bronchi.**

7. Observe the **lungs,** and if available, observe a prepared slide of lung tissue.

8. If so directed by your instructor, remove a portion of the trachea, the bronchi, and the lungs, keeping them all in one piece. Place this specimen in a small container of water. Holding the trachea with your forceps, gently but firmly stroke the lung repeatedly with the blunt wooden base of one of your probes. If you work carefully, the alveolar tissue will be fragmented and rubbed away, leaving the branching system of air tubes and blood vessels.

Figure 15.10   Air and food passages in the fetal pig.
A probe can pass from the mouth to the larynx to the esophagus.

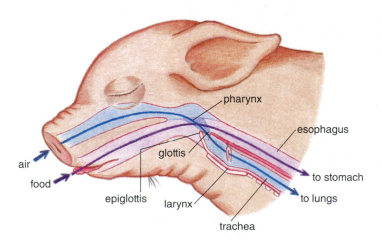

Observation: Digestive System in Pigs

1. Using these terms (esophagus, large intestine, small intesting, stomach) trace the path of food from the mouth to the anus:

   mouth_____   _____

   _____   _____

   _____   anus_____

2. Open the **mouth,** and insert a blunt probe into the esophagus (see Fig. 15.10). Then trace the **esophagus** to the stomach.

3. Open one side of the **stomach,** and examine its interior surface. Does it appear smooth or rough? _____

4. Find the pyloric sphincter, the muscle that surrounds the entrance to the duodenum, the first part of the **small intestine.** Record the length of the small intestine. _____

   If you have not done so before, find the bile duct that empties into the duodenum. The bile duct comes from the _____.

5. Find the **cecum,** a projection where the small intestine enters the large intestine. How does the appearance of the pig's large intestine differ from that of a human?

   _____

6. Carefully cut the mesenteries holding the colon of the **large intestine** in place, and uncoil the large intestine. Record the length of the large intestine. _____ How does the length of the large intestine compare with that of the small intestine? _____

7. Locate again the liver, pancreas, and gallbladder, three accessory organs of digestion.

Observation: Cardiovascular System in Pigs

Heart

1. Trace the path of blood through the heart, starting with the vena cava and ending with the aorta. Mention all the chambers of the heart and the valves (see Fig. 14.2).

To the heart:	From the lungs:

   vena cavae   _____

   _____   _____

   _____ valve   _____ valve

   _____   _____

   _____ valve   _____ valve

   _____   aorta

   _____

2. Keeping the heart inside the pig, cut the pericardial sac (the tissue that surrounds the heart).
3. Look for and identify the vessels attached to the heart.
4. Section the heart, and look for its four chambers. Remnants of the atrioventricular valves can be seen as thin sheets of whitish tissue attached to fine, white, tendinous strands.
5. With your blunt probe, find the oval opening in the wall between the two atrial chambers. Recall that this is a shunt that allows blood to bypass lung circulation prior to birth.

Blood Vessels

1. In general, arteries take blood _____ the heart, and veins take blood _____ the heart.
2. Locate the following blood vessels in your pig. State their origin and destination.

   a. coronary artery _____

   b. cardiac vein _____

   c. carotid artery _____

   d. jugular vein _____

   e. subclavian artery _____

   f. subclavian vein _____

   g. renal artery _____

   h. renal vein _____

   i. iliac artery _____

   j. iliac vein _____

Hepatic Portal System and Associated Vessels

A **portal system** goes from one capillary bed to another without passing through the heart. For example, the hepatic portal vein takes blood from the intestinal capillaries to capillaries in the liver. The liver plays an important role in processing and storing materials absorbed from the intestine. The hepatic veins take blood from the liver to the inferior (posterior) vena cava.

1. In your pig, the hepatic portal vein is dorsal to the bile duct and will not be blue if the latex did not enter it. To find the hepatic portal vein, break the mesenteries in the region of the bile duct (Fig. 15.11).
2. The **hepatic veins** consist of three or four vessels from the liver to the inferior vena cava. To see them, scrape away the liver with the blunt side of the scalpel until liver material has been removed and only a mass of cords remains.
3. Identify the **umbilical vein** leading into the liver and, on its posterior surface, the **venous duct** which is the main channel through the liver in the fetus (see Table 14.2).

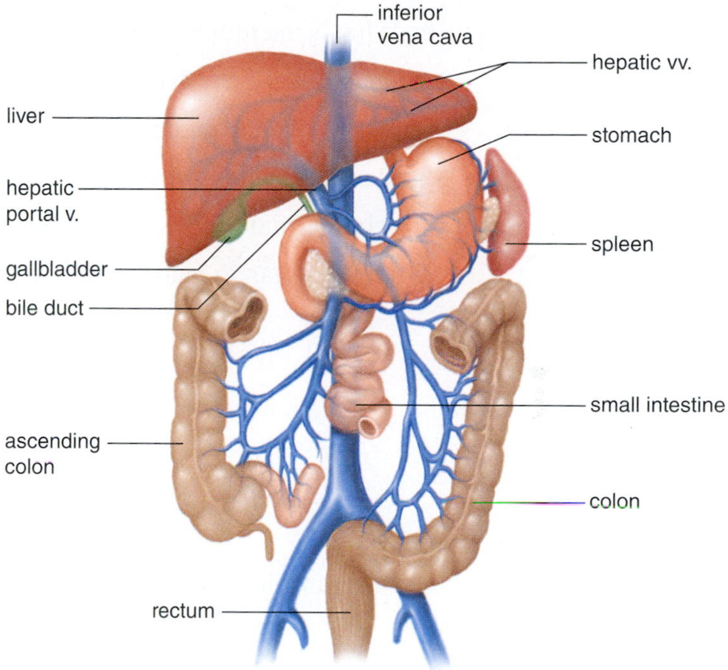

Figure 15.11   Hepatic portal system.
The hepatic portal vein lies between the digestive tract and the liver. The hepatic veins enter the inferior vena cava. (v. = vein; vv. = veins)

Storage of Pigs

1. Before leaving the laboratory, place your pig in the plastic bag provided.
2. Expel excess air from the bag, and tie it shut.
3. Write your name and section on the tag provided, and attach it to the bag. Your instructor will indicate where the bags are to be stored until the next laboratory period.
4. Clean the dissecting tray and tools, and return them to their proper location.
5. Wipe off your goggles and wash your hands.

Laboratory Review 15

_____ 1. Which structures in the urinary system carry urine to the bladder?

_____ 2. Which structure in the urinary system receives urine from the bladder?

_____ 3. What is the outer portion of the kidney proper called?

_____ 4. Where are the testes located in human males?

_____ 5. What is the function of the vas deferens?

_____ 6. What is the function of the prostate gland?

_____ 7. Where are the ovaries located?

_____ 8. What is the function of the uterus?

_____ 9. What is the function of the ovaries?

_____ 10. The vas deferens in males compares best with which structure in females?

_____ 11. Which type of mammal, a pig or a human, has uterine horns?

_____ 12. Name two glands that add fluid to semen after sperm reach the urethra.

_____ 13. A urogenital sinus is present only in female (pigs or humans)?

_____ 14. Which organ in males carries urine or sperm?

Thought Questions

15. On the basis of anatomy, explain why the urethra is part of both the urinary and the reproductive systems in males.

16. A vasectomy is a procedure in which the vas deferens are severed. Why would such a procedure cause sterility?

17. Explain why it would help students if the pulmonary arteries were colored with blue latex instead of red latex in fetal pigs.

18. Pigs and humans have a hepatic portal system. Of what benefit is this system?

16

Homeostasis

Introduction

Homeostasis refers to the dynamic equilibrium of the body's internal environment. The **internal environment** consists of blood and tissue fluid. The body's cells take nutrients from tissue fluid and return their waste molecules to it. Tissue fluid, in turn, exchanges molecules with the blood. This is called capillary exchange. All internal organs contribute to homeostasis, but this laboratory specifically examines the contributions of the blood, lungs, and kidneys (Fig. 16.1).

Figure 16.1   Contributions of organs to homeostasis.
The lungs exchange gases with blood; the kidneys remove nitrogenous wastes from blood; and the intestinal tract adds nutrients as regulated by the liver to blood.

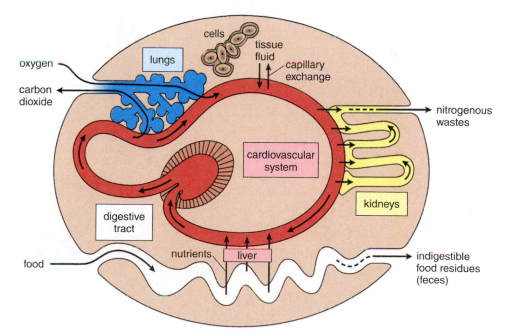

16.1　Heartbeat and Blood Flow

Recall that the cardiovascular system consists of the heart, blood vessels, and blood (Fig. 16.2). Arteries carry blood away from the heart while veins transport blood toward the heart. Arteries branch into smaller vessels called arterioles that enter capillary beds. Capillary beds are present throughout the organs and tissues of the body. An exchange of gases takes place across the thin walls of **pulmonary capillaries**. In the lungs, CO_2 leaves the blood and O_2 enters the blood. An exchange of gases and nutrients for metabolic wastes takes place across the thin walls of **systemic capillaries**. In the body tissues, O_2 and nutrients exit the blood, while CO_2 and metabolic wastes enter the blood.

We will see that the heart is vital to homeostasis because its contraction (called the **heartbeat**) keeps the blood moving in the arteries and arterioles which take blood to the capillaries. The exchanges that take place across capillaries help maintain homeostasis.

Liver

The hepatic portal vein lies between the

_____ and

the _____ .

This placement allows the liver to regulate what molecules enter the blood from the digestive tract. For example, if the hormone insulin is present, the liver removes excess glucose and stores it as glycogen. Later, the liver breaks down glycogen to glucose to keep the glucose concentration constant.

Figure 16.2　The circulatory system.
The heart provides the pumping action that transports the blood through the arteries, capillary beds, and veins.

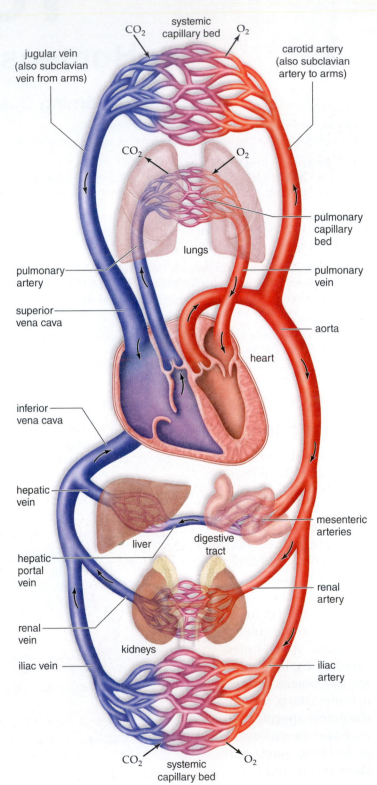

Heartbeat

During a heartbeat, first the atria contract and then the ventricles contract. When a chamber contracts, it is called **systole**, and when a chamber relaxes, it is called **diastole**. The atria and ventricles take turns being in systole.

Time	Atria	Ventricles
0.15 sec	Systole	Diastole
0.30 sec	Diastole	Systole
0.40 sec	Diastole	Diastole

Usually, there are two heart sounds with each heartbeat (Fig. 16.3). The first sound (*lub*) is low and dull and lasts longer than the second sound. It is caused by the closure of valves following atrial systole. The second sound (*dub*) follows the first sound after a brief pause. The sound has a snapping quality of higher pitch and shorter duration. The *dub* sound is caused by the closure of valves following ventricle systole.

Figure 16.3   The heartbeat sounds.
a. When the atria contract (are in systole), the ventricles fill with blood. **b.** Closure of the valves between atria and ventricles results in a **lub** sound. When the ventricles contract, blood enters the attached arteries (aorta and pulmonary trunk). **c.** Closure of valves between ventricles and arteries results in a **dub** sound. Blood enters the heart from the attached veins (vena cavas and pulmonary veins) once more.

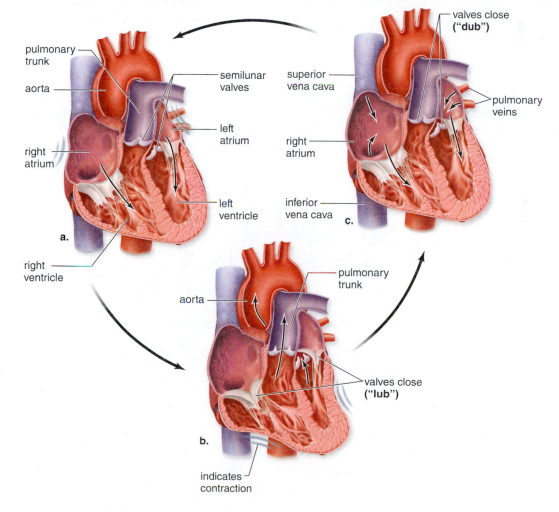

In the following procedure, you will work with a partner and use a stethoscope to listen to the heartbeat. It will not be necessary for you to count the number of beats per minute.

1. Obtain a stethoscope, and properly position the earpieces. They should point forward. Place the bell of the stethoscope on the left side of your partner's chest between the fourth and fifth ribs. This is where the apex (tip) of the heart is closest to the body wall.

2. Which of the two sounds (**lub** or **dub**) is louder? _____

3. Now switch, and your partner will determine your heartbeat.

Blood Pressure

Blood pressure is highest just after ventricular systole (contraction) and it is lowest during ventricular

diastole (relaxation). Why? _____

We would expect a person to have lower blood pressure readings at rest than after exercise. Why?

_____

A number of different types of digital blood pressure monitors are available, and your instructor will instruct you on how to use the type you will be using for this Experimental Procedure. The resting blood pressure readings for an individual are displayed on the monitor shown in Figure 16.4. A blood pressure reading at or below 120/80 (systolic/diastolic) is considered normal.

Figure 16.4 Measurement of blood pressure and pulse.
There are many different types of digital blood pressure/pulse monitors now available. The one shown here uses a cuff to be placed on the arm. Others use a cuff for the wrist.

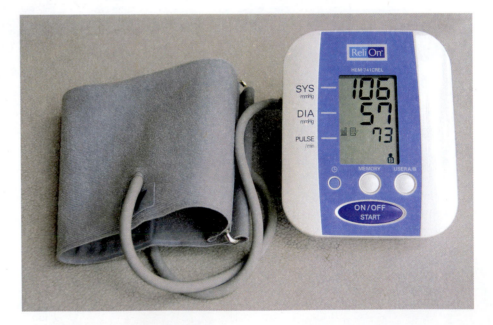

During this experimental procedure you may work with a partner or by yourself. If working with a partner, each of you will assist the other in taking blood pressure readings. When you note the blood pressure readings, also note the pulse reading.

Blood Pressure and Pulse at Rest

1. Reduce your activity as much as possible.
2. Use the blood pressure monitor to obtain several blood pressure readings, average them, and record your results in Table 16.1. Also note the pulse rates and average. Record in Table 16.1.

Blood Pressure and Pulse After Exercise

1. Run in place for 1 minute.
2. Immediately use the blood pressure monitor to obtain a blood pressure reading, and record it in Table 16.1. Also note the pulse rate and record in Table 16.1.

Table 16.1 Blood Pressure				
	Rates at Rest		Rates After Exercise	
	Blood Pressure	Pulse	Blood Pressure	Pulse
Partner				
Yourself				

Conclusions: Blood Pressure

- Knowing that exercise increases the heart rate, offer an explanation for your results. _____

  _____

- Under what conditions in everyday life would you expect the heart rate and the blood pressure to increase, even though you were not exercising? _____

  _____

  When might this be an advantage? _____

  A disadvantage? _____

16.2 Blood Flow and Systemic Capillary Exchange

We associate death with lack of a heartbeat, but the real problem is lack of blood flow to the capillaries.

Blood Flow

The beat of the heart moves blood into the aorta, which divides into arterioles, and then arterioles divide into capillaries. Venules, which receive blood from capillaries, combine to form veins, which take blood back to the heart.

Experimental Procedure: Blood Flow

1. Observe blood flow through arterioles, capillaries, and venules, either in the tail of a goldfish or in the webbed skin between the toes of a frog, as prepared by your instructor.
2. Examine under low and high power of the microscope.
3. Watch the pulse and the swiftly moving blood in the arterioles.
4. Contrast this with the more slowly moving blood that circulates in the opposite direction in the venules. Many criss-crossing capillaries are visible.
5. Look for blood cells floating in the bloodstream. Don't confuse blood cells with chromatophores, irregular black patches of pigment that may be visible in the skin.

Systemic Capillary Exchange

The beat of the heart creates blood pressure, and blood pressure is necessary to capillary exchange. Blood pressure acts to move water out of a capillary, while osmotic pressure (created by the presence of proteins in the blood) acts to move water into a capillary (Fig. 16.5). Blood pressure is higher than osmotic pressure at the arteriole end of a capillary and water moves out of a capillary. But blood pressure lessens as the blood moves through a capillary bed. This means that osmotic pressure is higher than blood pressure at the venule end of a capillary and water moves back into capillary.

Figure 16.5 Systemic capillary exchange.
At a systemic capillary, an exchange takes place across the capillary wall. Between the arterial end and the venule end, molecules follow their concentration gradient. Oxygen and nutrients move out of a capillary while carbon dioxide and wastes move into the capillary.

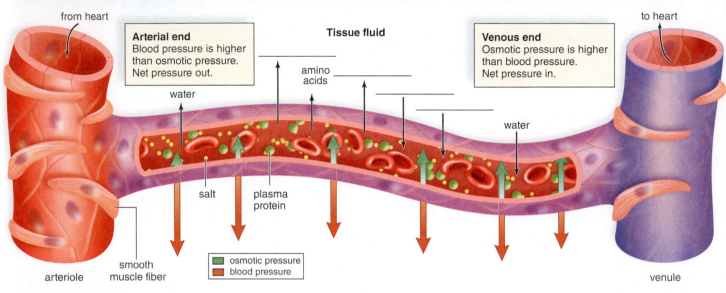

Conclusions: Systemic Capillary Exchange

- What generates blood pressure? _____
- Why are tissue cells always in need of glucose and oxygen? _____

  Add glucose at the end of an appropriate arrow in Figure 16.5. Do the same for oxygen.

- Why are tissue cells always producing carbon dioxide? _____

  Add carbon dioxide at the start of an appropriate arrow in Figure 16.5. Do the same for metabolic wastes.

16.3 Lung Structure and Human Respiratory Volumes

The right and left lungs lie in the thoracic cavity on either side of the heart. Air moves from the nasal passages to the trachea, bronchi, bronchioles, and finally, lungs.

Lung Structure

A **lung** is a spongy organ consisting of irregularly shaped air spaces called **alveoli** (sing., alveolus) (Fig. 16.6a). The alveoli are surrounded by a rich network of tiny blood vessels called pulmonary capillaries. *In Fig. 16.6a, use a labeled arrow to show oxygen entering blood from an alveoli and another labeled arrow to show carbon dioxide entering alveoli from the blood.*

1. Observe a prepared slide of a stained section of a lung (Fig. 16.6*b*). In stained slides, the nuclei of the cells forming the thin alveolar walls appear purple or dark blue.
2. Look for areas that show red or orange disc-shaped **erythrocytes.** These are the red blood cells that contain hemoglobin which takes up oxygen and transports it to the tissues. When these appear in strings, you are looking at capillary vessels in side view.
3. In some part of the slide, you may even observe an artery. Thicker, circular or oval structures with a lumen (cavity) are cross sections of **bronchioles,** tubular pathways through which air reaches the air spaces.
4. In Figure 16.6*c*, note that in emphysema, alveoli have burst. In smokers, small bronchioles collapse and trapped air in alveoli causes them to burst. Now gas exchange is minimal.

Figure 16.6 Healthy lung tissue versus emphysema.
a. The lungs normally contain many air sacs called alveoli where gas exchange occurs. **b.** Micrograph of normal lung tissue. **c.** In smokers, emphysema can occur; the alveoli burst and gas exchange is inadequate.

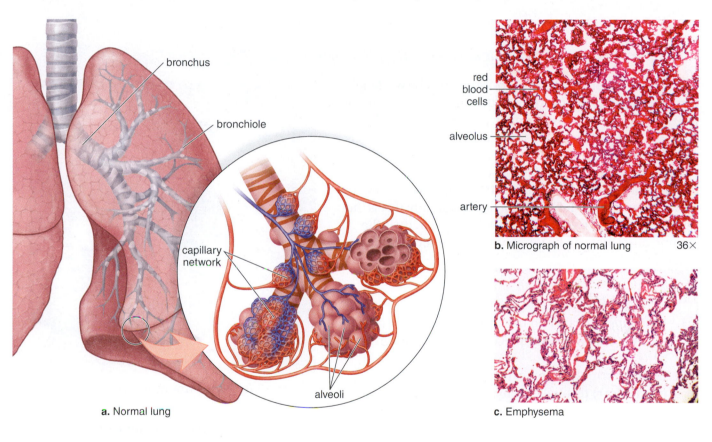

bronchus

bronchiole

capillary network

alveoli

a. Normal lung

red blood cells

alveolus

artery

b. Micrograph of normal lung   36×

c. Emphysema

Human Respiratory Volumes

Breathing in, called **inspiration** or inhalation, is the active part of breathing because that's when contraction of rib cage muscles causes the rib cage to move up and out, and contraction of the diaphragm causes the diaphragm to lower. Due to an enlarged thoracic cavity, the lungs expand and air is drawn into them. Breathing out, called **expiration** or exhalation, occurs when relaxation of these same muscles causes the thoracic cavity to resume its original capacity. Now air is pushed out of the lungs (Fig. 16.7).

Figure 16.7 Inspiration and expiration.

a. Inspiration occurs after the rib cage moves up and out and the diaphragm moves down. Air rushes into lungs because they expand as the thoracic cavity expands. **b.** Expiration occurs as the rib cage moves down and in and the diaphragm moves up. As the thoracic cavity and lungs get smaller, air is pushed out.

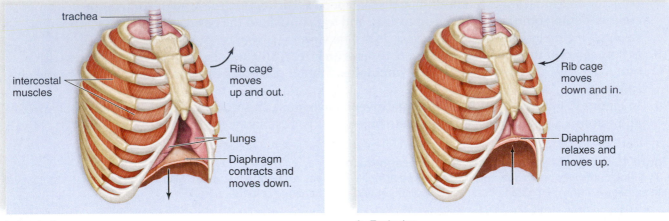

a. Inspiration

b. Expiration

Experimental Procedure: Human Respiratory Volumes

During this Experimental Procedure you will be working with a spirometer, an instrument that measures the amount of exhaled air (Fig. 16.8). Normally, about 500–600 ml of air move into and out of the lungs with each breath. This is called the **tidal volume** (**TV**). You can inhale deeply after a normal breath and more air will enter the lungs; this is the **inspiratory reserve volume** (**IRV**). You can also force more air out of your lungs after a normal breath; this is the **expiratory reserve volume** (**ERV**). **Vital capacity** is the volume of air that can be forcibly exhaled after forcibly inhaling.

Figure 16.8 Nine-liter student wet spirometer.

Tidal Volume (TV)

1. When it's your turn to use the spirometer, install a new disposable mouthpiece and set the spirometer to zero.
2. Inhale normally, then exhale normally (with *no* extra effort) through the mouthpiece of the spirometer. Record your measurement in Table 16.2.
3. Three readings are needed, so twice more set the spirometer to zero and repeat the same procedure. Record your measurements in Table 16.2.
4. Later, if necessary, change your readings to milliliters (ml), and calculate your average TV in ml.

   In your own words, what is tidal volume? _____

   _____

Expiratory Reserve Volume (ERV)

1. Make sure the spirometer is set to zero.
2. Inhale and exhale normally and then force as much air out as possible into the spirometer. Record your measurement in Table 16.2.

3. Three readings are needed, so twice more, set the spirometer to zero and repeat the same procedure. Record your measurements in Table 16.2.
4. Later, if necessary, change your readings to ml, and calculate your average ERV.

In your own words, what is expiratory reserve volume? _____

_____

Vital Capacity (VC)

1. Make sure the spirometer is set to zero.
2. Inhale as much as possible and then exhale as much as possible into the spirometer.
3. Three readings are needed, so twice more, set the spirometer to zero and repeat the same procedure. Record your measurements in Table 16.2.
4. Later, if necessary, change your readings to ml, and calculate your average VC.

In your own words, what is vital capacity? _____

_____

Inspiratory Reserve Volume (IRV)

It will be necessary for us to calculate IRV because a spirometer measures only exhaled air, not inhaled air.

Explain. _____

From having measured vital capacity (VC) you can see that VC = TV + IRV + ERV. To calculate IRV, simply subtract the average TV + the average ERV from the value you recorded for the average VC:

$$IRV = VC - (TV + ERV) = \underline{\hspace{2cm}} \text{ ml. Record your IRV in Table 16.2.}$$

Table 16.2 Measurements of Lung Volumes

Tidal Volume (TV)	Expiratory Reserve Volume (ERV)	Vital Capacity (VC)	Inspiratory Reserve Volume (IRV)
1st	1st	1st	_____
2nd	2nd	2nd	_____
3rd	3rd	3rd	_____
Average ____ ml	Average ____ ml	Average ____ ml	Calculated value = ____ ml

Conclusions: Human Respiratory Volumes

- Vital capacity varies with age, sex, and height; however, typically for men, vital capacity is about 5,200 ml and for women, it is about 4,000 ml. How does your vital capacity compare to the typical values for your gender? _____ If smaller than normal, are you a smoker or is there any health reason why it would be smaller? If larger than normal, are you a sports enthusiast or do you play a musical instrument that involves inhaling and exhaling deeply? _____

- Diffusion alone accounts for pulmonary gas exchange. Therefore, how does good lung ventilation assist gas exchange?_____

_____

16.4 Kidneys

The **kidneys** are bean-shaped organs that lie along the dorsal wall of the abdominal cavity.

Kidney Structure

Figure 16.9 shows the structure of a kidney, macroscopic and microscopic. The macroscopic structure of a kidney is due to the placement of over 1 million **nephrons.** Nephrons are tubules that do the work of producing urine.

Figure 16.9   Longitudinal section of a kidney.
a. The kidneys are served by the renal artery and renal vein. **b.** Macroscopically, a kidney has three parts: renal cortex, renal medulla, and renal pelvis. **c.** Microscopically, each kidney contains over a million nephrons.

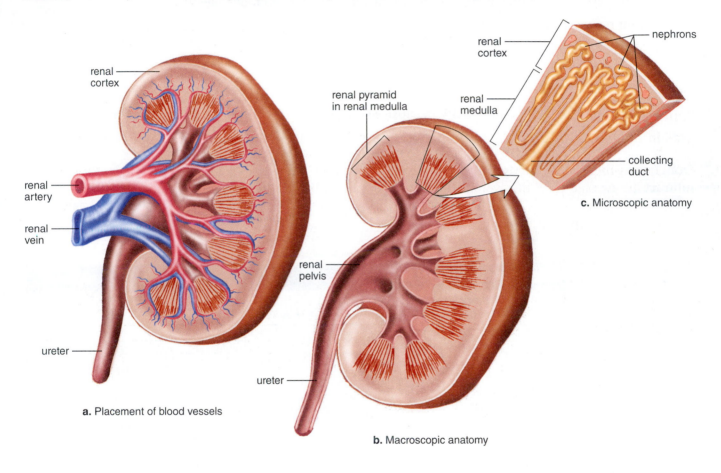

Observation: Kidney Model

Study a model of a kidney, and with the help of Figure 16.9, locate the following:

1. **Renal cortex:** a granular region
2. **Renal medulla:** contains the renal pyramids
3. **Renal pelvis:** where urine collects

Observation: Nephron Structure

Study a nephron model and, with the help of Figure 16.10, identify the following parts of a nephron:

1. **Glomerular capsule:** (Bowman's capsule): closed end of the nephron pushed in on itself to form a cuplike structure; the inner layer has pores that allow **glomerular filtration** to occur; substances move from the blood to inside the nephron.
2. **Proximal convoluted tubule:** The inner layer of this region has many microvilli that allow **tubular reabsorption** to occur; substances move from inside the nephron to the blood.
3. **Loop of the nephron:** Nephron narrows to form a U-shaped portion which functions in water reabsorption.
4. **Distal convoluted tubule:** second convoluted section that lacks microvilli and functions in **tubular secretion**; substances move from blood to inside nephron.

Several nephrons enter one collecting duct. The **collecting ducts** also function in water reabsorption, and they conduct urine to the pelvis of a kidney.

Observation: Circulation About a Nephron

Study a nephron model and, with the help of Figure 16.10 and Table 16.3, trace the path of blood from the renal artery to the renal vein:

1. **Afferent arteriole:** small vessel that conducts blood from the renal artery to a nephron.
2. **Glomerulus:** capillary network that exists inside the glomerular capsule; small molecules move from inside the capillary to the inside of the glomerulus during glomerular filtration.
3. **Efferent arteriole:** small vessel that conducts blood from the glomerulus to the peritubular capillary network.
4. **Peritubular capillary network:** surrounds the proximal convoluted tubule, the loop of the nephron, and the distal convoluted tubule.
5. **Venule:** takes blood from the peritubular capillary network to the renal vein.

Table 16.3 Blood Vessels Serving the Nephron	
Name of Structure	**Significance**
Afferent arteriole	Brings arteriolar blood to the glomerulus
Glomerulus	Capillary tuft enveloped by glomerular capsule
Efferent arteriole	Takes arteriolar blood away from the glomerulus
Peritubular capillary network	Capillary bed that envelops the rest of the nephron
Venule	Takes venous blood away from the peritubular capillary network

Kidney Function

The kidneys produce urine and in doing so help maintain homeostasis in several ways. Urine formation requires three steps: **glomerular filtration**, **tubular reabsorption**, and **tubular secretion** (see Fig. 16.10).

Figure 16.10 Nephron structure and blood supply.
The three main processes in urine formation are described in boxes and color coded to arrows that show the movement of molecules out of or into the nephron at specific locations. In the end, urine is composed of the substances within the collecting duct (see brown arrow).

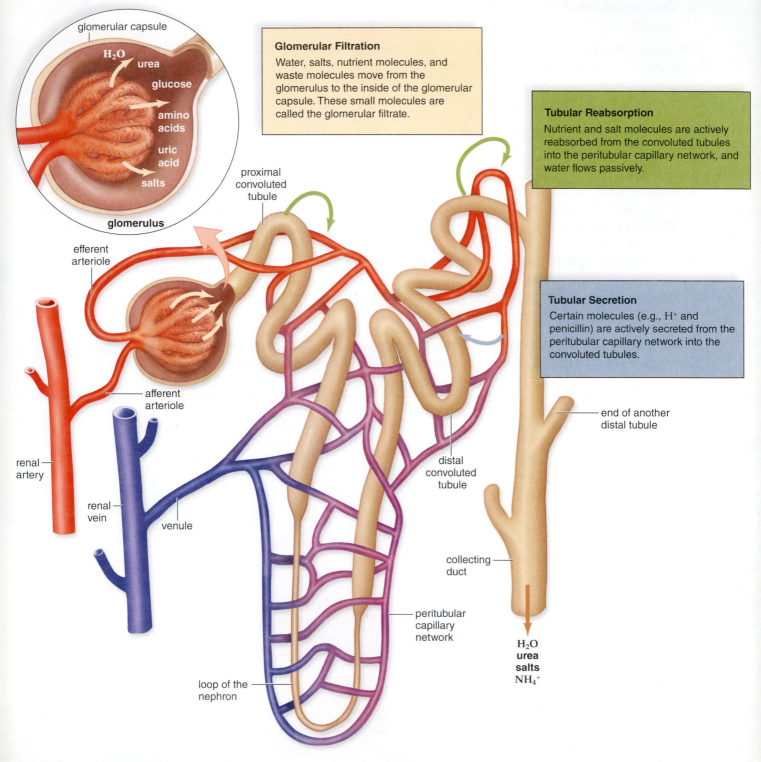

Glomerular Filtration

Water, salts, nutrient molecules, and waste molecules move from the glomerulus to the inside of the glomerular capsule. These small molecules are called the glomerular filtrate.

Tubular Reabsorption

Nutrient and salt molecules are actively reabsorbed from the convoluted tubules into the peritubular capillary network, and water flows passively.

Tubular Secretion

Certain molecules (e.g., H^+ and penicillin) are actively secreted from the peritubular capillary network into the convoluted tubules.

glomerular capsule

H_2O
urea
glucose
amino acids
uric acid
salts

glomerulus

efferent arteriole

proximal convoluted tubule

afferent arteriole

renal artery

renal vein

venule

loop of the nephron

distal convoluted tubule

peritubular capillary network

end of another distal tubule

collecting duct

H_2O
urea
salts
NH_4^+

Glomerular Filtration

1. Blood entering the glomerulus contains blood cells, proteins, glucose, amino acids, salts, urea, and water. Blood cells and proteins are too large to pass through the glomerular wall and enter the filtrate.
2. Blood pressure causes small molecules of glucose, amino acids, salts, urea, and water to exit the blood and enter the glomerular capsule. The fluid in the glomerular capsule is called the **filtrate.**
3. In the list that follows, draw an arrow from left to right for the small molecules that leave the glomerulus and become part of the filtrate.

     Glomerulus   **Glomerular (Filtrate)**
     Cells
     Proteins
     Glucose
     Amino acids
     Urea
     Water and salts

4. Complete the second column in Table 16.4, p. 226. Use an X to indicate that the substance is at the locations noted.

Tubular Reabsorption

1. When the filtrate enters the proximal convoluted tubule, it contains glucose, amino acids, urea, water, and salts. Some water and salts remain in the nephron, but enough are *passively* reabsorbed into the peritubular capillary to maintain blood volume and blood pressure. Use this information to state a way kidneys help maintain homeostasis. _____

   _____

2. The cells that line the proximal convoluted tubule are also engaged in active transport and usually completely reabsorb nutrients (glucose and amino acids) into the peritubular capillary. What would happen to cells if the body lost all its nutrients by way of the kidneys? _____

   _____

3. Which of the filtrate substances is reabsorbed the least and will become a part of urine? _____

   Urea is a nitrogenous waste. State here another way kidneys contribute to homeostasis. _____

   _____

4. In the list that follows, draw an arrow from left to right for all those molecules passively reabsorbed into the blood of the peritubular capillary. Use darker arrows for those that are reabsorbed completely by active transport.

     Proximal Convoluted Tubule   **Peritubular Capillary**
     Water and salts
     Glucose
     Amino acids
     Urea

Tubular Secretion

1. During tubular secretion, certain substances—for example, penicillin and histamine—are actively secreted from the peritubular capillary into the fluid of the tubule. Also, hydrogen ions (H^+) and ammonia (NH_3) are secreted as NH_4^+ as necessary. Complete the last column of Table 16.4. Check your entries against Figure 16.10.

2. The blood is buffered, but only the kidneys can excrete H^+. The excretion of H^+ by the kidneys raises the pH of the blood. Use this information to state a third way the kidneys contribute to homeostasis. _____

_____

Table 16.4 Urine Constituents		
In Glomerulus	**In Filtrate**	**In Urine**
Blood cells		
Proteins		
Glucose		
Amino acids		
Urea		
Water and salts		
NH_4^+		

Urinalysis: A Diagnostic Tool

Urinalysis can indicate whether the kidneys are functioning properly or whether an illness such as diabetes mellitus is present. The procedure is easily performed with a Chemstrip test strip, which has indicator spots that produce specific color reactions when certain substances are present in urine.

Experimental Procedure: Urinalysis

A urinalysis has been ordered, and you are to test the urine for a possible illness. (In this laboratory, you will be testing simulated urine.)

Assemble Supplies

1. Obtain three Chemstrip urine test strips each of which tests for leukocytes, pH, protein, glucose, ketones, and blood, as noted in Figure 16.11.
2. The color key on the diagnostic color chart or on the Chemstrip vial label will explain what any color changes mean in terms of the pH level and amount of each substance present in the urine sample. You will use these color blocks to read the results of your test.
3. Obtain three "specimen containers of urine" marked 1 through 3. Among them are a normal specimen and two that indicate the patient has an illness. Have a piece of absorbent paper ready to use.

Test the Specimen

1. Briefly (no longer than 1 second) dip a test strip into the first specimen of urine. Be sure the chemically treated patches on the test strip are totally immersed.
2. Draw the edge of the strip along the rim of the specimen container to remove excess urine.
3. Turn the test strip on its side, and tap once on a piece of absorbent paper to remove any remaining urine and to prevent the possible mixing of chemicals.
4. After 60 seconds, read the results as follows: Hold the strip close to the color blocks on the diagnostic color chart (Fig. 16.11) or vial label, and match carefully, ensuring that the strip is properly oriented to the color chart. Enter the test results in Figure 16.11. Use a negative symbol (−) for items that are not present in the urine, a plus symbol (+) for those that are present, and a number for the pH.
5. Test the other two specimens.

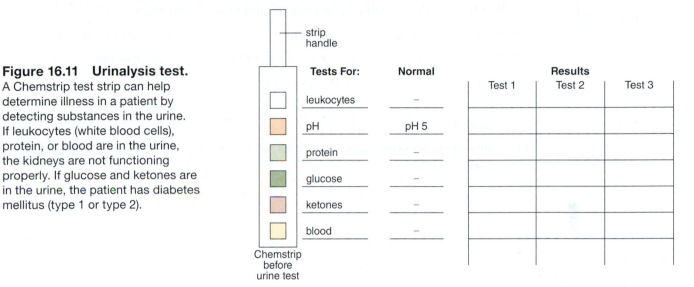

Figure 16.11 Urinalysis test.
A Chemstrip test strip can help determine illness in a patient by detecting substances in the urine. If leukocytes (white blood cells), protein, or blood are in the urine, the kidneys are not functioning properly. If glucose and ketones are in the urine, the patient has diabetes mellitus (type 1 or type 2).

Tests For:	Normal	Test 1	Test 2	Test 3
leukocytes	−			
pH	pH 5			
protein	−			
glucose	−			
ketones	−			
blood	−			

Chemstrip before urine test

Conclusion: Urinalysis

- State below if the urinalysis is normal or indicates a urinary tract infection (leukocytes, blood, and possibly protein in the urine) or that the patient has diabetes mellitus.

Test strip 1 _____

Test strip 2 _____

Test strip 3 _____

- The hormone insulin promotes the uptake of glucose by cells. When glucose is in the urine, either the pancreas is not producing insulin (diabetes mellitus type 1) or cells are resistant to insulin (diabetes mellitus type 2). Ketones (acids) are also in the urine because the cells are metabolizing fat instead of glucose. Explain why. _____

_____

Why is the pH of urine lower than normal? _____

_____

- If urinalysis shows that proteins are excreted instead of retained in the blood, would capillary exchange in the tissues (see Fig. 16.5) be normal? _____ Why or why not? _____

_____

Laboratory Review 16

_____ **1.** Systole of what chamber causes a systemic blood pressure reading of 120?

_____ **2.** Body cells receive their oxygen and nutrients by way of which type vessels?

_____ **3.** The presence especially of proteins in the blood account for what pressure that draws water back into the systemic capillaries?

_____ **4.** What blood vessel takes nutrients absorbed from the digestive system to the liver?

_____ **5.** Emphysema due to smoking causes the _____ of lung tissue to burst.

_____ **6.** What is another name for erythrocytes that transport oxygen in the blood?

_____ **7.** When we exhale, the diaphragm relaxes and moves in what direction?

_____ **8.** When measuring tidal volume, should a student exhale normally or maximally?

_____ **9.** Vital capacity is expected to have a (higher or lower) volume than tidal volume.

_____ **10.** Where does urine collect before exiting the kidney?

_____ **11.** When molecules leave the glomerular capsule, they enter what portion of a nephron?

_____ **12.** Name the process by which molecules move from the proximal convoluted tubule into the blood.

_____ **13.** Name a substance that is normally in the blood, the filtrate, and the urine.

_____ **14.** Glucose in the urine indicates that a person may have what condition?

Thought Questions

15. In your own words, what is homeostasis?

16. We studied three ways the kidneys contribute to homeostasis. Name two of these.

17. How do pulmonary capillaries contribute to homeostasis?

18. How does the heartbeat (i.e., cardiac contraction) contribute to homeostasis?

17

Nervous System and Senses

Learning Outcomes

17.1 Central Nervous System
- Identify the parts of the brain studied, and state the functions of each part.   230–33
- Give examples to show that the parts of the brain work together.   232–33
- Describe the anatomy of the spinal cord and tell how the cord functions as a relay station.   234

17.2 Peripheral Nervous System
- Distinguish between cranial nerves and spinal nerves on the basis of location and function.   234
- Describe the anatomy and physiology of a spinal reflex arc.   234–36

17.3 The Human Eye
- Identify the parts of the eye and state a function for each part.   236–37
- Explain the occurrences of the blind spot and the process of accommodation.   238–39

17.4 The Human Ear
- Using photographs, other images, or models, identify the parts of the ear and state a function for each part.   239–41

17.5 Sensory Receptors in Human Skin
- Describe the anatomy of the human skin and explain the distribution and function of sensory receptors.   241–42
- Relate the abundance of touch receptors to the ability to distinguish between two different touch points.   242

17.6 Human Chemoreceptors
- Relate the ability to distinguish foods to the senses of smell and taste.   243

Introduction

The nervous system has two major divisions: the central nervous system (CNS), consisting of the brain and spinal cord, and the peripheral nervous system (PNS), which contains cranial nerves and spinal nerves (Fig. 17.1). Sensory receptors detect changes in environmental stimuli, and nerve impulses move along sensory nerve fibers to the brain and the spinal cord. The brain and spinal cord sum up the data before sending impulses via motor nerve fibers to effectors (muscles and glands) so a response to stimuli is possible. Nervous tissue consists of neurons; whereas the brain and spinal cord contain all parts of neurons, nerves contain only axons.

Figure 17.1   The nervous system.
The central nervous system (CNS) is in the midline of the body, and the peripheral nervous system (PNS) is outside the CNS.

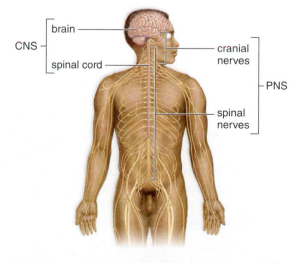

17.1 Central Nervous System

The brain is the enlarged anterior end of the spinal cord; it contains centers that receive input from and can command other regions of the nervous system.

> ⚠️ **Latex gloves** Wear protective latex gloves when handling preserved animal organs. Use protective eyewear and exercise caution when using sharp instruments during this laboratory. Wash hands thoroughly upon completion of this laboratory.

Preserved Sheep Brain

The sheep brain (Fig. 17.2) is often used to study the brain. It is easily available and large enough that individual parts can be identified.

Observation: Preserved Sheep Brain

Examine the exterior and a midsagittal (longitudinal) section of a preserved sheep brain or a model of the human brain, and with the help of Figure 17.2, identify the following.

1. **Ventricles:** interconnecting spaces that produce and serve as a reservoir for cerebrospinal fluid, which cushions the brain. Toward the anterior, note the lateral ventricle (on one longitudinal section) and similarly a lateral ventricle (on the other longitudinal section). Trace the second ventricle to the third and then the fourth ventricles.

2. **Cerebrum:** most developed area of the brain; responsible for higher mental capabilities. The cerebrum is divided into the right and left **cerebral hemispheres,** joined by the **corpus callosum,** a broad sheet of white matter. The outer portion of the cerebrum is highly convoluted and divided into the following surface lobes (see Fig. 17.4):

    a. **Frontal lobe:** controls motor functions and permits voluntary muscle control; it also is responsible for abilities to think, problem solve, speak, and smell.

    b. **Parietal lobe:** receives information from sensory receptors located in the skin and also the taste receptors in the mouth. A groove called the **central sulcus** separates the frontal lobe from the parietal lobe.

    c. **Occipital lobe:** interprets visual input and combines visual images with other sensory experiences. The optic nerves split and enter opposite sides of the brain at the optic chiasma, located in the diencephalon.

    d. **Temporal lobe:** has sensory areas for hearing and smelling. The olfactory bulb contains nerve fibers that communicate with the olfactory cells in the nasal passages and take nerve impulses to the temporal lobe.

3. **Diencephalon:** portion of the brain where the third ventricle is located. The hypothalamus and thalamus are also located here.

    a. **Thalamus:** two connected lobes located in the roof of the third ventricle. The thalamus is the highest portion of the brain to receive sensory impulses before the cerebrum. It is believed to control which received impulses are passed on to the cerebrum. For this reason, the thalamus sometimes is called the "gatekeeper to the cerebrum."

    b. **Hypothalamus:** forms the floor of the third ventricle and contains control centers for appetite, body temperature, blood pressure, and water balance. Its primary function is homeostasis. The hypothalamus also has centers for pleasure, reproductive behavior, hostility, and pain.

4. **Cerebellum:** located just posterior to the cerebrum as you observe the brain dorsally, the cerebellum's two lobes make it appear rather like a butterfly. In cross section, the cerebellum has an internal pattern that looks like a tree. The cerebellum coordinates equilibrium and motor activity to produce smooth movements.

Figure 17.2   The sheep brain.

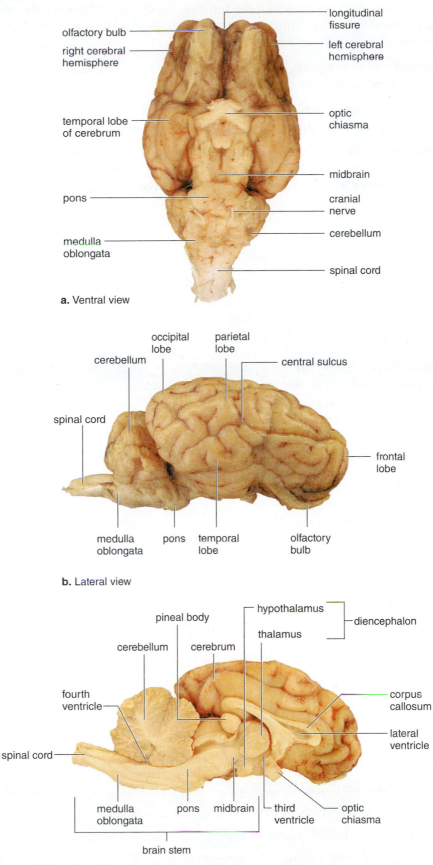

longitudinal fissure

left cerebral hemisphere

olfactory bulb

right cerebral hemisphere

temporal lobe of cerebrum

optic chiasma

midbrain

pons

cranial nerve

cerebellum

medulla oblongata

spinal cord

a. Ventral view

occipital lobe

parietal lobe

cerebellum

central sulcus

spinal cord

frontal lobe

medulla oblongata

pons

temporal lobe

olfactory bulb

b. Lateral view

hypothalamus

diencephalon

pineal body

thalamus

cerebellum

cerebrum

fourth ventricle

corpus callosum

lateral ventricle

spinal cord

medulla oblongata

pons

midbrain

third ventricle

optic chiasma

brain stem

c. Longitudinal section

5. **Brain stem:** part of the brain that connects with the spinal cord. Because it includes the pons and medulla oblongata, it contains centers for the functioning of internal organs; because of its location, it serves as a relay station for nerve impulses passing from the cord to the brain. Therefore, it helps keep the rest of the brain alert and functioning.

   a. **Midbrain:** anterior to the pons, the midbrain serves as a relay station for sensory input and motor output. It also contains a reflex center for eye muscles.

   b. **Pons:** the ventral, bulblike enlargement on the brain stem. It serves as a passageway for nerve impulses running between the medulla and the higher brain regions.

   c. **Medulla oblongata** (or simply **medulla**): the most posterior portion of the brain stem. It controls internal organs; for example, blood pressure, cardiac, and breathing control centers are present in the medulla. Nerve impulses pass from the spinal cord through the medulla to and from higher brain regions.

The Human Brain

Based on your knowledge of the sheep brain, complete Table 17.1 by stating the major functions of each part of the brain listed. *Also label Figure 17.3.*

Table 17.1   Summary of Brain Functions

Part	Major Functions
Cerebrum	
Cerebellum	
Diencephalon Thalamus	
Hypothalamus	
Brain stem Midbrain	
Pons	
Medulla oblongata	

Which parts of the brain work together to achieve the following?

1. Good eye–hand coordination _____

2. Concentrating on homework when TV is playing _____

3. Avoiding dark alleys while walking home at night _____

4. Keeping the blood pressure 17.4 within the normal range. _____

Figure 17.3 The human brain (longitudinal section).
The cerebrum is larger in humans than in sheep. Label where indicated.

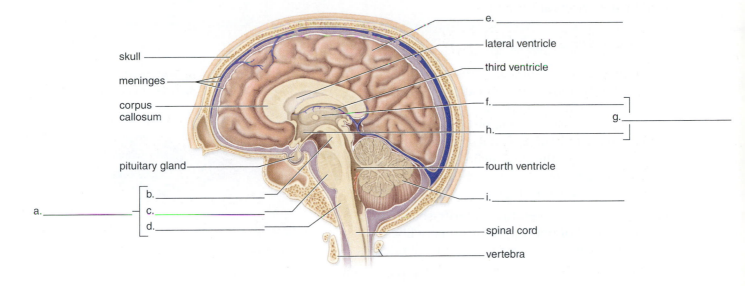

skull

meninges

corpus callosum

pituitary gland

a. _____

b. _____

c. _____

d. _____

e. _____

lateral ventricle

third ventricle

f. _____

g. _____

h. _____

fourth ventricle

i. _____

spinal cord

vertebra

Cerebral Lobes

As stated previously, the outer portion of the cerebrum is highly convoluted and divided into lobes as illustrated in Figure 17.4. The various sense organs send nerve impulses to a particular lobe where the nerve impulses are integrated to give us our senses of vision, hearing, smell, taste, and touch. Although not stated in Figure 17.4, the frontal lobe helps us remember smells of some significance to us. *Write the name of a sense next to the appropriate lobe in Figure 17.4.*

Figure 17.4 The cerebral lobes.
Each lobe has centers for integrating nerve impulses received from a particular type of sense organ. Our five senses result from this activity of the brain.

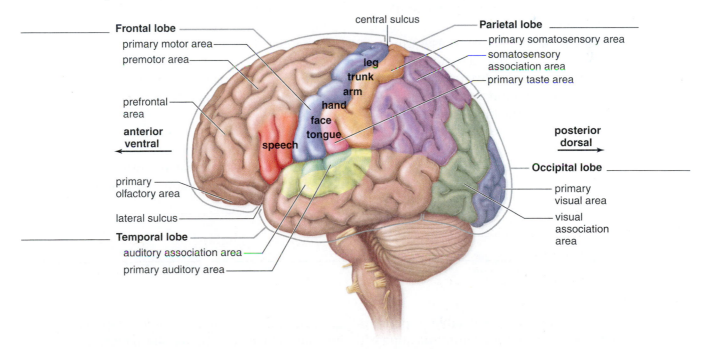

Frontal lobe _____

primary motor area

premotor area

prefrontal area

anterior ventral ←

primary olfactory area

lateral sulcus

Temporal lobe _____

auditory association area

primary auditory area

central sulcus

leg
trunk
arm
hand
face
tongue

speech

Parietal lobe _____

primary somatosensory area

somatosensory association area

primary taste area

posterior dorsal →

Occipital lobe _____

primary visual area

visual association area

The Spinal Cord

The spinal cord is a part of the central nervous system. It lies in the middorsal region of the body and is protected by the vertebral column.

Observation: The Spinal Cord

1. Examine a prepared slide of a cross section of the spinal cord under the lowest magnification possible. For example, some microscopes are equipped with a short scanning objective that enlarges about 3.5×, with a total magnification of 35×. If a scanning objective is not available, observe the slide against a white background with the naked eye.

2. Identify the following with the help of Figure 17.5:

   a. **Gray matter:** a central, butterfly-shaped area composed of masses of short nerve fibers, interneurons, and motor neuron cell bodies.

   b. **White matter:** masses of long fibers that lie outside the gray matter and carry impulses up and down the spinal cord. In living animals, white matter appears white because an insulating myelin sheath surrounds long fibers.

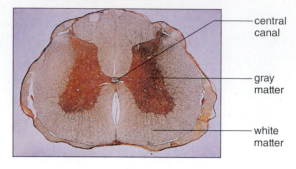

Figure 17.5   The spinal cord.
Photomicrograph of spinal cord cross section.

- central canal
- gray matter
- white matter

17.2   Peripheral Nervous System

The peripheral nervous system contains the cranial nerves and the spinal nerves. Twelve pairs of cranial nerves project from the inferior surface of the brain. The cranial nerves are largely concerned with nervous communication between the head, neck, and facial regions of the body and the brain. The 31 pairs of spinal nerves emerge from either side of the spinal cord (Fig. 17.6).

Spinal Nerves

Each spinal nerve contains long fibers of sensory neurons and long fibers of motor neurons. In Figure 17.6, identify the following:

1. **Sensory neuron:** takes nerve impulses from a sensory receptor to the spinal cord. The cell body of a sensory neuron is in the dorsal root ganglion.

2. **Interneuron:** lies completely within the spinal cord. Some interneurons have long fibers and take nerve impulses to and from the brain. The interneuron in Figure 17.6 transmits nerve impulses from the sensory neuron to the motor neuron.

3. **Motor neuron:** takes nerve impulses from the spinal cord to an effector—in this case, a muscle. Muscle contraction is one type of response to stimuli.

Suppose you were walking barefoot and stepped on a prickly sandbur. Describe the pathway of information, starting with the pain receptor in your foot, that would allow you to both feel and respond to this unwelcome stimulus. _____

_____

_____

Spinal Reflexes

A **reflex** is an involuntary and predictable response to a given stimulus that allows a quick response to environmental stimuli without communicating with the brain. When you touch a sharp tack, you

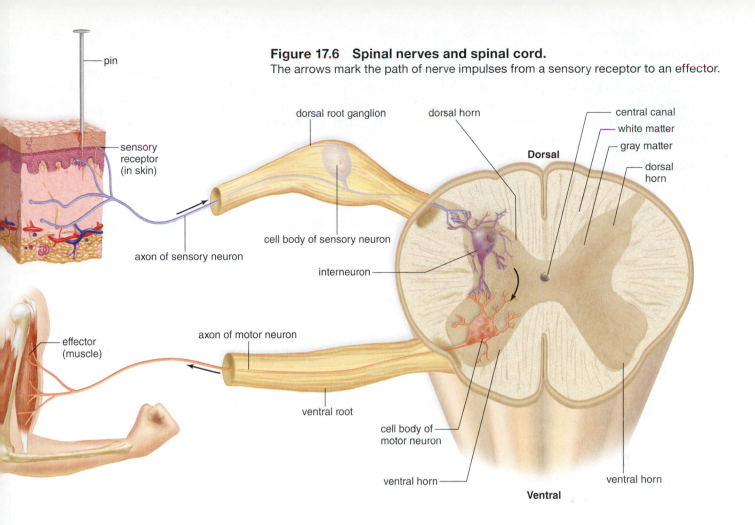

Figure 17.6 Spinal nerves and spinal cord.

The arrows mark the path of nerve impulses from a sensory receptor to an effector.

pin

sensory receptor (in skin)

axon of sensory neuron

dorsal root ganglion

cell body of sensory neuron

interneuron

dorsal horn

Dorsal

central canal

white matter

gray matter

dorsal horn

effector (muscle)

axon of motor neuron

ventral root

cell body of motor neuron

ventral horn

ventral horn

Ventral

immediately withdraw your hand (see Fig. 17.6). When a spinal reflex occurs, a sensory receptor is stimulated and generates nerve impulses that pass along the three neurons mentioned earlier—the sensory neuron, interneuron, and motor neuron—until the effector responds. In the spinal reflexes that follow, a receptor detects the tap, and sensory neurons conduct nerve impulses to interneurons in the spinal cord. The interneurons send a message via motor neurons to the effectors, muscles in the leg or foot. These reflexes are involuntary because the brain is not involved in formulating the response. *Consciousness* of the stimulus lags behind the response because information must be sent up the spinal cord to the brain before you can become aware of the tap.

Experimental Procedure: Spinal Reflex

Although many reflexes occur in the body, only a tendon reflex is investigated in this Experimental Procedure. One easily tested tendon reflex involves the **patellar tendon.** When this tendon is tapped with a reflex hammer (Fig. 17.7) or, in this

Figure 17.7 Knee-jerk reflex.

The quick response when the patellar tendon is stimulated by tapping with a reflex hammer indicates that a reflex has occurred.

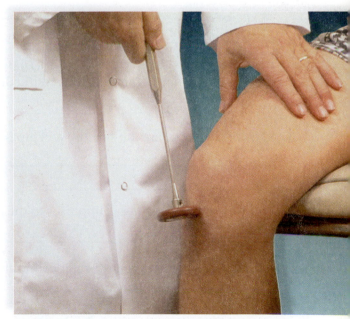

Knee-jerk (patellar) reflex

experiment, with a meter stick, the attached muscle is stretched. This causes a receptor to generate nerve impulses which are transmitted along sensory neurons to the spinal cord. Nerve impulses from the cord then pass along motor neurons and stimulate the muscle, causing it to contract. As the muscle contracts, it tugs on the tendon, causing movement of a bone opposite the joint. Receptors in other tendons, such as the Achilles tendon, respond similarly. Such reflexes help the body automatically maintain balance and posture.

Knee-Jerk (Patellar) Reflex

1. Have the subject sit on a table so that his or her legs hang freely.
2. Sharply tap one of the patellar tendons just below the patella (kneecap) with a meter stick.
3. In this relaxed state, does the leg flex (move toward the buttocks) or extend (move away from the buttocks)? _____

17.3 The Human Eye

The human eye is responsible for sight. Light rays enter the eye and strike the **rod cells** and **cone cells,** the photoreceptors for sight. The rods and cones generate nerve impulses that go to the brain via the optic nerve.

Observation: The Human Eye

1. Examine a human eye model, and identify the structures listed in Table 17.2 and depicted in Figure 17.8.
2. Trace the path of light from outside the eye to the retina.

   _____

   _____

   _____

3. During **accommodation,** the lens rounds up to aid in viewing near objects or flattens to aid in viewing distant objects. Which structure holds the lens and is involved in accommodation?

   _____

   _____

4. **Refraction** is the bending of light rays so that they can be brought to a single focus. Which of the structures listed in Table 17.2 aid in refracting and focusing light rays?

   _____

5. Specifically, what are the sensory receptors for sight, and where are they located in the eye?

   _____

   _____

6. What structure takes nerve impulses to the brain from the rod cells and cone cells?

   _____

7. Which cerebral lobe processes nerve impulses from an eye? _____

Table 17.2   Parts of the Human Eye

Part	Location	Function
Sclera	Outer layer of eye	Protects and supports eyeball
Cornea	Transparent portion of sclera	Refracts light rays
Choroid	Middle layer of eye	Absorbs stray light rays
Retina	Inner layer of eye	Contains receptors for sight
Rod cells	In retina	Make black-and-white vision possible
Cone cells	Concentrated in fovea centralis	Make color vision possible
Fovea centralis	Special region of retina	Makes acute vision possible
Lens	Interior of eye between cavities	Refracts and focuses light rays
Ciliary body	Extension from choroid	Holds lens in place; functions in accommodation
Iris	More anterior extension of choroid	Regulates light entrance
Pupil	Opening in middle of iris	Admits light
Humors (aqueous and vitreous)	Fluid media in anterior and posterior compartments, respectively, of eye	Transmit and refract light rays; support eyeball
Optic nerve	Extension from posterior of eye	Transmits impulses to occipital lobe of brain

Figure 17.8   Anatomy of the human eye.
The sensory receptors for vision are the rod cells and cone cells present in the retina of the eye.

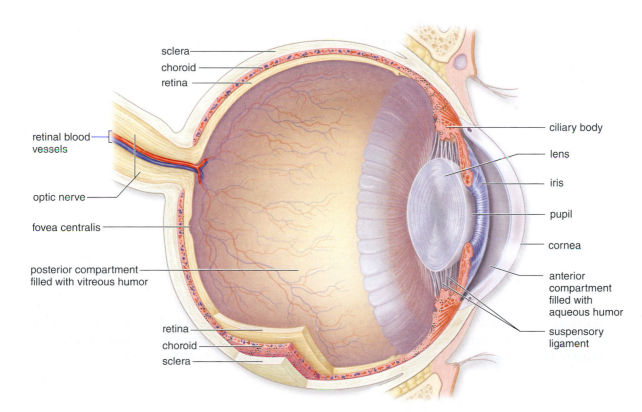

The Blind Spot of the Eye

The **blind spot** occurs where the optic nerve fibers exit the retina. No vision is possible at this location because of the absence of rod cells and cone cells.

Experimental Procedure: Blind Spot of the Eye

This Experimental Procedure requires a laboratory partner. Figure 17.9 shows a small circle and a cross several centimeters apart.

Figure 17.9   Blind spot.
This dark circle (or cross) will disappear at one location because there are no rod cells or cone cells at each eye's blind spot, where vision does not occur.

Left Eye

1. Hold Figure 17.9 approximately 30 cm from your eyes. The cross should be directly in front of your left eye. If you wear glasses, keep them on.
2. Close your right eye.
3. Stare only at the cross with your left eye. You should also be able to see the circle in the same field of vision. Slowly move the paper toward you until the circle disappears.
4. Repeat the procedure as many times as needed to find the blind spot.
5. Then slowly move the paper closer to your eyes until the circle reappears. Because only your left eye is open, you have found the blind spot of your left eye.
6. With your partner's help, measure the distance from your eye to the paper when the circle first

   disappeared. Left eye: _____ cm

Right Eye

1. Hold Figure 17.9 approximately 30 cm from your eyes. The circle should be directly in front of your right eye. If you wear glasses, keep them on.
2. Close your left eye.
3. Stare only at the circle with your right eye. You should also be able to see the cross in the same field of vision. Slowly move the paper toward you until the cross disappears.
4. Repeat the procedure as many times as needed to find the blind spot.
5. Then slowly move the paper closer to your eyes until the cross reappears. Because only your right eye is open, you have found the blind spot of your right eye.
6. With your partner's help, measure the distance from your eye to the paper when the cross first

   disappeared. Right eye: _____ cm

Why are you unaware of a blind spot under normal conditions? Although the eye detects patterns of light and color, it is the brain that determines what we visually perceive. The brain interprets the visual input based in part on past experiences. In this exercise, you created an artificial situation in which you became aware of how your perception of the world is constrained by the eye's anatomy.

Accommodation of the Eye

When the eye accommodates to see objects at different distances, the shape of the lens changes. The lens shape is controlled by the ciliary muscles attached to it. When you are looking at a distant object, the lens is in a flattened state. When you are looking at a closer object, the lens becomes more rounded. The elasticity of the lens determines how well the eye can accommodate. Lens elasticity decreases with increasing age, a condition called **presbyopia.** Presbyopia is the reason many older people need bifocals to see near objects.

Experimental Procedure: Accommodation of the Eye

This Experimental Procedure requires a laboratory partner. It tests accommodation of either your left or your right eye.

1. Hold a pencil upright by the eraser and at arm's length in front of whichever of your eyes you are testing (Fig. 17.10).
2. Close the opposite eye.
3. Move the pencil from arm's length toward your eye.
4. Focus on the end of the pencil.
5. Move the pencil toward you until the end is out of focus. Measure the distance (in centimeters)

   between the pencil and your eye: _____ cm

6. At what distance can your eye no longer

   accommodate for distance? _____ cm

7. If you wear glasses, repeat this experiment without your glasses, and note the accommodation distance of your eye without glasses:

   _____ cm. (Contact lens wearers need not make these determinations, and they should write the words *contact lens* in this blank.)

8. The "younger" lens can easily accommodate for closer distances. The nearest point at which the end of the pencil can be clearly seen is called the **near point.** The more elastic the lens,

   the "younger" the eye (Table 17.3). How "old" is the eye you tested? _____

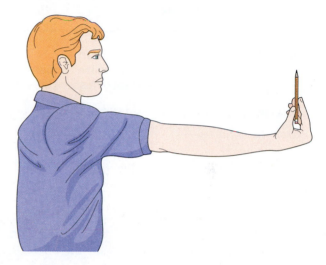

Figure 17.10 Accommodation.
When testing the ability of your eyes to accommodate to see a near object, always keep the pencil in this position.

Table 17.3 Near Point and Age Correlation						
Age (years)	10	20	30	40	50	60
Near point (cm)	9	10	13	18	50	83

17.4 The Human Ear

The human ear, whose parts are listed and depicted in Table 17.4 and Figure 17.11, serves two functions: hearing and balance.

Observation: The Human Ear

Examine a human ear model, and find the structures depicted in Figure 17.11 based on the information given in Table 17.4.

Table 17.4 Parts of the Human Ear

Part	Medium	Function	Mechanoreceptor
Outer ear	Air		
Pinna		Collects sound waves	—
Auditory canal		Filters air	—
Middle ear	Air		
Tympanic membrane and ossicles		Amplify sound waves	—
Auditory tube		Equalizes air pressure	—
Inner ear	Fluid		
Semicircular canals		Rotational equilibrium	Stereocilia embedded in cupula
Vestibule (contains utricle and saccule)		Gravitational equilibrium	Stereocilia embedded in otolithic membrane
Cochlea (spiral organ)		Hearing	Stereocilia embedded in tectorial membrane

Figure 17.11 Anatomy of the human ear.
The outer ear extends from the pinna to the tympanic membrane. The middle ear extends from the tympanic membrane to the oval window. The inner ear encompasses the semicircular canals, the vestibule, and the cochlea.

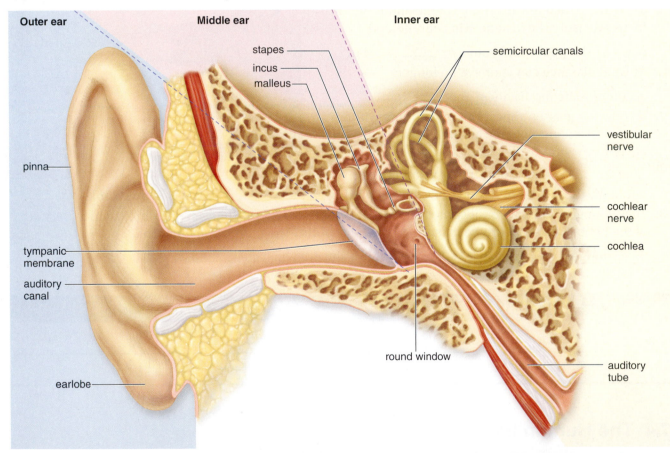

When you hear, sound waves are picked up by the **tympanic membrane** and amplified by the **malleus, incus,** and **stapes.** This creates pressure waves in the canals of the **cochlea** that lead to stimulation of **hair cells,** the receptors for hearing. Hair cells in the utricle and saccule of the vestibule and in semicircular canals are receptors for equilibrium (i.e., balance). Nerve impulses from the cochlea travel by way of the cochlear nerve and the vestibular nerve to the brain and eventually are interpreted by the _____ lobe of the brain as sound.

Humans locate the direction of sound according to how fast it is detected by either or both ears. A difference in the hearing ability of the two ears can lead to a mistaken judgment about the direction of sound. You and a laboratory partner should perform this Experimental Procedure on each other. Enter the data for *your* ears, not your partner's ears, in the spaces provided.

1. Ask the subject to be seated, with eyes closed. Then strike a tuning fork or rap two spoons together at the five locations listed in number 2. Use a random order.
2. Ask the subject to give the exact location of the sound in relation to his or her head. Record the subject's perceptions when the sound is

   a. directly below and behind the head. _____

   b. directly behind the head. _____

   c. directly above the head. _____

   d. directly in front of the face. _____

   e. to the side of the head. _____

3. Is there an apparent difference in hearing between your two ears?

17.5 Sensory Receptors in Human Skin

The sensory receptors in human skin respond to touch, pain, temperature, and pressure (Fig. 17.12). There are individual sensory receptors for each of these stimuli, as well as free nerve endings able to respond to pressure, pain, and temperature.

Figure 17.12 Sensory receptors in the skin.
Each type of receptor shown responds primarily to a particular stimulus.

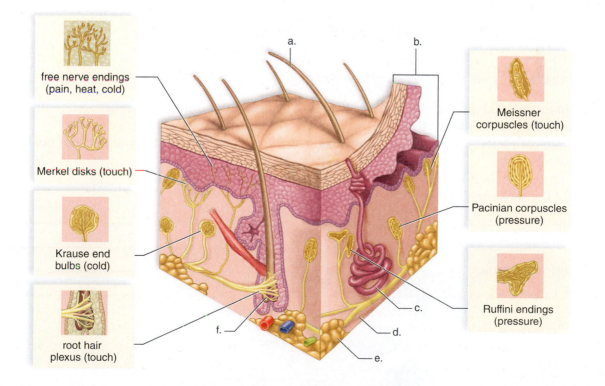

free nerve endings (pain, heat, cold)

Merkel disks (touch)

Krause end bulbs (cold)

root hair plexus (touch)

Meissner corpuscles (touch)

Pacinian corpuscles (pressure)

Ruffini endings (pressure)

Sense of Touch

The dermis of the skin contains touch receptors, whose concentration differs in various parts of the body.

Experimental Procedure: Sense of Touch

You will need a laboratory partner to perform this Experimental Procedure. Enter *your* data, not the data of your partner, in the spaces provided.

1. Ask the subject to be seated, with eyes closed.
2. Then test the subject's ability to discriminate between the two points of a hairpin or a pair of scissors at the four locations listed in number 5.
3. Hold the points of the hairpin or scissors on the given skin area, with both of the points simultaneously and gently touching the subject.
4. Ask the subject whether the experience involves one or two touch sensations.
5. Record the shortest distance between the hairpin or scissor points for a two-point discrimination.

   a. Forearm: _____ mm

   b. Back of the neck: _____ mm

   c. Index finger: _____ mm

   d. Back of the hand: _____ mm

6. Which of these areas apparently contains the greatest density of touch receptors? _____

   Why is this useful? _____

7. Do you have a sense of touch at every point in your skin? _____ Explain. _____

   _____

8. What specific part of the brain processes nerve impulses from touch and pain receptors?

   _____

Sense of Heat and Cold

Temperature receptors respond to a change in temperature.

Experimental Procedure: Sense of Heat and Cold

1. Obtain three 1,000 ml beakers, and fill one with *ice water,* one with *tap water* at room temperature, and one with *warm water* (45°–50°C).
2. Immerse your left hand in the ice-water beaker and your right hand in the warm-water beaker for 30 seconds.
3. Then place both hands in the beaker with room-temperature tap water.
4. Record the sensation in the right and left hands.

   a. Right hand: _____

   b. Left hand: _____

5. Explain your results: _____

   _____

17.6 Human Chemoreceptors

The taste receptors, called _____, located in the mouth, and the smell receptors, called
_____, located in the nasal cavities, are the chemoreceptors that respond to molecules
in the air and water. Nerve impulses from taste receptors go to the _____ lobe of the
brain, while those from smell receptors go to the _____ lobe of the brain.

Experimental Procedure: Sense of Taste and Smell

You will need a laboratory partner to perform the following procedures. It will not be necessary for all tests to
be performed on both partners. You should take turns being either the subject or the experimenter. Dispose of
used cotton swabs in a hazardous waste container or as directed by your instructor.

Taste and Smell

1. Students work in groups. Each group has one experimenter and several subjects.
2. The experimenter should obtain a LifeSavers candy from the various flavors available, without letting
   the subject know what flavor it is.
3. The subject closes both eyes and holds his or her nose.
4. The experimenter gives the LifeSavers candy to the subject, who places it on his or her tongue.
5. The subject, while still holding his or her nose, guesses the flavor of the candy. The experimenter
   records the guess in Table 17.5.
6. The subject releases his or her nose and guesses the flavor again. The experimenter records the guess
   and the actual flavor in Table 17.5.

Table 17.5 Taste and Smell Experiment

Subject	Actual Flavor	Flavor While Holding Nose	Flavor After Releasing Nose
1			
2			
3			
4			
5			

Conclusions: Sense of Taste and Smell

- From your results, how would you say that smell affects the taste of LifeSavers candy?

  _____

- What do you conclude about the effect of smell on your sense of taste?

  _____

  _____

  _____

Laboratory Review 17

1. What portion of the brain is largest in humans?
2. What portion of the brain directly controls muscular coordination?
3. What is the most inferior portion of the brain stem?
4. What structures protect the spinal cord?
5. Are motor neuron cell bodies located in the gray or white matter of the spinal cord?
6. What type of neuron is found completely within the central nervous system?
7. Which neuron's cell body is in the dorsal root ganglion?
8. What layer of the eye contains the sensory receptors for sight?
9. Where on the retina is the blind spot located?
10. What do you call the outer layer of the eye?
11. What part of the ear contains the sensory receptors for hearing?
12. Where in relation to the head is it most difficult to detect the location of a sound?
13. In which portion of the ear are the malleus, incus, and stapes located?
14. What layer of the skin contains sensory receptors?
15. Are touch receptors distributed evenly or unevenly in the skin?
16. What senses are dependent on chemoreceptors?
17. What is the benefit of the fact that the spinal cord and spinal nerves function below the level of consciousness?
18. Identify the type of neuron responsible for transmitting nerve impulses from the spinal cord to an effector.

Thought Questions

19. Trace the path of sound waves in the human ear—from the tympanic membrane to the sensory receptors for hearing.

20. Trace the path of light in the human eye—from the exterior to the retina.

21. In a drag race, drivers must wait until the green light is illuminated before they can move their vehicle. Explain why a time delay exists, based on the information presented in this exercise.

18

Musculoskeletal System

Introduction

The term **musculoskeletal** system recognizes that contraction of muscles causes the bones to move. The skeletal system consists of the bones and joints, along with the cartilage and ligaments that occur at the joints. The muscular system contains three types of muscles: smooth, cardiac, and skeletal. The skeletal muscles are most often attached to bones after crossing a joint (Fig. 18.1). In humans, the biceps brachii muscle has two **origins**, and the triceps brachii has three origins on the humerus and scapula. Find the biceps brachii **insertion** on the anterior surface of your elbow after contracting your biceps muscle. Locate the ulna at your posterior elbow. The triceps brachii tendon inserts on the ulna.

Muscles work in antagonistic pairs. For example, when the biceps brachii contracts, the bones of the forearm are pulled upward, and the triceps brachii relaxes; and when the triceps brachii contracts, the bones of the forearm are pulled downward, and the biceps brachii relaxes.

Figure 18.1   Muscular action.
Muscles, such as these muscles of the arm, cause bones to move. The origins and insertion of the triceps are noted.

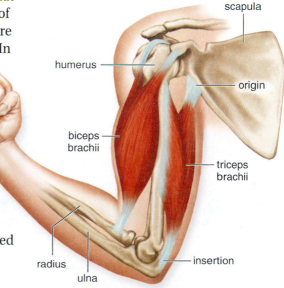

scapula

humerus

origin

biceps brachii

triceps brachii

radius

ulna

insertion

18.1 Anatomy of a Long Bone

Although the bones of the skeletal system vary considerably in shape as well as in size, a long bone, such as the human femur, illustrates the general principles of bone anatomy (Fig. 18.2).

Figure 18.2   Anatomy of a long bone from the macroscopic to the microscopic level.
A long bone is encased by the periosteum except at the ends, where it is covered by hyaline (articular) cartilage (see micrograph, *top left*). Spongy bone located in each epiphysis may contain red bone marrow. The medullary cavity contains yellow bone marrow and is bordered by compact bone, which is shown in the enlargement and micrograph (*right*). (hyaline cartilage: Magnification ×250; compact bone: Magnification ×100; spongy bone: Magnification ×4700)

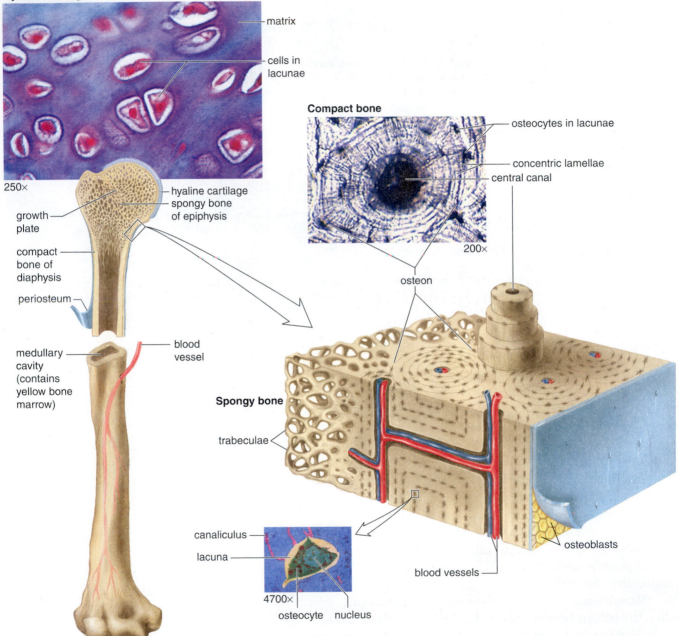

Hyaline cartilage
matrix
cells in lacunae
250×
growth plate
hyaline cartilage
spongy bone of epiphysis
compact bone of diaphysis
periosteum
blood vessel
medullary cavity (contains yellow bone marrow)

Compact bone
osteocytes in lacunae
concentric lamellae
central canal
200×
osteon

Spongy bone
trabeculae
canaliculus
lacuna
4700×
osteocyte   nucleus
blood vessels
osteoblasts

Observation: Anatomy of a Long Bone

Examine the exterior and a longitudinal section of a long bone or a model of a long bone, and with the help of Figure 18.2, identify the following:

1. **Periosteum:** tough, fibrous connective tissue covering continuous with the ligament and tendons that anchor bones; the periosteum allows blood vessels to enter the bone and service its cells.
2. Expanded portions at each end of the bone (*epiphysis*) that contain spongy bone.
3. Extended portion, or shaft (*diaphysis*) of a long bone that lies between the epiphyses; walls of diaphysis are compact bone.
4. **Hyaline (articular) cartilage:** layer of cartilage where the bone articulates with (meets) another bone; decreases friction between bones during movement.
5. **Medullary cavity:** cavity located in the diaphysis that stores yellow marrow, which contains a large amount of fat.

    Label the diaphysis and the epiphysis (twice) in Figure 18.2. Which of these contains the growth line

    where a long bone can grow in length? _____

Observation: Tissues of a Long Bone

The medullary cavity is bounded at the sides by **compact bone** and at the ends by **spongy bone.** Beyond a thin shell of compact bone is the layer of articular cartilage. **Red marrow,** a specialized tissue that produces all types of blood cells, occurs in the spongy bone of the skull, ribs, sternum, and vertebrae and in the ends of the long bones.

1. Examine a prepared slide of compact bone, and with the help of Figure 18.2, identify
    a. **Osteons:** cylindrical structural units
    b. **Lamellae:** concentric rings of matrix
    c. **Matrix:** nonliving material maintained by osteocytes; contains mineral salts (notably calcium salts) and protein
    d. **Lacunae:** cavities between the lamellae that contain osteocytes (bone cells)
    e. **Central canal:** canal in the center of each osteon; Figure 18.2 shows that there are

        _____ in a central canal.
    f. **Canaliculi:** tiny channels that contain the processes of cells; these processes allow nutrients to pass between the osteocytes. (sing., *canaliculus*)

    Describe how an osteocyte located near a central canal can pass nutrients to osteocytes located far

    from the central canal. _____

    _____

2. Examine a prepared slide of spongy bone, and with the help of Figure 18.2, identify
    a. **Trabeculae:** bony bars and plates made of mineral salts and protein
    b. **Lacunae:** cavities scattered throughout the trabeculae that contain osteocytes
    c. **Red bone marrow** within large spaces separated by the trabeculae

    What activity occurs in red bone marrow? _____

3. Examine a prepared slide of hyaline cartilage, and with the help of Figure 18.2, identify
    a. **Lacunae:** cavities in twos and threes scattered throughout the matrix, which contain chondrocytes (cells that maintain cartilage)
    b. **Matrix:** material more flexible than bone because it consists primarily of protein

    Seniors tend to have joints that creak. What might be the matter? _____

18.2 The Skeleton

The human skeleton is divided into axial and appendicular components. The **axial skeleton** is the main longitudinal portion and includes the skull (Fig. 18.3), the vertebral column, the sternum, and the ribs. The **appendicular skeleton** includes the bones of the appendages and their supportive pectoral and pelvic (shoulder and hip) girdles.

Figure 18.3   Three views of skull.

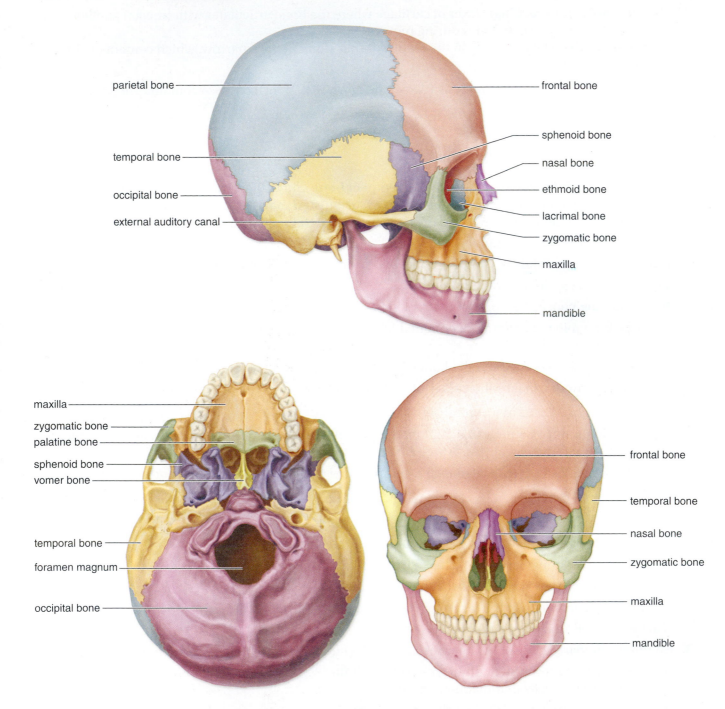

Observation: Axial Skeleton

Examine a human skeleton, and with the help of Figures 18.3 and 18.4, identify the **foramen magnum,** a large opening through which the spinal cord passes, and the following bones.

1. The **skull** is composed of many bones fused together at fibrous joints called **sutures.** Note the following in the cranium (brain case):

    a. **Frontal bone:** forms forehead

    b. **Parietal bones:** extend to sides of skull

    c. **Occipital bone:** curves to form base of skull

Figure 18.4   Human skeletal system.
a. Anterior view. **b.** Posterior view. The axial skeleton appears in blue, while the appendicular skeleton is shown in tan.

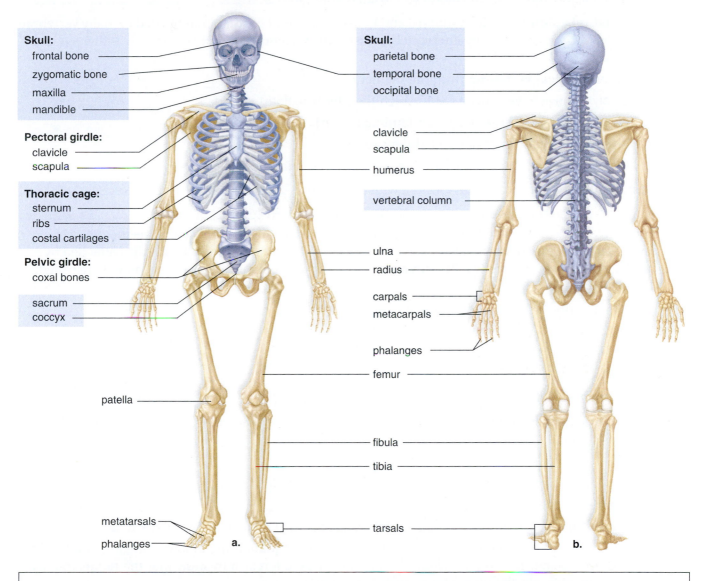

Skull:
    frontal bone
    zygomatic bone
    maxilla
    mandible

Pectoral girdle:
    clavicle
    scapula

Thoracic cage:
    sternum
    ribs
    costal cartilages

Pelvic girdle:
    coxal bones

    sacrum
    coccyx

patella

metatarsals
phalanges

a.

Skull:
    parietal bone
    temporal bone
    occipital bone

clavicle
scapula
humerus

vertebral column

ulna
radius

carpals
metacarpals

phalanges

femur

fibula

tibia

tarsals

b.

NOTE: According to accepted terminology, the upper limb is composed of the humerus in the **arm** and the radius and ulna in the **forearm.** The lower limb is composed of the femur in the **thigh** and the tibia and fibula in the **leg.**

d. **Temporal bones:** located on sides of skull

e. **Sphenoid bone:** helps form base and sides of skull, as well as part of the orbits

Which of these bones contribute to forming the face? _____

Which could best be associated with wearing glasses? _____

2. Note the facial bones:

a. **Mandible:** the lower jaw

b. **Maxillae:** the upper jaw and anterior portion of the hard palate

c. **Palatine bones:** posterior portion of hard palate and floor of nasal cavity

d. **Zygomatic bones:** cheekbones

e. **Nasal bones:** bridge of nose

Which of these is movable and allows you to chew your food? _____

3. The **vertebral column** provides support and houses the **spinal cord.** It is composed of many vertebrae separated from one another by intervertebral disks. The vertebral column customarily is divided into five series:

a. Seven **cervical vertebrae** (forming the neck region)

b. Twelve **thoracic vertebrae** (with which the ribs articulate)

c. Five **lumbar vertebrae** (in the abdominal region)

d. Five fused sacral vertebrae, called the **sacrum**

e. Four fused caudal vertebrae forming the **coccyx** in humans

Which of the vertebrae can be associated with the chest? _____

Which of the vertebrae could be the cause of your aching back? _____

4. The twelve pairs of **ribs** and their associated muscles form a bony case that supports the thoracic cavity wall. The ribs connect posteriorly with the thoracic vertebrae, and some are also attached by cartilage directly or indirectly to the sternum. Those ribs without any anterior attachment are called **floating ribs.**

Which of the ribs help form the protective part of the rib cage? _____

After studying Figure 18.4, give several reasons why the axial skeleton is critical to life.

Observation: Appendicular Skeleton

Examine a human skeleton, and with the help of Figure 18.4, identify the following bones.

1. The **pectoral girdles,** which support the upper limbs, are composed of the **clavicle** (collarbone) and **scapula** (shoulder bone).

Why is the shoulder apt to become dislocated? _____

2. The upper limb (arm plus the forearm) is composed of the following:

a. **Humerus:** the large long bone of the arm

b. **Radius:** the long bone of the forearm, with a pivot joint at the elbow that allows rotational motion

c. **Ulna:** the other long bone of the forearm, with a hinge joint at the elbow that allows motion in only one plane. Take hold of your elbow, and twist the forearm to show that the radius rotates over the ulna but the ulna doesn't move during this action.

d. **Carpals:** a group of small bones forming the wrist

e. **Metacarpals:** slender bones forming the palm

f. **Phalanges:** the bones of the fingers

Which of these bones would you use to pick up a teacup? _____

3. The **pelvic girdle** forms the basal support for the lower limbs and is composed of two **coxal** (hip) **bones.** Each coxal bone consists of the *ilium* (which is superior to the other two), the *pubis,* and the *ischium.* (The pubis is ventral to the ischium.)
   The female pelvis is much broader and shallower than that of the male. The angle between the pubic bones looks like a U in females and a V in males.
   How is it advantageous for the female pelvis to be broader and more shallow than that of

   a male? _____

4. The lower limb (the thigh plus the leg) is composed of a series of loosely articulated bones, including the following:

   a. **Femur:** the long bone of the thigh

   b. **Patella:** kneecap

   c. **Tibia:** the larger of the two long bones of the leg; feel for the bump on the inside of the ankle

   d. **Fibula:** the smaller of the two long bones of the leg; feel for the bump on the outside of the ankle

   e. **Tarsals:** a group of small bones forming the ankle

   f. **Metatarsals:** slender anterior bones of the foot

   g. **Phalanges:** the bones of the toes

   Which of these bones would you use to kick a soccer ball? _____

Hip Replacement

Find the hip joint in Figure 18.4. The terminology "hip replacement" is a misnomer because no part of the coxal bone is replaced. Instead, the hip joint is replaced. The head of the femur is replaced with a metal ball attached to a metal stem. The stem is inserted into the femur. The inside of the socket on the coxal bone is bolstered with a plastic (or plastic and metal) cup. After the replacement, no part of the joint is living tissue.

18.3 The Skeletal Muscles

This laboratory is concerned with skeletal muscles—those muscles that make up the bulk of the human body. With the help of Figure 18.5 and Tables 18.1 and 18.2, identify the major muscles of the body.

Naming Muscles

Muscles are named for various characteristics, as shown in the following list:

1. **Size:** The gluteus *maximus* is the largest muscle, and it forms the buttocks.
2. **Shape:** The *deltoid* is shaped like a Greek letter delta, or triangle.
3. **Direction of fibers:** The *rectus abdominis* is a longitudinal muscle of the abdomen (*rectus* means "straight").
4. **Location:** The *frontalis* overlies the frontal bone.
5. **Number of attachments:** The *biceps brachii* has two attachments, or origins.
6. **Action:** The *extensor digitorum* extends the fingers, or digits.

Figure 18.5 Human superficial skeletal muscles.
All the muscles of the human body are named in accordance with the structure and/or function of the underlying bone. The muscles highlighted here are those noted in the observation on page 255.

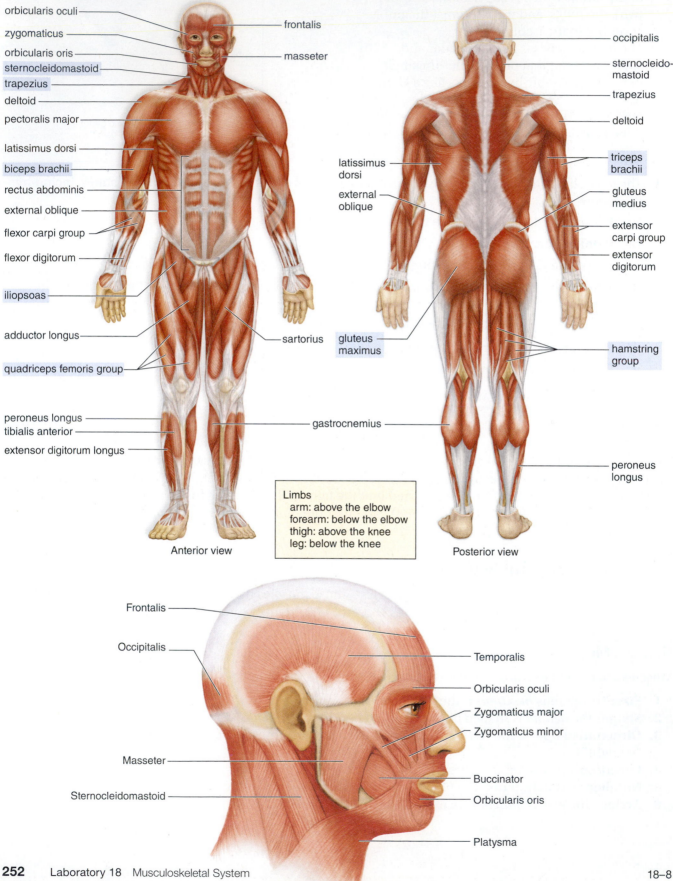

orbicularis oculi

zygomaticus

orbicularis oris

sternocleidomastoid

trapezius

deltoid

pectoralis major

latissimus dorsi

biceps brachii

rectus abdominis

external oblique

flexor carpi group

flexor digitorum

iliopsoas

adductor longus

quadriceps femoris group

peroneus longus

tibialis anterior

extensor digitorum longus

frontalis

masseter

sartorius

gluteus maximus

gastrocnemius

Anterior view

Limbs
arm: above the elbow
forearm: below the elbow
thigh: above the knee
leg: below the knee

occipitalis

sternocleido-mastoid

trapezius

deltoid

triceps brachii

gluteus medius

extensor carpi group

extensor digitorum

latissimus dorsi

external oblique

hamstring group

peroneus longus

Posterior view

Frontalis

Occipitalis

Masseter

Sternocleidomastoid

Temporalis

Orbicularis oculi

Zygomaticus major

Zygomaticus minor

Buccinator

Orbicularis oris

Platysma

Table 18.1 Muscles (Anterior View)

Name	Action
Head and Neck	
Frontalis	Wrinkles forehead and lifts eyebrows
Orbicularis oculi	Closes eye (winking)
Zygomaticus	Raises corner of mouth (smiling)
Masseter	Closes jaw (chewing)
Orbicularis oris	Closes and protrudes lips (kissing)
Upper Limb and Trunk	
External oblique	Compresses abdomen; rotates trunk
Rectus abdominis	Flexes spine
Pectoralis major (pulls arm across chest)	Flexes and adducts shoulder and arm ventrally
Deltoid	Abducts and raises arm at shoulder joint
Biceps brachii	Flexes forearm and supinates hand
Flexor carpi group	Flexes wrist
Flexor digitorum	Flexes fingers
Lower Limb	
Adductor longus	Adducts and flexes thigh
Iliopsoas	Flexes thigh at hip joint
Sartorius	Rotates thigh (sitting cross-legged)
Quadriceps femoris group	Extends leg
Peroneus longus	Everts foot
Tibialis anterior	Dorsiflexes and inverts foot
Flexor digitorum longus	Flexes toes (not shown)
Extensor digitorum longus	Extends toes

Table 18.2 Muscles (Posterior View)

Name	Action
Head and Neck	
Occipitalis	Moves scalp backward
Sternocleidomastoid	Turns head to side; flexes neck and head
Trapezius	Extends head; raises and adducts shoulders dorsally (shrugging shoulders)
Upper Limb and Trunk	
Latissimus dorsi	Extends and adducts shoulder and arm dorsally (pulls arm across back)
Deltoid	Abducts and raises arm at shoulder joint
External oblique	Rotates trunk
Triceps brachii	Extends forearm
Extensor carpi group	Extends wrist
Extensor digitorum	Extends fingers
Buttocks and Lower Limb	
Gluteus medius	Abducts thigh
Gluteus maximus	Extends thigh (forms buttock)
Hamstring group	Flexes leg and extends thigh at hip joint
Gastrocnemius	Flexes leg and foot (tiptoeing)

Match these muscles to these functions:

Doing a pelvic tilt and curving the spine _____

Swinging movements of arms during walking and swimming _____

Pulling forearm toward you when rowing _____

Helping maintain the trunk in an erect posture _____

Extending and separating the fingers _____

Wrinkling the skin of the forehead _____

Antagonistic Pairs

Skeletal muscles are attached to the skeleton, and their contraction causes the movement of bones at a joint. Because muscles shorten when they contract, they can only pull; they cannot push. Therefore, muscles work in **antagonistic pairs.** Usually, contraction of one member of the pair causes a bone to move in one direction, and contraction of the other member of the pair causes the same bone to move in an opposite direction.

Figure 18.6 demonstrates the following types of joint movements:

Flexion	Moving jointed body parts toward each other
Extension	Moving jointed body parts away from each other
Adduction	Moving a part toward a vertical plane running through the longitudinal midline of the body
Abduction	Moving a part away from a vertical plane running through the longitudinal midline of the body
Rotation	Moving a body part around its own axis; **circumduction** is moving a body part in a wide circle

Figure 18.6 Joint movements.
a. Flexion and extension. **b.** Adduction and abduction. **c.** Rotation and circumduction. **d.** Inversion and eversion. Red dots indicate pivot points.

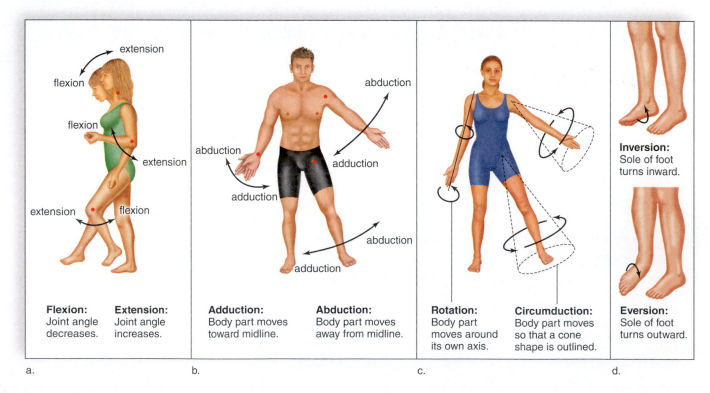

a. b. c. d.

| Inversion | A movement of the foot in which the sole is turned inward |
| Eversion | A movement of the foot in which the sole is turned outward |

These terms describe the action of the muscles listed in Tables 18.1 and 18.2.

Observation: Antagonistic Pairs

Locate the following antagonistic pairs in Figure 18.5*a*. In each case, state their opposing actions by inserting one of these functions: *flexes, extends, raises, lowers, adducts,* or *abducts*.

1. The biceps brachii _____ the forearm.

   The triceps brachii _____ the forearm.

2. The sternocleidomastoid _____ the head.

   The trapezius _____ the head.

3. The iliopsoas _____ the thigh.

   The gluteus maximus _____ the thigh.

4. The quadriceps femoris group _____ the leg.

   The hamstring group _____ the leg.

Isotonic and Isometric Contractions

A muscle contains many muscle fibers. When a muscle contracts, usually some fibers undergo isotonic contraction, and others undergo isometric contraction. When the tension of muscle fibers is sufficient to lift a load, many fibers change length as they lift the load. The muscle contraction is said to be **isotonic** (same tension). In contrast, when the tension of muscle fibers is used only to support rather than to lift a load, the muscle contraction is said to be **isometric** (same length). The length of many fibers remains the same, but their tension still changes.

Experimental Procedure: Isotonic and Isometric Contractions

Isotonic Contraction

1. Start with your left forearm resting on a table. Watch the anterior surface of your left arm while you slowly bend your elbow and bring your left forearm toward the arm. An isotonic contraction of the biceps brachii produces this movement.

2. If a muscle contraction produces movement, is this an isometric or isotonic contraction? _____

Isometric Contraction

1. Place the palm of your left hand underneath a tabletop. Push up against the table while you have your right hand cupped over the anterior surface of your left arm so that you can feel the muscle there undergo an isometric contraction.

2. Is the biceps brachii or the triceps brachii located on the anterior surface of the arm? _____

3. What change did you notice in the firmness of this muscle as it contracted? _____

4. Did your hand or forearm move as you pushed up against the table? _____

5. Given your answer to question 4, did this muscle's fibers shorten as you pushed up against

   the tabletop? _____

18.4   Mechanism of Muscle Fiber Contraction

A whole skeletal muscle is made up of many cells called **muscle fibers** and muscle fibers contain many myofibrils (Fig. 18.7). Muscle fibers are striated—that is, they have alternating light and dark bands as seen in a light micrograph of muscle fibers in longitudinal section.

Electron microscopy has shown that striations are due to the placement of protein filaments of **myosin** and **actin.** During contraction, units of the muscle, called **sarcomeres,** shorten because myosin filaments attach to actin filaments by cross-bridges that pull the actin filaments inward. ATP serves as the immediate energy source for sarcomere contraction. Potassium (K^+) and magnesium (Mg^{2+}) ions are cofactors for the breakdown of ATP by myosin.

Observation: Skeletal Muscle

Examine a prepared slide of skeletal muscles, and identify long, multinucleated fibers arranged in a parallel fashion. Note that the muscle fibers are striated (Fig. 18.7). See also Section 11.1, page 153 for a photomicrograph of skeletal muscle.

Experimental Procedure: Muscle Fiber Contraction

1. Label two slides "slide 1" and "slide 2." Mount a strand of glycerinated muscle fibers in a drop of glycerol on each slide. Place each slide on a millimeter ruler, and measure the length of the strand. Record these lengths in the first row in Table 18.3.
2. If there is more than a small drop of glycerol on the slides, soak up the excess on a piece of lens paper held at the edge of the glycerol farthest from the fiber strand.
3. To slide 1, add a few drops of a salt solution containing potassium (K^+) and magnesium (Mg^{2+}) ions, and measure any change in strand length. Record your results in Table 18.3.
4. To slide 2, add a few drops of ATP solution, and measure any change in strand length. Record your results in Table 18.3.
5. Now add ATP solution to slide 1. Measure any change in strand length, and record your results in Table 18.3.
6. To slide 2, add a few drops of the K^+/Mg^{2+} salt solution, and measure any change in strand length. Record your results in Table 18.3.

Table 18.3   Glycerinated Muscle Contraction

Solution	Length (mm)	
	Slide 1	Slide 2
Glycerol alone		
K^+/Mg^{2+} salt solution alone		—
ATP alone	—	
Both salt solution and ATP		

7. To demonstrate that you understand the requirements for contraction, state the function of each of the substances listed in Table 18.4.

Table 18.4   Summary of Muscle Fiber Contraction

Substance	Function
Myosin	
Actin	
K^+/Mg^{2+} salt solution	
ATP	

Figure 18.7 Microscopic structure of a skeletal muscle fiber.

The striations you see in a sarcomere are due to the placement of two types of myofilaments: actin filaments and myosin filaments. When a sarcomere contracts, the myosin filaments pull on the actin filaments so that they move toward the center and the H zone disappears.

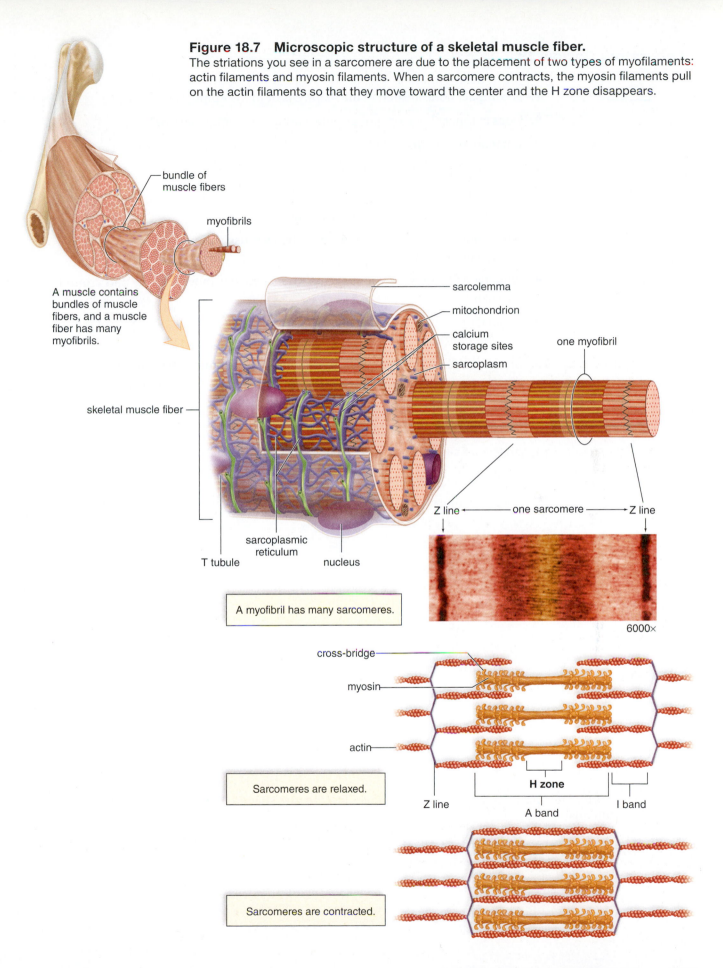

A muscle contains bundles of muscle fibers, and a muscle fiber has many myofibrils.

A myofibril has many sarcomeres.

6000×

Sarcomeres are relaxed.

Sarcomeres are contracted.

_____ 1. Is compact bone located in the diaphysis or in the epiphyses?

_____ 2. Does compact bone or spongy bone contain red bone marrow?

_____ 3. What are bone cells called?

_____ 4. What are the vertebrae in the neck region called?

_____ 5. Name the strongest bone in the lower limb.

_____ 6. What bones are part of a pectoral girdle?

_____ 7. What type of joint movement occurs when a muscle moves a limb toward the midline of the body?

_____ 8. What type of joint movement occurs when a muscle moves a body part around its own axis?

_____ 9. Skeletal muscle is voluntary, and its appearance is _____ because of the placement of actin and myosin filaments.

_____ 10. Glycerinated muscle requires the addition of what molecule to supply the energy for muscle contraction?

_____ 11. Actin and myosin are what type of molecule?

_____ 12. Does the quadriceps femoris group flex or extend the leg?

_____ 13. Does the biceps brachii flex or extend the forearm?

_____ 14. What muscle forms the buttocks?

_____ 15. Name the muscle group that is antagonistic to the quadriceps femoris group.

Thought Questions

16. Aside from support and movement, what important function might the axial skeleton have? (*Hint*: Think of the organs in the vicinity of these bones.)

17. When you see glycerinated muscle shorten, what is happening microscopically?

18. Which girdle best identfies the gender? Explain how.

19. Both hands and feet have metatarsals and phalanges but they are shaped differently. How does this benefit humans?

19

Development

Learning Outcomes

19.1 Embryonic Development
- Identify the cellular stages of development with reference to slides of early sea star development.   260
- Identify the tissue stages of development with reference to slides of frog development.   261–62
- Associate the germ layers with the development of various organs.   262–63
- Identify which organs develop first in a vertebrate embryo (e.g., frog, chick, and human).   264–67
- Compare embryonic feature of the sea star, frog, and chick.   268

19.2 Extraembryonic Membranes, the Placenta, and the Umbilical Cord
- Distinguish between and give a function for the extraembryonic membranes, the placenta, and the umbilical cord.   269
- Trace the development of the extraembryonic membranes in humans and compare their functions in a human and chick.   269-71

19.3 Fetal Development
- Trace the main events of human fetal development.   272–73

Introduction

The early development of animals is quite similar, regardless of the species. The fertilized egg, or zygote, undergoes successive divisions by cleavage, forming a mulberry-shaped ball of cells called a morula and then a hollow ball of cells called a blastula. The fluid-filled cavity of the blastula is the blastocoel. Later, some of the surface cells fold inward, or invaginate, eventually forming a double-walled structure. The outer layer is called the ectoderm, and the inner layer is the endoderm. Between these layers, a middle layer, or mesoderm, arises. The embryo is now called a gastrula. In particular, the presence of yolk (nutrient material) influences how the gastrula comes about.

All later development can be associated with the three **germ layers** that give rise to different tissues and systems: (1) The **ectoderm** forms the nervous system and the skin plus its accessory structures, such as hair and nails; (2) the **endoderm** forms the lining of the digestive system and respiratory system; and (3) the **mesoderm** gives rise to the cardiovascular, muscular, reproductive, and skeletal systems and to connective tissue. **Embryonic development** comes to a close when all the basic organs have formed. Refinements and an increase in size occur during **fetal development**.

zygote

embryo at one week; implants in uterine wall

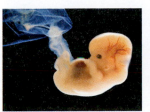
embryo at eight weeks

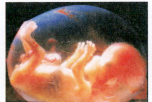
fetus at three months

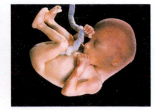
fetus at five months

Development occurs in stages.

19.1 Embryonic Development

We will divide embryonic development into three stages: cellular, tissue layer, and organ development. In human beings, it takes two months to complete embryonic development. It is impossible for us to view the stages of embryonic development in a human being, so we will use the sea star, frog, and chick as our observational material.

Cellular Stages of Development

The cellular stages of development include the following:

- **Zygote formation:** A single sperm fertilizes an egg and the result is a zygote, the first cell of the new individual.
- **Morula formation:** Zygote divides into a number of smaller cells until there is a cluster of 16–32 cells called a morula.
- **Blastula formation:** The morula becomes a blastula, a hollow ball of cells.

Observation: Cellular Stages of Development in the Sea Star

The cellular stages of development are remarkably similar in all animals. Therefore, we can view slides of sea star development to study the cellular stages of human development (Fig. 19.1). A sea star is an invertebrate that develops in the ocean and, therefore, will develop easily in the laboratory where it can be observed.

Obtain slides or view a model of sea star development and note the following:

1. **Zygote.** Both plants and animals begin life as a single cell, a zygote. A zygote contains chromosomes from each parent. Explain. _____

   _____

2. **Cleavage.** View slides showing various numbers of cells due to the process of cleavage, cell division without growth until the morula stage. Is the morula about the same size as the zygote? _____
   Explain. _____

   _____

3. **Blastula.** The cavity of a blastula is called the blastocoel. Label the blastocoel in Figure 19.1. The formation of a hollow cavity is important to the next stage of development.

Figure 19.1   Starfish development.
All animals, including starfish and humans, go through the same cellular stages from cleavage to blastula. (Magnification 75×)

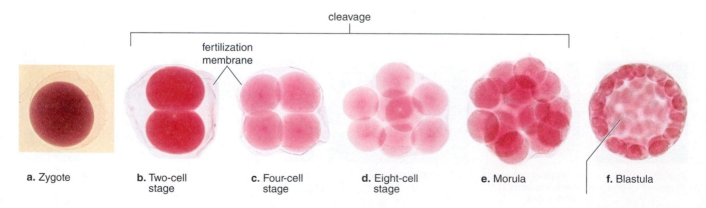

a. Zygote   b. Two-cell stage   c. Four-cell stage   d. Eight-cell stage   e. Morula   f. Blastula

Tissue Stages of Development

The tissue stages of development include the following:

- **Early gastrula stage.** This stage begins when certain cells begin to push or invaginate into the blastocoel, creating a double layer of cells. The outer layer is called the **ectoderm,** and the inner layer is called the **endoderm.**
- **Late gastrula stage.** Gastrulation is not complete until there are three layers of cells. The third layer, called **mesoderm,** occurs between the other two layers already mentioned.

Observation: Tissue Stages of Development in a Frog

It is traditional to view frog gastrulation. A frog is a vertebrate, and so its development is expected to be closer to that of a human than is the development of a sea star. In Figure 19.2, note that the yellow (vegetal pole) cells are heavily laden with yolk, and the blue (animal pole) cells are the ones that invaginate into the blastocoel, forming the early gastrula.

1. **Early gastrula stage.** Obtain a cross section of a frog gastrula. Most likely, your slide is the equivalent of Figure 19.2b, number 3, in which case you will see two cavities, the old blastocoel and newly forming *archenteron,* which forms once the animal pole cells have invaginated. The archenteron will become the digestive tract.

Figure 19.2   Drawings of frog developmental stages.
a. During cleavage, the number of cells increases but overall size remains the same. **b.** During gastrulation, three tissue layers form. **c.** During neurulation, the notochord and neural tube form.

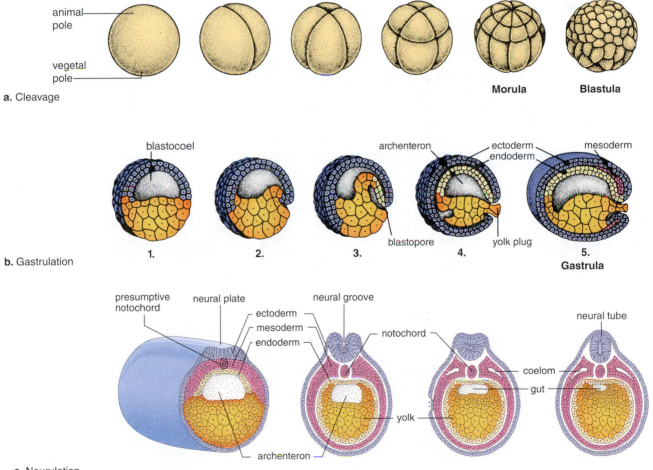

2. **Late gastrula stage:** The moderate amount of yolk also influences the formation of the mesoderm. This germ layer develops by invagination of cells at the lateral and ventral lips of the blastopore.

Compare formation of the mesoderm in the frog to that in the sea star. _____

_____

_____

3. **Neurula stage:** During neurulation in the frog, two folds of ectoderm grow upward as the neural folds with a groove between them (Fig. 19.3). The flat layer of ectoderm between them is the **neural plate.** The tube resulting from closure of the folds is the **neural tube,** which will become the spinal (nerve) cord and brain. An examination of the neurula in cross section shows that the nervous system develops directly above the **notochord,** a structure that arises from invaginated cells in the middorsal region (see Fig. 19.2c). *Draw a series of sketches that shows how the neural tube develops.*

How does the ectoderm form the neural tube? _____

_____

The notochord is said to **induce** the formation of the nervous system. Experiments have shown that if contact with notochord tissue is prevented, no neural plate is formed. Even more dramatic are experiments in which presumptive (soon-to-be) notochord is transplanted under an area of ectoderm not in the dorsal midline. This ectoderm then is induced to differentiate into neural plate tissue, something it would not normally do. **Induction** is believed to be one means by which development is usually orderly. The part of the embryo that induces the formation of an adjacent organ is said to be an **organizer** and is believed to carry out its function by releasing one or more chemical substances.

Figure 19.3   Photographs of frog during neural stage.

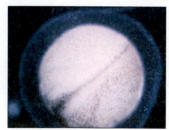  

Neural plate   Neural groove   `neural folds   Advanced neurula

Observation: Tissue Stages of Development in a Human

In a model of human development, observe the same stages of development already observed in frog slides. After implantation, gastrulation in humans turns the inner cell mass into the **embryonic disk.** Figure 19.4 shows the embryonic disk, which has the three layers of cells we have been discussing: the ectoderm, mesoderm, and endoderm. Figure 19.4 also shows the significance of these layers, often called the **germ layers.** The future organs of an individual can be traced back to one of the germ layers.

Figure 19.4 Embryonic disk.

The embryonic disk has three germ layers called ectoderm, mesoderm, and endoderm. Organs and tissues can be traced back to a particular germ layer as indicated in this illustration.

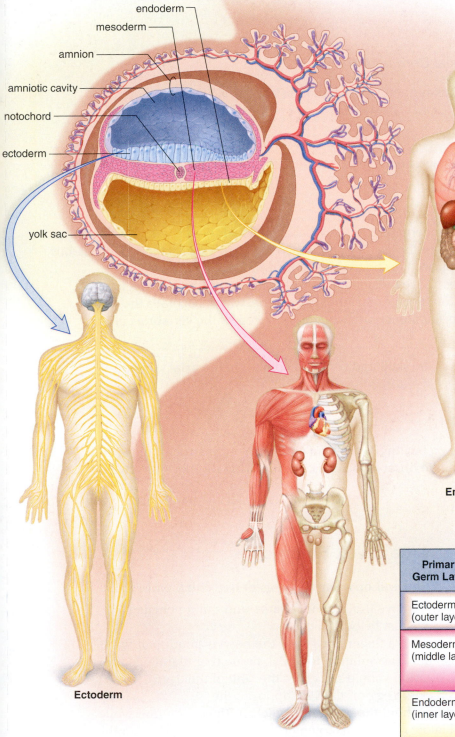

endoderm
mesoderm
amnion
amniotic cavity
notochord
ectoderm
yolk sac

Ectoderm

Mesoderm

Endoderm

Primary Germ Layer	Human Adult Structures
Ectoderm (outer layer)	Epidermis of skin; epithelial lining of oral cavity and rectum; nervous system
Mesoderm (middle layer)	Skeleton; muscular system; dermis of skin; cardiovascular system; urinary system; reproductive system; outer layers of respiratory and digestive systems
Endoderm (inner layer)	Epithelial lining of digestive tract and respiratory tract; associated glands of these systems; epithelial lining of urinary bladder

Organ Stages of Development

As soon as all three embryonic tissue layers (ectoderm, endoderm, and mesoderm) are established, the organ level of development begins. It continues until development is complete. The first organs to develop are the

- Digestive tract. You have already observed the start of the archenteron during gastrulation.
- Spinal cord and brain
- Heart

Observation: Development of the Spinal Cord and Brain

One of the first systems to form is the nervous system. Why might it be beneficial for the nervous system to begin development first? _____

_____

1. Obtain a cross section of a frog neurula stage, and match it to one of the drawings in Figure 19.2c and 19.3.

   Which drawing seems to best match your slide? _____
   Your instructor will confirm your match for you.

2. Recall that a neural tube develops from ectoderm (Fig. 19.2c). When neural folds rise up and fuse, the neural tube has formed. The neural tube, which runs the length of the embryo, is the first sign of the central nervous system. The nerve cord, also called the spinal cord, and the brain both develop from the neural tube.
       Notice how the neural tube develops above the notochord, a dorsal supporting rod that later becomes the vertebral column. Why would you expect the neural tube, which becomes the spinal cord, to develop in the same vicinity as the notochord, which is replaced by the vertebral column?

   _____

   If you are uncertain, review the functions of the skull and vertebral column in the axial skeleton (see page 250).

Observation: Development of the Heart

A chick embryo offers an opportunity to view a beating heart in an embryo. Your instructor may show you various stages. In particular you will want to observe the chick embryo from the 48-hour stage up to the 96-hour stage.

Observing Live Chick Embryos

Use the following procedure for selecting and opening the eggs of live chick embryos:

1. Choose an egg of the proper age to remove from the incubator, and put a penciled × on the uppermost side. The embryo is just below the shell.
2. Add warmed chicken Ringer solution to a finger bowl until the bowl is about half full. (Chicken Ringer solution is an isotonic salt solution for chick tissue that maintains the living state.) The chicken Ringer solution should not cover the yolk of the egg.
3. On the edge of the dish, gently crack the egg on the side opposite the ×.
4. With your thumbs placed over the ×, hold the egg in the chicken Ringer solution while you pry it open from below and allow its contents to enter the solution. If you open the egg too slowly or too quickly, the shell may damage the delicate membranes surrounding the embryo.

1. Follow the standard procedure (see page 264) for selecting and opening an egg containing a 48-hour chick embryo.
2. The embryo has turned so that the head region is lying on its side. Refer to Figure 19.5, and identify the following:

   a. **Shape of the embryo,** which has started to bend. The head is now almost touching the heart.

   b. **Heart,** contracting and circulating blood. Can you make out a ventricle, an atrium, and the aortic arches in the region below the head? Later, only one aortic arch will remain.

   c. **Vitelline arteries** and **veins,** which extend over the yolk. The vitelline veins carry nutrients from the yolk sac to the embryo.

   d. **Brain** with several distinct regions.

   e. **Eye,** which has a developing lens.

   f. **Margin (edge) of the amnion,** which can be seen above the vitelline arteries (see next section for amnion).

   g. **Somites,** blocks of developing muscle tissue that differentiate from mesoderm, which now number 24 pairs.

   h. **Caudal fold** of the amnion. The embryo will be completely enveloped when the head fold and caudal fold meet the margin of the amnion.

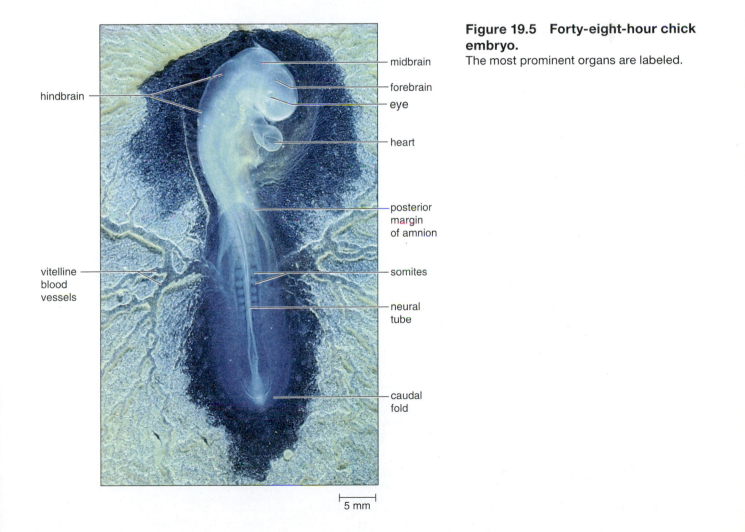

Figure 19.5   Forty-eight-hour chick embryo.
The most prominent organs are labeled.

As a chick embryo continues to grow, various organs differentiate further (Fig. 19.6). The neural tube closes along the entire length of the body and is now called the spinal cord. The allantois, an extraembryonic membrane, is seen as a sac extending from the ventral surface of the hindgut near the tail bud. The digestive system forms specialized regions, and there are both a mouth and an anus. The yolk sac, the extraembryonic membrane that encloses the yolk, is attached to the ventral wall, but when the yolk is used up, the ventral wall closes.

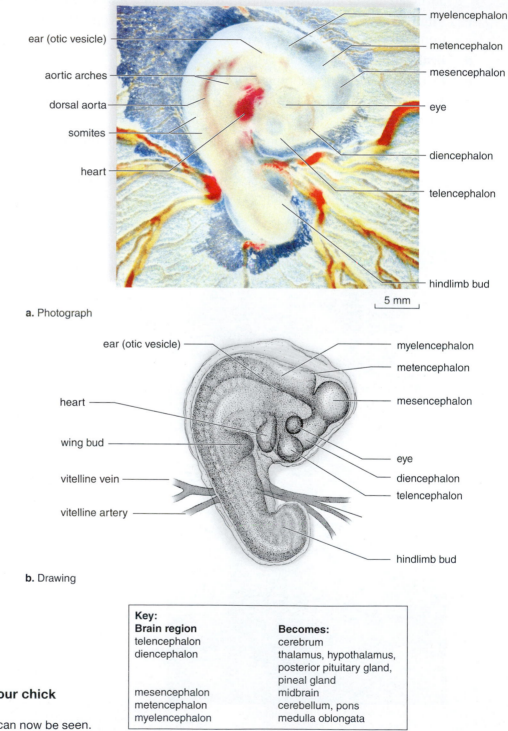

a. Photograph

5 mm

b. Drawing

Figure 19.6 Ninety-six-hour chick embryo.
Brain regions listed in the key can now be seen.

Key:	
Brain region	**Becomes:**
telencephalon	cerebrum
diencephalon	thalamus, hypothalamus, posterior pituitary gland, pineal gland
mesencephalon	midbrain
metencephalon	cerebellum, pons
myelencephalon	medulla oblongata

Study models or other study aids available that show the development of the nervous system and the heart in human beings and/or show models of human embryos of different ages. Also view Figure 19.7, which depicts the external appearance of the embryo from the fourth to the seventh week of development.

During the embryonic period of development, the growing baby is susceptible to environmental influences, including the following:

- Drugs, such as alcohol; certain prescriptions; and recreational drugs. These can cause birth defects.
- Infections such as rubella, also called German measles, and other viral infections.
- Nutritional deficiencies.
- X-rays or radiation therapy.

Figure 19.7 External appearance of the human embryo.
a. Weeks 4 to 5 and (**b**) weeks 6 to 7.

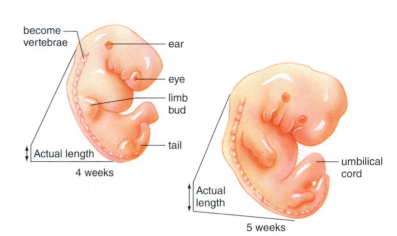

a. Weeks 4 and 5
- Head dominant, but body getting longer.
- Limb buds are visible.
- Eyes and ears begin to form.
- Tissue for vertebrae extend into tail.

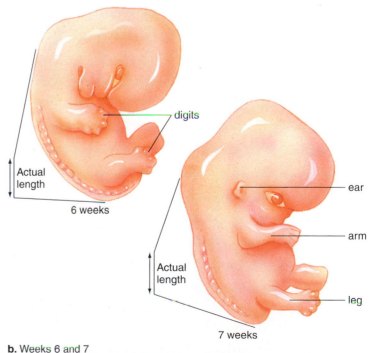

b. Weeks 6 and 7
- Head still dominant, but tail has disappeared.
- Facial features continue to develop.
- Hands and feet have digits.
- All organs are more developed.

Comparison of Embryonic Features of a Developing Sea Star, Frog, and Chick

Complete Table 19.1 by placing an X in the appropriate square if the feature pertains to the organism.

Table 19.1 Comparison of Embryonic Features of a Developing Sea Star, Frog, and Chick			
Feature	Sea Star	Frog	Chick
Has the most yolk			
Blastula is a circular cavity			
Germ layers are present			
Primitive streak is present			
Notochord is present			
Waste is deposited in water			

Comparison of Chick and Human Development

As illustrated in Figure 19.8, the early stages of human development are quite similar to those of the chick. Differences become marked only as development proceeds.

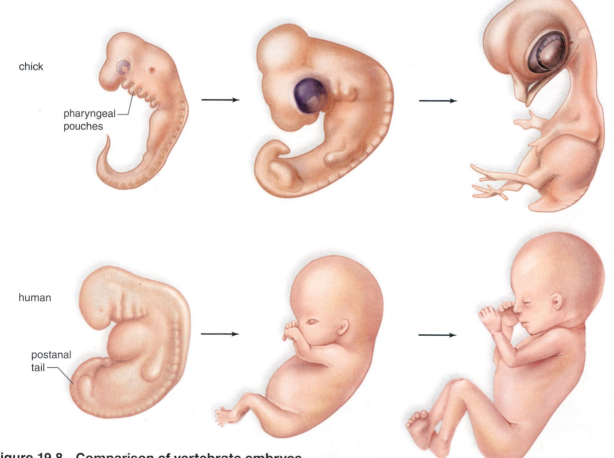

Figure 19.8 Comparison of vertebrate embryos.
Successive stages in the development of chick and human. Early stages (far left) are similar; differences become apparent as development continues.

19.2 Extraembryonic Membranes, the Placenta, and the Umbilical Cord

- The **extraembryonic membranes** take their name from the observation that they are not part of the embryo proper. They are outside the embryo, and therefore they are "extra."
- In humans, the **placenta** is the structure that provides the embryo with nutrient molecules and oxygen and takes away its waste molecules, such as carbon dioxide. The fetal half of the placenta contains the fetal capillaries. The maternal half of the placenta is the uterine wall where maternal blood vessels meet the fetal capillaries.
- The **umbilical cord** is a tubular structure that contains two of the extraembryonic membranes (the allantois and the yolk sac) and also the **umbilical blood vessels.** The umbilical blood vessels bring fetal blood to and from the placenta. When a baby is born and begins to breathe on its own, the umbilical cord is cut and the remnants become the navel. *In this drawing, label the umbilical cord, the umbilical blood vessels and the placenta, which contains the maternal blood vessels.*

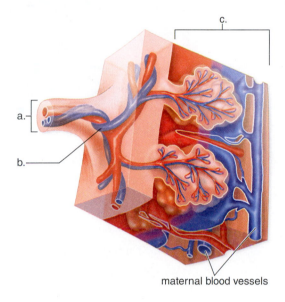

maternal blood vessels

Observation: The Extraembryonic Membranes

In a model, and in Figure 19.9, trace the development of the extraembryonic membranes. Also, note the development of the placenta and the umbilical cord. The extraembryonic membranes are as follows:

- **Chorion.** The chorion is the outermost membrane, and in chicks it lies just below the porous shell, where it functions in gas exchange. In humans, an outer layer of cells surrounding the inner cell mass at the blastocyst stage becomes the chorion. Notice in Figure 19.9 that the treelike **chorionic villi** are a part of the chorion.
- **Amnion.** The amnion forms the amniotic cavity, which envelops the embryo and contains the amniotic fluid that cushions and protects the developing offspring (Fig. 19.10). All animals, whether the sea star, the frog, the chick, or the human, develop in an aqueous environment. Birth of a human is imminent when "the water breaks," and the amniotic fluid is lost.
- **Allantois.** The allantois serves as a storage area for metabolic waste in the chick. In humans, the allantois extends into the umbilical cord. It accumulates the small amount of urine produced by the fetal kidneys and later contributes to formation of the urinary bladder. Its blood vessels become the umbilical blood vessels.
- **Yolk sac.** The yolk sac is the first embryonic membrane to appear. In the chick, the yolk sac does contain yolk, food for the developing embryo. In humans, the yolk sac contains plentiful blood vessels and is the first site of blood cell formation.

Figure 19.9 Development of extraembryonic membranes in humans.

a. At first, no organs are present in the embryo, only tissues. The amniotic cavity is above the embryonic disk, and the yolk sac is below. The chorionic villi are present. **b, c.** The allantois and yolk sac, two more extraembryonic membranes, are positioned inside the body stalk as it becomes the umbilical cord. **d.** At 35+ days, all membranes are present, and the umbilical cord takes blood vessels between the embryo and the chorion (placenta).

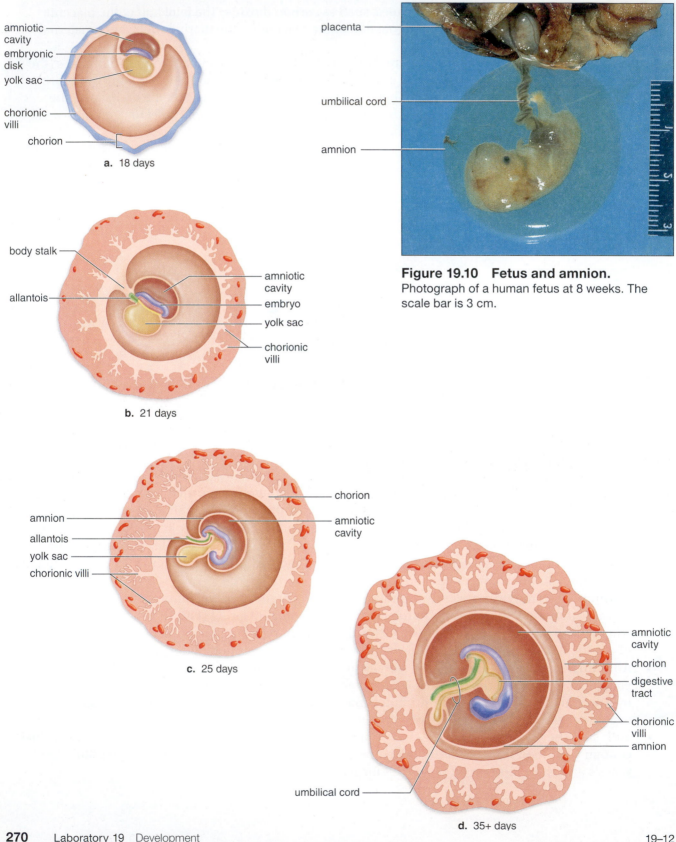

a. 18 days

amniotic cavity
embryonic disk
yolk sac
chorionic villi
chorion

b. 21 days

body stalk
allantois
amniotic cavity
embryo
yolk sac
chorionic villi

c. 25 days

amnion
allantois
yolk sac
chorionic villi
chorion
amniotic cavity

d. 35+ days

amniotic cavity
chorion
digestive tract
chorionic villi
amnion
umbilical cord

placenta
umbilical cord
amnion

Figure 19.10 Fetus and amnion.
Photograph of a human fetus at 8 weeks. The scale bar is 3 cm.

Comparison of Extraembryonic Membranes in Chick and Human

Consult Figure 19.11 and use the information on page 269 to complete Table 19.2, which compares the function of the extraembryonic membranes in the chick and in the human. Reptiles, which we now know include birds, were the first animals to have extraembryonic membranes. These membranes allowed reptiles to develop on land. They also allow mammals, including humans, to develop inside the uterus of the mother.

Table 19.2 Functions of Extraembryonic Membranes in Chick and Human		
Membrane	**Chick**	**Human**
Amnion		
Allantois		
Chorion		
Yolk sac		

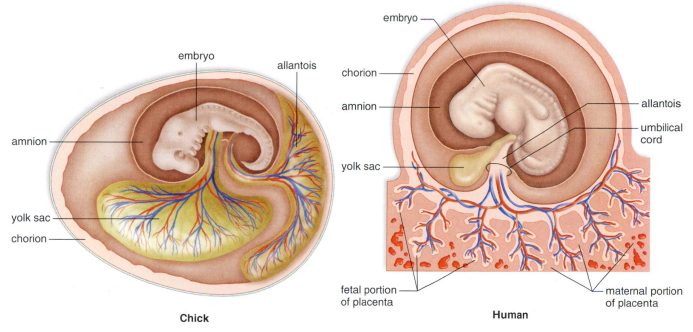

Chick

Human

Figure 19.11 Extraembryonic membranes.
The chick and a human have the same extraembryonic membranes, but except for the amnion, they have different functions.

19.3 Fetal Development

During fetal development (last seven months), the skeleton becomes ossified (bony), reproductive organs form, arms and legs develop fully, and the fetus enlarges in size and gains weight.

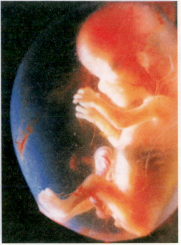 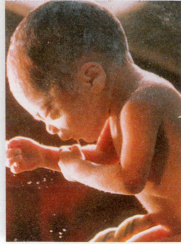

Three- to four-month-old fetus   Seven- to eight-month-old fetus

Observation: Fetal Development

1. Using Table 19.3 and Figure 19.12 to assist you, examine models of fetal development.
2. In Table 19.3, note the following.

   a. **External genitals:** About the third month, it is possible to tell male from female if an ultrasound is done.

   b. **Quickening:** Fetal movement is felt during the fourth or fifth months.

   c. **Vernix caseosa:** Beginning with the fifth month, the skin is covered with a cheesy coating called vernix caseosa.

   d. **Lanugo:** During the sixth and seventh months, the body is covered with fine, downy hair termed lanugo.

Table 19.3 Fetal Development

Month	Events for Mother	Events for Baby
Third month	Uterus is the size of a grapefruit.	Possible to distinguish sex. Fingernails appear.
Fourth month	Fetal movement is felt by those who have been previously pregnant. Heartbeat is heard by stethoscope.	Bony skeleton visible. Hair begins to appear. 150 mm (6 in.), 170 g (6 oz.).
Fifth month	Fetal movement is felt by those who have not been previously pregnant. Uterus reaches up to level of umbilicus and pregnancy is obvious.	Protective cheesy coating, called vernix caseosa, begins to be deposited. Heartbeat can be heard.
Sixth month	Doctor can tell where baby's head, back, and limbs are. Breasts have enlarged, nipples and areolae are darkly pigmented, and colostrum is produced.	Body is covered with fine hair called lanugo. Skin is wrinkled and reddish.
Seventh month	Uterus reaches halfway between umbilicus and rib cage.	Testes descend into scrotum. Eyes are open. 300 mm (12 in.), 1,350 g (3 lb).
Eighth month	Weight gain is averaging about a pound a week. Difficulty in standing and walking because center of gravity is thrown forward.	Body hair begins to disappear. Subcutaneous fat begins to be deposited.
Ninth month	Uterus is up to rib cage, causing shortness of breath and heartburn. Sleeping becomes difficult.	Ready for birth. 530 mm (20½ in.), 3,400 g (7½ lb).

Figure 19.12 Human development.
Changes occurring from the fifth week to the eighth month.

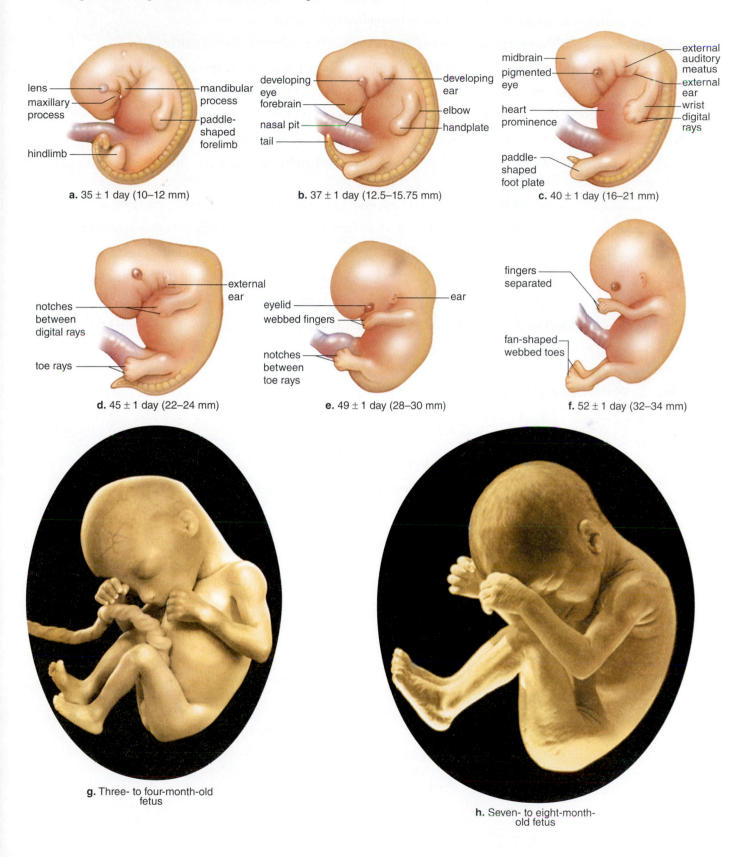

a. 35 ± 1 day (10–12 mm)

lens
maxillary process
hindlimb
mandibular process
paddle-shaped forelimb

b. 37 ± 1 day (12.5–15.75 mm)

developing eye
forebrain
nasal pit
tail
developing ear
elbow
handplate

c. 40 ± 1 day (16–21 mm)

midbrain
pigmented eye
heart prominence
paddle-shaped foot plate
external auditory meatus
external ear
wrist
digital rays

d. 45 ± 1 day (22–24 mm)

notches between digital rays
toe rays
external ear

e. 49 ± 1 day (28–30 mm)

eyelid
webbed fingers
notches between toe rays
ear

f. 52 ± 1 day (32–34 mm)

fingers separated
fan-shaped webbed toes

g. Three- to four-month-old fetus

h. Seven- to eight-month-old fetus

Laboratory Review 19

_____ 1. By what process does the embryo divide with no increase in size?

_____ 2. Name the stage of development when the embryo is a solid ball of cells.

_____ 3. Name the stage of development when the embryonic germ layers are forming.

_____ 4. The nervous system develops from which embryonic germ layer?

_____ 5. The intestinal tract develops from which embryonic germ layer?

_____ 6. What structure in an embryo induces formation of the nervous system?

_____ 7. Which group has four extraembryonic membranes—aquatic animals or land animals?

_____ 8. Which two organ systems appear before the others during chick development?

_____ 9. Is the human embryo or the fetus most likely to resemble that of other animals?

_____ 10. What term is used to refer to the last seven months of human development?

_____ 11. How does the morula stage of sea star development differ from the blastula stage?

_____ 12. What is the function of the amnion in both chicks and humans?

_____ 13. The allantois blood vessels become the _____ blood vessels in humans.

Thought Questions

14. What chiefly causes a frog's morula, blastula, and gastrula to look different than those of a sea star?

15. For each scenario, hypothesize whether a neural tube will develop.

   a. Notochord is removed. Does ectoderm in this location become a neural tube? _____
   Explain.

   b. Ectoderm above notochord is replaced with belly ectoderm. Does belly ectoderm become a neural tube? _____ Explain.

16. Why does the sea star embryo have no need of the yolk associated with the chick embryo?

20

Patterns of Inheritance

Introduction

Gregor Mendel, sometimes called the "father of genetics," formulated the basic laws of genetics examined in this laboratory. He determined that individuals have two alternate forms of a gene (now called **alleles**) for each trait in their body cells. Today, we know that alleles are on the chromosomes. An individual can be **homozygous dominant** (two dominant alleles, *GG*), **homozygous recessive** (two recessive alleles, *gg*), or **heterozygous** (one dominant and one recessive allele, *Gg*). **Genotype** refers to an individual's genes, while **phenotype** refers to an individual's appearance (Fig. 20.1). Homozygous dominant and heterozygous individuals show the dominant phenotype; homozygous recessive individuals show the recessive phenotype.

Figure 20.1   Genotype versus phenotype.
Only with homozygous recessive do you immediately know the genotype.

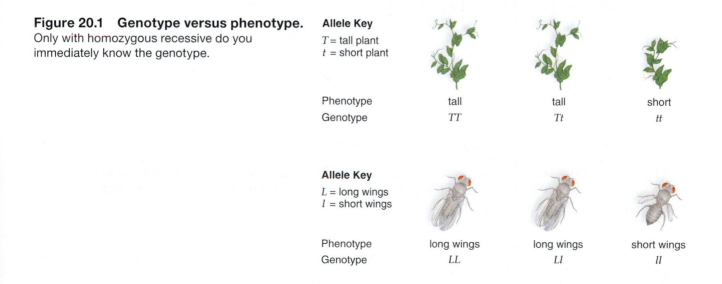

Allele Key
T = tall plant
t = short plant

Phenotype	tall	tall	short
Genotype	*TT*	*Tt*	*tt*

Allele Key
L = long wings
l = short wings

Phenotype	long wings	long wings	short wings
Genotype	*LL*	*Ll*	*ll*

Punnett Squares

Punnett squares, named after the man who first used them, allow you to easily determine the results of a cross between individuals whose genotypes are known. Consider that when fertilization occurs, two gametes, such as a sperm and an egg, join together. Whereas individuals have two alleles for every trait, gametes have only one allele because alleles are on the chromosomes and homologues separate during meiosis. Heterozygous parents with the genotype *Aa* produce two types of gametes: 50% of the gametes contain an *A* and 50% contain an *a*. A Punnett square allows you to vertically line up all possible types of sperm and to horizontally line up all possible types of eggs. Every possible combination of gametes occurs within the squares and these combinations indicate the genotypes of the offspring. In Figure 20.2, one offspring is *AA* = homozygous dominant, two are *Aa* = heterozygous, and one is *aa* = homozygous recessive. Therefore, three of the offspring will have the dominant phenotype and one individual will have the recessive phenotype. This is said to be a **phenotypic ratio** of 3:1.

A Punnett square can be used for any cross regardless of the trait(s) and the genotypes of the parents. All you need to do is use the correct letters for the particular trait(s), and make sure you have given the parents the correct genotypes and correct proportion of each type gamete. Then you can determine the genotypes of the offspring and the resulting phenotypic ratio among the offspring.

Figure 20.2 What are the expected results of a cross?

A Punnett square allows you to determine the expected phenotypic ratio for a cross.

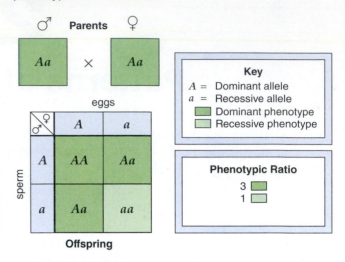

Virtual Lab **Punnett Squares** A virtual lab called Punnett Squares is available on the *Inquiry into Life* website **www.mhhe.com/mader Inquiry14**. It will allow you to practice filling in Punnett squares and determining your results.

20.1   One-Trait Crosses

A single pair of alleles is involved in one-trait crosses. Mendel found that reproduction between two heterozygous individuals *(Aa)*, called a **monohybrid cross**, results in both dominant and recessive phenotypes among the offspring. In Figure 20.2, the expected phenotypic ratio among the offspring is 3:1. Three offspring have the dominant phenotype for every one that has the recessive phenotype.

Mendel realized that these results are obtainable only if the alleles of each parent segregate (separate from each other) during meiosis. Therefore, Mendel formulated his first law of inheritance:

Law of Segregation

Each organism contains two alleles for each trait, and the alleles segregate during the formation of gametes. Each gamete (egg or sperm) then contains only one allele for each trait. When fertilization occurs, the new organism has two alleles for each trait, one from each parent.

Inheritance is a game of chance. Just as there is a 50% probability of heads or tails when tossing a coin, there is a 50% probability that a sperm or egg will have an *A* or an *a* when the parent is *Aa*. The chance of an equal number of heads or tails improves as the number of tosses increases. In the same way, the chance of an equal number of gametes with *A* and *a* improves as the number of gametes increases. Therefore, the 3:1 ratio among offspring is more likely the more offspring you count for the same type cross.

Color of Tobacco Seedlings

In tobacco plants, a dominant allele *(C)* for chlorophyll gives the plants a green color, and a recessive allele *(c)* for chlorophyll causes a plant to appear white. If a tobacco plant is homozygous for the recessive allele *(c)*, it cannot manufacture chlorophyll and thus appears white (Fig. 20.3).

Figure 20.3 Monohybrid cross.
These tobacco seedlings are growing on an agar plate. The white plants cannot manufacture chlorophyll.

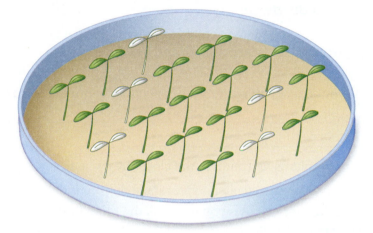

Experimental Procedure: Color of Tobacco Seedlings

1. Obtain a numbered agar plate on which tobacco seedlings are growing. They are the offspring of a cross between heterozygous parents: the cross *Cc* × *Cc*. Complete the Punnett square to determine the expected phenotypic ratio.

   What is the expected phenotypic ratio? _____

2. Record the plate number and using a stereomicroscope, view the seedlings. Count the number that are green and the number that are white. Record your results in Table 20.1.
3. Repeat this procedure for two additional plates. Total the number that are green and the number that are white.
4. Complete Table 20.1 by recording the class data. Total the number that are green and the number that are white per class.

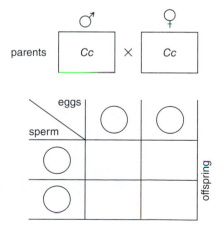

Key:
C = green
c = white

Table 20.1 Color of Tobacco Seedlings

| | Number of Offspring | | |
	Green Color	White Color	Phenotypic Ratio
Plate # _____			
Plate # _____			
Plate # _____			
Totals			
Class data			

Conclusions: Color of Tobacco Seedlings

- In the last column of Table 20.1, record the actual phenotypic ratio per observed plate; per total number of green versus white plants you counted; and per the entire class. To determine the actual phenotypic ratio, divide the number of green color seedlings in a plate by the number of white color seedlings in a plate. Do your results differ from the expected phenotypic ratio? Explain. _____

_____

_____

- Mendel found that the more plants he counted, the closer he came to the expected phenotypic ratio. Was your class data closer to the expected phenotypic ratio than your individual data? _____ This is expected, because the more crosses you observe, the more likely it is that all types of sperm and eggs will have a chance to come together.

Color of Corn Kernels

In corn plants, the allele for purple kernel (P) is dominant over the allele for yellow kernel (p) (Fig. 20.4).

Figure 20.4 Monohybrid cross.
Two types of kernels are seen on an ear of corn following a monohybrid cross: purple and yellow.

1. Obtain an ear of corn from the supply table. You will be examining the results of the cross *Pp* × *pp*. Complete the Punnett square to determine the expected phenotypic ratio. Note that when one parent has only one possible type of gamete, only one column is needed in the Punnett square.

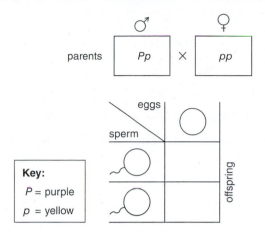

What is the expected phenotypic ratio among the offspring? _____

2. Record the sample number and then count the number of kernels that are purple and the number that are yellow. As before, use two more samples, and record your results in Table 20.2.

Table 20.2 Color of Corn Kernels

	Number of Kernels		
	Purple	**Yellow**	**Phenotypic Ratio**
Sample # _____			
Sample # _____			
Sample # _____			
Totals			
Class data			

Conclusions: Color of Corn Kernels

- In the last column of Table 20.2, record the actual phenotypic ratio per observed sample; per total number of purple versus yellow kernels you counted; and per the entire class. Do your results differ from the expected phenotypic ratio? Explain. _____

  _____

- Was your class data closer to the expected phenotypic ratio than your individual data? _____ Explain. _____

  _____

Practice Problems

1. In pea plants, purple flowers (*P*) is dominant and white flowers (*p*) is recessive. What is the genotype of pure-breeding white plants? Pure-breeding means that they produce plants with only one phenotype. _____ If pure-breeding purple plants are crossed with these white plants, what phenotype is expected? _____

2. In pea plants, tall (*T*) is dominant and short (*t*) is recessive. A heterozygous tall plant is crossed with a short plant. What is the expected phenotypic ratio? _____

3. Unexpectedly to the farmer, two tall plants have some short offspring. What is the genotype of the parent plants and the short offspring? parent _____ offspring _____

4. In horses, two trotters are mated to each other and produce only trotters; two pacers are mated to each other and produce only pacers. When one of these trotters is mated to one of the pacers, all the horses are trotters. Create a key and show the cross. key _____ cross _____

5. A brown dog is crossed with two different black dogs. The first cross produces only black dogs and the second cross produces equal numbers of black and brown dogs. What is the genotype of the brown dog? _____ the first black dog _____ the second black dog _____

6. In pea plants, green pods (*G*) is dominant and yellow pods (*g*) is recessive. When two pea plants with green pods are crossed, 25% of the offspring have yellow pods. What is the genotype of all plants involved? plants with green pods _____ plants with yellow pods _____

7. A breeder wants to know if a dog is homozygous black or heterozygous black. If the dog is heterozygous, which cross is more likely to produce a brown dog, *Bb* × *bb* or *Bb* × *Bb*? Explain.

_____

_____

8. If the cross in #6 produces 220 plants, how many offspring have green pods and how many have yellow pods? _____ If the cross in #2 produces 220 plants, how many offspring are tall and how many are short? _____

20.2 Two-Trait Crosses

Two-trait crosses involve two pairs of alleles. Mendel found that during a **dihybrid cross,** when two dihybrid individuals *(AaBb)* reproduce, the phenotypic ratio among the offspring is 9:3:3:1, representing four possible phenotypes. He realized that these results could be obtained only if the alleles of the parents segregated independently of one another when the gametes were formed. From this, Mendel formulated his second law of inheritance:

Law of Independent Assortment

Members of an allelic pair segregate (assort) independently of members of another allelic pair. Therefore, all possible combinations of alleles can occur in the gametes.

The FOIL method is a way to determine the gametes. FOIL stands for *First* two alleles from each trait; *Outer* two alleles from each trait; *Inner* two alleles from each trait; *Last* two alleles from each trait. Here is how the FOIL method can help you determine the gametes for the genotype *PpSs:*

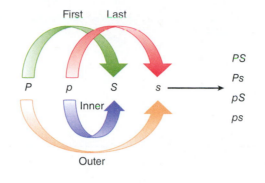

Color and Texture of Corn

In corn plants, the allele for purple kernel *(P)* is dominant over the allele for yellow kernel *(p),* and the allele for smooth kernel *(S)* is dominant over the allele for rough kernel *(s)* (Fig. 20.5).

Figure 20.5 Dihybrid cross.
Four types of kernels are seen on an ear of corn following a dihybrid cross: purple smooth, purple rough, yellow smooth, and yellow rough.

20 mm

Experimental Procedure: Color and Texture of Corn

1. Obtain an ear of corn from the supply table. You will be examining the results of the cross *PpSs × PpSs.*

2. Do the Punnett square on page 282, in order to state the expected phenotypic ratio among the offspring. _____

3. Count the number of kernels of each possible phenotype listed in Table 20.3. Record the sample number and your results in Table 20.3. Use three samples, and total your results for all samples. Also record the class data, i.e., the number of kernels that are the four phenotypes per class.

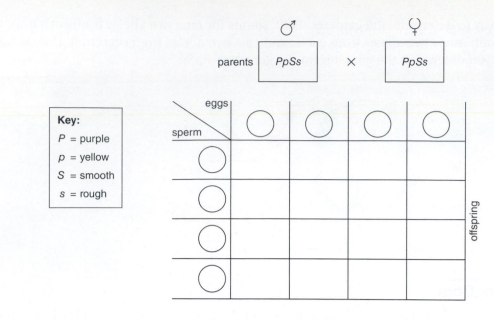

parents $PpSs$ × $PpSs$

Key:
P = purple
p = yellow
S = smooth
s = rough

eggs

sperm

offspring

Table 20.3 Color and Texture of Corn

	Number of Kernels				
	Purple Smooth	Purple Rough	Yellow Smooth	Yellow Rough	Phenotypic Ratio
Sample # _____					
Sample # _____					
Sample # _____					
Totals					
Class data					

Conclusions: Color and Texture of Corn

- Calculate the actual phenotypic ratios based on the data and record in Table 20.3. Do the results differ from the expected ratio per individual data? _____ Per class data? _____ Explain. _____
_____
_____

Wing Length and Body Color in *Drosophila*

Drosophila are the tiny flies you often see flying around ripe fruit; therefore, they are called fruit flies. If a culture bottle of fruit flies is on display, take a look at it. Because so many flies can be grown in a small culture bottle, fruit flies have contributed substantially to our knowledge of genetics. If you were to examine *Drosophila* flies under the stereomicroscope, they would appear like this:

In *Drosophila*, long wings *(L)* are dominant over short (vestigial) wings *(l)*, and gray body *(G)* is dominant over black (ebony) body *(g)*. Consider the cross *LlGg* × *llgg* and complete this Punnett square:

> **Key:**
> *L* = long wing
> *l* = short (vestigial) wing
> *G* = gray body
> *g* = ebony (black) body

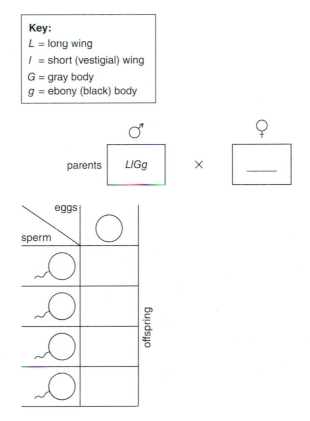

What is the expected phenotypic ratio for this cross? _____

Experimental Procedure: Wing Length and Body Color in Drosophila

If your instructor has frozen flies available, cross out the numbers in Table 20.4 and use the stereomicroscope or a hand lens to count the flies of each type given in Table 20.4. Otherwise simply use the data supplied for you in Table 20.4.

Table 20.4	Wing Length and Body Color in *Drosophila**				
	Phenotypes				
	Long Gray	Long Ebony	Short Gray	Short Ebony	Phenotypic Ratio
Number of offspring	28	32	28	30	
Class data	128	120	120	120	

*Wings and body are understood in this table.

Conclusions: Wing Length and Body Color in Drosophila

- Calculate the actual phenotypic ratio based on the data and record in Table 20.4. Do the results differ from the expected ratio per individual data? _____ Per class data? _____ Explain. _____

_____

Practice Problems

1. In tomatoes, tall is dominant and short is recessive. Red fruit is dominant and yellow fruit is recessive. Choose a key for height _____ for color of fruit _____ What is the genotype of a plant heterozygous for both traits? _____ What are the possible gametes for this plant?

_____

2. Using words, what are the likely parental genotypes if the results of a two-trait problem are 1:1:1:1 among the offspring? _____ × _____

3. In horses, black (*B*) and a trotting gait (*T*) are dominant, while brown (*b*) and a pacing gait (*t*) are recessive. If a black trotter (homozygous for both traits) is mated to a brown pacer, what phenotypic ratio is expected among the offspring? _____

4. Two black trotters have a brown pacer offspring. What is the genotype of all horses involved? black trotter parents _____ brown pacer offspring _____

5. The phenotypic ratio among the offspring for two corn plants producing purple and smooth kernels is 9:3:3:1. (See lab for the key.) What is the genotype of these plants? parental plants _____ the 9 offspring _____ 3 of the offspring _____ the other 3 _____ and the 1 offspring? _____

6. Which matings could produce at least some fruit flies heterozygous in both traits? Write yes or no beside each. (You do not need a key.)

ggLl × *Ggll* _____   *GGLl* × *ggLl* _____   *GGLL* × *ggll* _____

Explain. _____

_____

7. State two new crosses that could not produce fruit flies heterozygous in both traits.

_____ × _____   _____ × _____

8. Chimpanzees are not deaf if they inherit both an allele *E* and an allele *G*. A cross between two deaf chimpanzees produces only chimpanzees that can hear. What are the genotypes of all chimpanzees involved? parents _____ × _____ offspring _____

20.3   X-Linked Crosses

In animals such as fruit flies, chromosomes differ between the sexes. All but one pair of chromosomes in males and females are the same; these are called **autosomes** because they do not actively determine sex. The pair that is different is called the **sex chromosomes**. In fruit flies and humans, the sex chromosomes in females are XX and those in males are XY.

Some alleles on the X chromosome have nothing to do with gender, and these genes are said to be X-linked. The Y chromosome does not carry these genes and indeed carries very few genes. Males with a normal chromosome inheritance are never heterozygous for X-linked alleles and if they inherit a recessive X-linked allele it will be expressed.

Red/White Eye Color in *Drosophila*

In fruit flies, red eyes (X^R) are dominant over white eyes (X^r). You will be examining the results of the cross $X^RY \times X^RX^r$. Complete this Punnett square and state the expected phenotypic ratio for this cross.

females _____   males _____

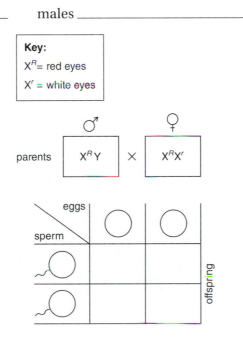

Key:
X^R = red eyes
X^r = white eyes

Experimental Procedure: Red/White Eye Color in Drosophila

If your instructor has frozen flies available, cross out the numbers in Table 20.5 and use the stereomicroscope or a hand lens to count the flies of each type given in Table 20.5. Use the art on page 283 to tell males from females and record male and female data separately. If frozen flies are not available, simply use the data supplied for you in Table 20.5.

Table 20.5 Red/White Eye Color in *Drosophila*

	Number of Offspring		
Your Data:	**Red Eyes**	**White Eyes**	**Phenotypic Ratio**
Males	16	17	
Females	63	0	
Class Data:			
Males	45	48	
Females	215	0	

Conclusions: Red/White Eye Color in Drosophila

- Calculate the phenotypic ratios based on the data for males and females separately and record in Table 20.5. Do the results differ from the expected ratio per individual data? _____ Per class data? _____ Explain. _____
  _____
  _____

Virtual Lab Sex-linked Traits A virtual lab called Sex-linked Traits is available on the *Inquiry into LIfe* website **www.mhhe.com/mader Inquiry14**. It will allow you to practice filling in Punnett squares and determining your results for X-linked traits.

- Using the Punnett square provided, calculate the expected phenotypic results for the cross $X^RY \times X^rX^r$. What is the expected phenotypic ratio among the offspring? males _____ females _____

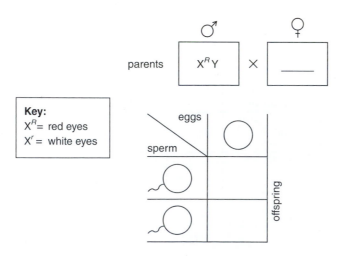

parents

Key:
X^R = red eyes
X^r = white eyes

Practice Problems

1. State the genotypes and gametes for each of these fruit flies:

	genotype	gamete(s)
white-eyed male	_____	_____
white-eyed female	_____	_____
red-eyed male	_____	_____
homozygous red-eyed female	_____	_____
heterozygous red-eyed female	_____	_____

2. What are the phenotypic ratios if a white-eyed female is crossed with a red-eyed male?

   males _____ females _____

3. Regardless of any type cross, do white-eyed males inherit the allele for white eyes from their father or their mother? _____ Explain. _____

4. In sheep, horns are sex linked; H = horns and h = no horns. Using symbols, what cross do you recommend if a farmer wants to produce hornless males? _____ × _____

5. In *Drosophila*, bar eye is sex linked; B = bar eye and b = no bar eye. What are the phenotypic ratios for these crosses?

   bar-eyed male × non–bar-eyed female _____ , _____

   bar-eyed male × heterozygous female _____ , _____

   non–bar-eyed male × heterozygous female _____ , _____

6. A female fruit fly has white eyes. What is the genotype of the father? _____ What could be the genotype of the mother? _____ or _____

7. In a cross between fruit flies, all the males have white eyes and the females are 1:1. What is the genotype of the parents? female parent _____ male parent _____

8. In a cross between fruit flies, a white-eyed male and a red-eyed female produce no offspring that have white eyes. What is the genotype of the parents? male parent _____ female parent _____

9. Make up a sex-linked genetic cross using words and parental genotypes. Create a key and show a Punnett Square and the phenotypic ratios.

_____ 1. A cross gives a 3:1 phenotypic ratio. What are the genotypes of the parents?

_____ 2. According to Mendel's law of segregation, parents who both have the genotype *Aa* would produce what gametes?

_____ 3. What is the genotype of a plant that produces green peas? Yellow peas are dominant.

_____ 4. To determine whether an animal with the dominant phenotype is heterozygous or homozygous, it is best to cross it with the _____ phenotype.

_____ 5. If you performed the *Drosophila* cross *LL* × *ll*, what phenotypic ratio would you expect among the offspring?

_____ 6. According to Mendel's law of independent assortment, how many different types of gametes would an *AaBb* parent have? An *AABb* parent?

_____ 7. What is the genotype of a homozygous long-winged fly that is heterozygous for gray body color?

_____ 8. What is the expected phenotypic ratio among offspring if both parents are heterozygous for both traits?

_____ 9. Why do you expect class data to be closer to the expected ratio than your individual data?

_____ 10. What is the genotype of a white-eyed male fruit fly? A white-eyed female fruit fly?

_____ 11. Which gender can have white eyes if the female parent is homozygous dominant for red eyes and the male parent has white eyes?

_____ 12. What does a father always give his male offspring in X-linked genetic crosses?

_____ 13. What is the phenotype of a tobacco plant with the genotype *Cc*?

Thought Questions

14. You count 73 long-winged flies and 27 short-winged flies from a cross between two heterozygous parents. How many flies did you expect to have long wings?

15. In horses, a trotting gait is dominant over a pacing gait and black is dominant over brown. What is the genotype of a black trotter whose mother was a brown pacer?

16. In the cross *AaBb* × *aaBb*, what are the gametes for *AaBb*? For *aaBb*? What are the genotypic results of the cross?

21

Human Genetics

Learning Outcomes

21.1 Determining the Genotype
- Determine the genotype by observation of the person and their relatives. 289–91

21.2 Determining Inheritance
- Do genetic problems involving autosomal dominant, autosomal recessive, and X-linked recessive alleles. 292–95
- Do genetic problems involving multiple allele inheritance and use blood type to help determine paternity. 296–97

21.3 Genetic Counseling
- Analyze a karyotype to determine if a person's chromosomal inheritance is as expected or whether a chromosome anomaly has occurred. 297–98
- Analyze a pedigree to determine if the pattern of inheritance is autosomal dominant, autosomal recessive, or X-linked recessive. 299–300
- Construct a pedigree to determine the chances of inheriting a particular phenotype when provided with generational information. 300–01

Introduction

In this laboratory, you will discover that the same principles of genetics apply to humans as they do to plants and fruit flies. A gene has two alternate forms, called **alleles,** for any trait, such as hairline, finger length, and so on. One possible allele, designated by a capital letter, is **dominant** over the **recessive** allele, designated by a lowercase letter. An individual can be **homozygous dominant** (two dominant alleles, *EE*), **homozygous recessive** (two recessive alleles, *ee*) or **heterozygous** (one dominant and one recessive allele, *Ee*). **Genotype** refers to an individual's alleles, and **phenotype** refers to an individual's appearance (Fig. 21.1). Homozygous dominant and also heterozygous individuals show the dominant phenotype; homozygous recessive individuals show the recessive phenotype.

Figure 21.1 Genotype versus phenotype.
Unattached earlobes (*E*) are dominant over attached earlobes (*e*). **a.** Homozygous dominant individuals have unattached earlobes. **b.** Homozygous recessive individuals have attached earlobes. **c.** Heterozygous individuals have unattached earlobes.

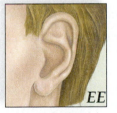
EE
a. Unattached earlobe

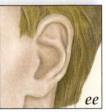
ee
b. Attached earlobe

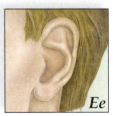
Ee
c. Unattached earlobe

21.1 Determining the Genotype

Humans inherit 46 chromosomes that occur in 23 pairs. Twenty-two of these pairs are called autosomes and one pair is the sex chromosomes. Autosomal traits are determined by alleles on the autosomal chromosomes.

Autosomal Dominant and Recessive Traits

Figure 21.2 shows a few human traits.

1. What is the homozygous dominant genotype for type of hairline? _____ What is the
   phenotype? _____

2. What is the homozygous recessive genotype for finger length? _____ What is the
   phenotype? _____

3. Why does the heterozygous individual *Ff* have freckles? _____

Figure 21.2   Commonly inherited traits in human beings.
The alleles indicate which traits are dominant and which are recessive.

   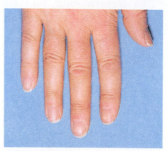

a. Widow's peak: *WW* or *Ww*   **b.** Straight hairline: *ww*   **e.** Short fingers: *SS* or *Ss*   **f.** Long fingers: *ss*

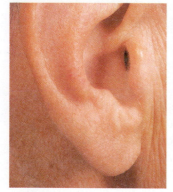  

c. Unattached earlobes: *EE* or *Ee*   **d.** Attached earlobes: *ee*   **g.** Freckles: *FF* or *Ff*   **h.** No freckles: *ff*

These genetic problems use the alleles from Figure 21.2 and Table 21.1.

4. Maria and the members of her immediate family have attached earlobes. What is Maria's genotype?
   _____ Her maternal grandfather has unattached earlobes. Deduce the genotype of her
   maternal grandfather. _____ Explain. _____
   _____

5. Moses does not have a bent little finger, but his parents do. Deduce the genotype of his parents.
   _____ of Moses. _____ Explain. _____
   _____

6. Manny is adopted. He has hair on the back of his hand. Could both of his parents have had hair on the
   back of the hand? _____ Could both of his parents have had no hair on the back of the hand?
   _____ Explain. _____
   _____

1. For this Experimental Procedure, you will need a lab partner to help you determine your phenotype for the traits listed in the first column of Table 21.1.
2. Determine your probable genotype. If you have the recessive phenotype, you know your genotype. If you have the dominant phenotype, you may be able to decide whether you are homozygous dominant or heterozygous by recalling the phenotype of your parents, siblings, or children. Circle your probable genotype in the second column of Table 21.1.
3. Your instructor will tally the class's phenotypes for each trait so that you can complete the third column of Table 21.1.
4. Complete Table 21.1 by calculating the percentage of the class with each trait. Are dominant phenotypes always the most common in a population? _____ Explain. _____

_____

Table 21.1   Autosomal Human Traits

Trait: d = Dominant r = Recessive	Probable Genotypes	Number in Class	Percentage of Class with Trait
Hairline:			
Widow's peak (d)	WW or Ww	_____	_____
Straight hairline (r)	ww	_____	_____
Earlobes:			
Unattached (d)	UU or Uu	_____	_____
Attached (r)	uu	_____	_____
Skin pigmentation:			
Freckles (d)	FF or Ff	_____	_____
No freckles (r)	ff	_____	_____
Hair on back of hand:			
Present (d)	HH or Hh	_____	_____
Absent (r)	hh	_____	_____
Thumb hyperextension—"hitchhiker's thumb":			
Last segment cannot be bent backward (d).	TT or Tt	_____	_____
Last segment can be bent back to 60° (r).	tt	_____	_____
Bent little finger:			
Little finger bends toward ring finger (d).	LL or Ll	_____	_____
Straight little finger (r)	ll	_____	_____
Interlacing of fingers:			
Left thumb over right (d)	II or Ii	_____	_____
Right thumb over left (r)	ii	_____	_____

21.2 Determining Inheritance

Recall that a Punnett square is a means to determine the genetic inheritance of offspring if the genotypes of both parents are known. In a **Punnett square**, all possible types of sperm are lined up vertically, and all possible types of eggs are lined up horizontally, or vice versa, so that every possible combination of gametes occurs within the square. Figure 21.3 shows how to construct a Punnett square when autosomal alleles are involved.

Figure 21.3   Punnett square.
In a Punnett square, all possible sperm are displayed vertically and all possible eggs are displayed horizontally, or vice versa. The genotypes of the offspring (in this case, also the phenotypes) are in the squares.

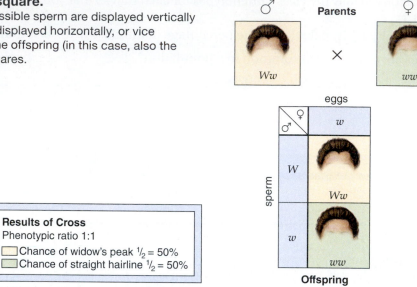

Results of Cross
Phenotypic ratio 1:1
☐ Chance of widow's peak ½ = 50%
☐ Chance of straight hairline ½ = 50%

Inheritance of Genetic Disorders

Figure 21.4 can be used to learn the chances of a particular phenotype.

In Figure 21.4*a*,

¼ of the offspring have the recessive phenotype = _____ % chance

¾ of the offspring have the dominant phenotype = _____ % chance

In Figure 21.4*b*,

½ of the offspring have the recessive or the dominant phenotype = _____ % chance

In these genetics problems, use letters to fill in the parentheses with the genotype of the parents.

1. **a.** With reference to Figure 21.4*a*, if a genetic disorder is recessive and both parents are

heterozygous (_____), what are the chances that an offspring will have the disorder? _____

b. With reference to Figure 21.4*a*, if a genetic disorder is dominant and both parents are heterozygous

(_____), what are the chances that an offspring will have the disorder? _____

2. **a.** With reference to Figure 21.4*b*, if the parents are heterozygous (_____) by homozygous

recessive (_____), and the genetic disorder is recessive, what are the chances that the offspring

will have the disorder? _____

b. With reference to Figure 21.4*b*, if the parents are heterozygous (_____) by homozygous

recessive (_____), and the genetic disorder is dominant, what are the chances that an offspring

will have the disorder? _____

Figure 21.4   Two common patterns of autosomal inheritance in humans.
a. Both parents are heterozygous. **b.** One parent is heterozygous and the other is homozygous recessive. The letter *A* stands for any trait that is dominant and the letter *a* stands for any trait that is recessive. Substitute the correct alleles for the problem you are working on. For example, *C* = normal; *c* = cystic fibrosis.

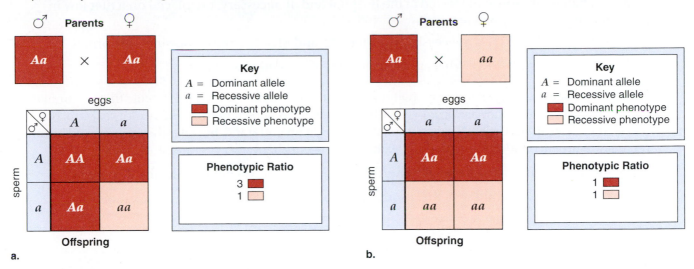

Autosomal Disorders

1. **Neurofibromatosis (NF),** sometimes called von Recklinghausen disease, is one of the most common genetic disorders. It affects roughly 1 in 3,000 people. It is seen equally in every racial and ethnic group throughout the world. At birth or later, the affected individual may have six or more large tan spots on the skin. Such spots may increase in size and number and become darker. Small benign tumors (lumps) called neurofibromas may occur under the skin or in the muscles. Neurofibromas are made up of nerve cells and other cell types.

   Neurofibromatosis is a dominant disorder. If a heterozygous woman reproduces with a homozygous

   normal man, what are the chances a child will have neurofibromatosis? _____

2. **Cystic fibrosis** is due to abnormal mucus-secreting tissues. At first, the infant may have difficulty regaining the birth weight despite good appetite and vigor. A cough associated with a rapid respiratory rate but no fever indicates lung involvement. Large, frequent, and foul-smelling stools are due to abnormal pancreatic secretions. Whereas children previously died in infancy due to infections, they now often survive because of antibiotic therapy.

   Cystic fibrosis is a recessive disorder. A **carrier** is an individual that appears to be normal but carries

   a recessive allele for a genetic disorder. A man and a woman are both carriers (_____) for cystic

   fibrosis. What are the chances a child will have cystic fibrosis? _____

3. **Huntington disease** does not appear until the 30s or early 40s. There is a progressive deterioration of the individual's nervous system that eventually leads to constant thrashing and writhing movements until insanity precedes death. Studies suggest that Huntington disease is due to a single faulty gene that has multiple effects, in which case there is now hope for a cure.
   People with Huntington disease seem to be more fertile than others. It is amazing that more than 1,000 of the cases in the United States in the past century can be traced to one man born in 1831.

   Huntington disease is a dominant disorder. Drina is 25 years old and as yet has no signs of

   Huntington disease. Her mother does have Huntington disease (_____), but her father is free

   (_____) of the disorder. What are the chances that Drina will develop Huntington disease?

   _____

4. **Phenylketonuria (PKU)** is characterized by severe mental retardation due to an abnormal accumulation of the common amino acid phenylalanine within cells, including neurons. The disorder takes its name from the presence of a breakdown product, phenylketone, in the urine and blood. Newborn babies are routinely tested at the hospital and, if necessary, are placed on a diet low in phenylalanine.

   Phenylketonuria (PKU) is a recessive disorder. Mr. and Mrs. Martinez appear to be normal, but they have a child with PKU. What are the genotypes of Mr. and Mrs. Martinez? _____

5. **Tay–Sachs disease** is caused by the inability to break down a certain type of fat molecule that accumulates around nerve cells until they are destroyed. Afflicted newborns appear normal and healthy at birth, but they do not develop normally. At first, they may learn to sit up and stand, but later they regress and become mentally retarded, blind, and paralyzed. Death usually occurs between ages three and four.

   Tay–Sachs is an autosomal recessive disorder. Is it possible for two individuals who do not have Tay–Sachs to have a child with the disorder? Explain. _____

   _____

X-Linked Disorders

The sex chromosomes designated X and Y carry genes just like the autosomal chromosomes. Some genes, particularly on the X chromosome, have nothing to do with gender inheritance and are said to be X-linked. **X-linked recessive disorders** are due to recessive genes carried on the X chromosomes. Males are more likely to have an X-linked recessive disorder than females because the Y chromosome is blank for this trait. Does a color-blind male give his son a recessive-bearing X or a Y that is blank for the recessive allele? _____

   The possible genotypes and phenotypes for an X-linked recessive disorder are as follows:

   Females
   $X^B X^B$ = normal vision
   $X^B X^b$ = normal vision (carrier)
   $X^b X^b$ = color blindness

   Males
   $X^B Y$ = normal vision
   $X^b Y$ = color blindness

An X-linked recessive disorder in a male is always inherited from his mother. Most likely, his mother is heterozygous and therefore does not show the disorder. She is designated a carrier for the disorder. Figure 21.5 shows how females can become carriers.

1. **a.** What is the genotype for a color-blind female?_____ How many recessive alleles does a female inherit to be color-blind? _____

   b. What is the genotype for a color-blind male? _____ How many recessive alleles does a male inherit to be color-blind? _____

2. **a.** With reference to Figure 21.5*a*, if the mother is a carrier (_____) and the father has normal vision (_____), what are the chances that a daughter will be color blind? _____

   b. A daughter will be a carrier? _____ **c.** A son will be color blind? _____

3. **a.** With reference to Figure 21.5*b*, if the mother has normal vision (_____) and the father is color blind (_____), what are the chances that a daughter will be color blind? _____

   b. A daughter will be a carrier? _____ **c.** A son will be color blind? _____

Figure 21.5 Two common patterns of X-linked inheritance in humans.
a. The sons of a carrier mother have a 50% chance of being color blind. **b.** A color-blind father has carrier daughters.

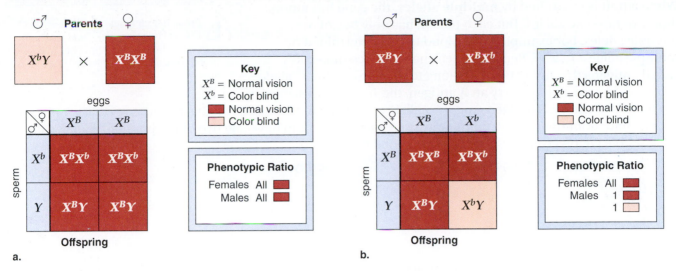

X-Linked Genetics Problems

For **color blindness,** there are two possible X-linked alleles involved. One affects the green-sensitive cones, whereas the other affects the red-sensitive cones. About 6% of men in the United States are color blind due to a mutation involving green perception, and about 2% are color blind due to a mutation involving red perception.

1. A woman with normal color vision (_____), whose father was color blind (_____), marries a man with normal color vision (_____). What genotypes could occur among their offspring?

_____

What genotypes could occur if it was the normal-visioned man's father who was color blind?

_____

2. Antonio's father is color blind (_____) but his mother is not color blind (_____).

Is Antonio necessarily color blind? _____ Explain. _____

Could he be color blind? _____ Explain. _____

Hemophilia is called the bleeder's disease because the affected person's blood is unable to clot. Although hemophiliacs do bleed externally after an injury, they also suffer from internal bleeding, particularly around joints. Hemorrhages can be checked with transfusions of fresh blood (or plasma) or concentrates of the clotting protein. The most common type of hemophilia is hemophilia A, due to absence or minimal presence of a particular clotting protein called factor VIII.

3. Make up a cross involving hemophilia that could be answered by a Punnett square, as in

Figure 21.5*a* or *b*. _____

_____

What is the answer to your genetics problem? _____

Multiple Alleles

When a trait is controlled by **multiple alleles,** the gene has more than two possible alleles. But each person has only two of the possible alleles. For example, ABO blood type is determined by multiple alleles: I^A, I^B, i. Red blood cells have surface molecules called antigens that indicate they belong to the person. The I^A allele causes red blood cells to carry an A antigen, the I^B allele causes red blood cells to carry a B antigen, and the i allele causes the red blood cells to have neither of these antigens. I^A and I^B are dominant to i. Remembering that each person can have any two of the possible alleles, these are possible genotypes and phenotypes for blood types.

> ⚠️ **Protective clothing** Wear protective laboratory clothing, latex gloves, and goggles. If the chemicals touch the skin, eyes, or mouth, wash immediately. If inhaled, seek fresh air.

Genotypes	Antigens on Red Cells	Blood Types
$I^A I^A$, $I^A i$	A	A
$I^B I^B$, $I^B i$	B	B
$I^A I^B$	A and B	AB
ii	none	O

Blood type also indicates whether the person is Rh positve or Rh negative. If the genotype is DD or Dd, the person is Rh positive and if the genotype is dd, the person is Rh negative. It is customary to simply attach a + or – superscript to the ABO blood type, as in A⁻.

Experimental Procedure: Using Blood Type to Help Determine Paternity

In this experimental procedure a mother, Wanda, is seeking support for her child, Sophia. We will use blood typing to decide which of three men could possibly be the father.

1. Obtain 3 testing plates, each of which contains three depressions; vials of blood from possible fathers 1, 2, and 3 respectively; vials of anti-A serum, anti-B serum, and anti-Rh serum. (All of these are synthetic.)

2. Using a wax pencil, number the plates so you know which plate is for possible father #1, #2, or #3. Look carefully at a plate and notice the wells are designated as A, B, or Rh.

3. Being sure to close the cap to each vial in turn, do the following using plate #1:

   Add a drop of father #1 blood to all three wells—close the cap.
   Add a drop of anti-A (blue) to the well designated A—close the cap.
   Add a drop of anti-B (yellow) to the well designated B—close the cap.
   Add a drop of anti-Rh (clear) to the well designated Rh—close the cap.

4. Stir the contents of each well with a mixing stick of the correct color. After a few minutes, examine the wells for agglutination, i.e., granular appearances that indicate the blood type. (Rh⁺ takes the longest to react.) If a person had AB⁺ blood, which wells would show agglutination? _____

   _____

5. Repeat steps 3 and 4 for plates #2 and #3.

6. Record the blood type results for each of the men in Table 21. 2.

Table 21.2   Blood Types of Involved Persons

	Mother*	Child*	Father?		
	Wanda	Sophia	#1	#2	#3
Blood type	B⁻	AB⁺			

*Your instructor may have you confirm these results.

Conclusion

1. Noting that only father #3 could have given Sophia the Rh antigen, from whom did she receive the I^B allele? _____ From which parent did she receive the I^A allele? _____ Is there any other possible interpretation to the results of blood typing? _____

Blood Typing Problems

1. A man with type A blood reproduces with a woman who has type B blood. Their child has blood type O. Using I^A, I^B, and i, give the genotype of all persons involved. man _____ woman _____ child _____

2. If a child has type AB blood and the father has type B blood, what could the genotype of the mother be? _____ or _____

3. If both mother and father have type AB blood, they cannot be the parents of a child who has what blood type? _____

4. What blood types are possible among the children if the parents are $I^A i \times I^B i$? (*Hint:* Do a Punnett square using the possible gametes for each parent.)

21.3   Genetic Counseling

Potential parents are becoming aware that many illnesses are caused by abnormal chromosomal inheritance or by gene mutations. Therefore they are seeking genetic counseling, which is available in many major hospitals. The counselor helps the couple understand the mode of inheritance for a condition of concern so that the couple can make an informed decision about how to proceed.

Determining Chromosomal Inheritance

If a genetic counselor suspects that a condition is due to a chromosome anomaly, he or she may suggest that the chromosomal inheritance be examined. It is possible to view the chromosomes of an individual because cells can be microscopically examined and photographed just before cell division occurs. A computer is then used to arrange the chromosomes by pairs. The resulting pattern of chromosomes is called a **karyotype.**

A trisomy occurs when the individual has three chromosomes instead of two chromosomes at one karyotype location. **Trisomy 21** (Down syndrome) is the most common autosomal trisomy in humans. Survival to adulthood is common. Characteristic facial features include an eyelid fold, a flat face, and a large fissured tongue. Some degree of mental retardation is common as is early-onset Alzheimer's disease. Sterility due to sexual underdevelopment may be present.

Observation: Sex Chromosome Anomalies

A female with **Turner syndrome** (XO) has only one sex chromosome, an X chromosome; the O signifies the absence of the second sex chromosome. Because the ovaries never become functional, these females do not undergo puberty or menstruation, and their breasts do not develop. Generally, females with Turner syndrome have a short build, folds of skin on the back of the neck, difficulty recognizing various spatial patterns, and normal intelligence. With hormone supplements, they can lead fairly normal lives.

When an egg having two X chromosomes is fertilized by an X-bearing sperm, an individual with **poly-X syndrome** results. The body cells have three X chromosomes and therefore 47 chromosomes. Although they tend to have learning disabilities, poly-X females have no apparent physical anomalies, and many are fertile and have children with a normal chromosome count.

Turner syndrome   XO

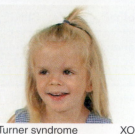
Poly-X syndrome   XXX

Klinefelter syndrome   XXY

Jacob syndrome   XYY

When an egg having two X chromosomes is fertilized by a Y-bearing sperm, a male with **Klinefelter syndrome** results. This individual is male in general appearance, but the testes are underdeveloped, and the breasts may be enlarged. The limbs of XXY males tend to be longer than average, muscular development is poor, body hair is sparse, and many XXY males have learning disabilities.

Jacob syndrome occurs in males who are usually taller than average, suffer from persistent acne, and tend to have speech and reading problems. At one time, it was suggested that XYY males were likely to be criminally aggressive, but the incidence of such behavior has been shown to be no greater than that among normal XY males.

Label each karyotype in Figure 21.6 as one of syndromes just discussed.

Figure 21.6   Sex chromosome anomalies.

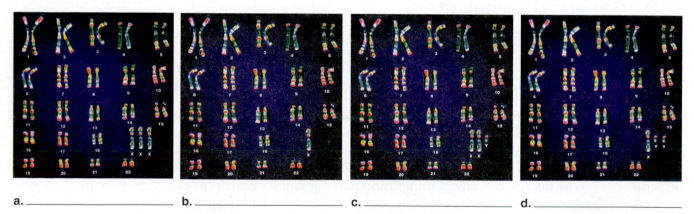

a. _____   b. _____   c. _____   d. _____

Determining the Pedigree

A pedigree shows the inheritance of a genetic disorder within a family and can help determine the inheritance pattern and whether any particular individual has an allele for that disorder. Then a Punnett square can be done to determine the chances of a couple producing an affected child.

The symbols used to indicate normal and affected males and females, reproductive partners, and siblings in a pedigree are shown in Figure 21.7.

For example, suppose you wanted to determine the inheritance pattern for straight hairline and you knew which members of a generational family had the trait (Fig. 21.8a).

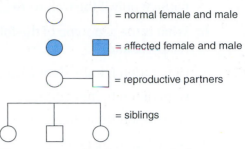

= normal female and male

= affected female and male

= reproductive partners

= siblings

Figure 21.7   Pedigree symbols.

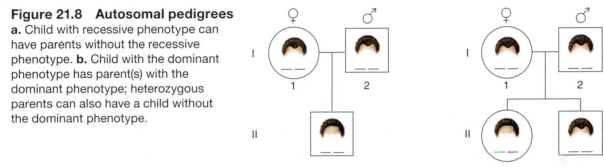

Figure 21.8   Autosomal pedigrees
a. Child with recessive phenotype can have parents without the recessive phenotype. **b.** Child with the dominant phenotype has parent(s) with the dominant phenotype; heterozygous parents can also have a child without the dominant phenotype.

a. Straight hairline is recessive.

b. Widow's peak is dominant.

A pedigree allows you to determine that straight hairline is autosomal recessive because two parents without this phenotype have a child with the phenotype. This can happen only if the parents are heterozygous and straight hairline is recessive. Similarly, a pedigree allows you to determine that widow's peak is autosomal dominant (Fig. 21.8b): a child with this phenotype has at least one parent with the dominant phenotype, but again, heterozygous parents can produce a child without widow's peak. *Give each person in Figure 21.8a and b a genotype.*

Not shown is an X-linked recessive pedigree. An X-linked recessive phenotypes occurs mainly in males and it skips a generation because a female who inherits a recessive allele for the condition from her father may have a son with the condition.

Pedigree Analysis

For each of the following pedigrees, determine how a genetic disorder is inherited. Is the pattern autosomal dominant, autosomal recessive, or X-linked recessive? Also decide the genotype of particular individuals in the pedigree.

1. Study the following pedigree:

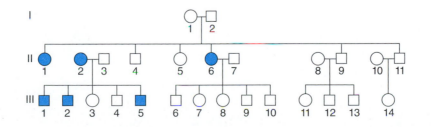

a. Notice that neither of the original parents is affected, but several children are affected. This could happen only if the trait were _____.

b. What is the genotype of the following individuals? Use *A* for the dominant allele and *a* for the recessive allele.

Generation I, individual 1: _____

Generation II, individual 1: _____

Generation III, individual 8: _____

2. Study the following pedigree:

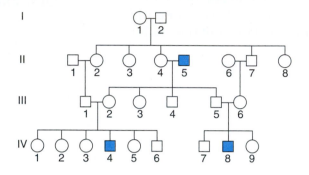

a. Notice that only males are affected. This could happen only if the trait were _____.

b. What is the genotype of the following individuals?

Generation I, individual 1: _____

Generation II, individual 8: _____

Generation III, individual 1: _____

Construction of a Pedigree

You are a genetic counselor who has been given the following information from which you will construct a pedigree.

1. Your data: Henry has a double row of eyelashes, which is a dominant trait. Both his maternal grandfather and his mother have double eyelashes. Their spouses are normal. Henry is married to Isabella and their first child, Polly, has normal eyelashes. The couple wants to know the chances of any child having a double row of eyelashes.

2. *Construct two pedigrees with symbols only for the underlined persons in #1. The pedigrees start with the maternal grandfather and grandmother and end with Polly.*

        Pedigree 1   **Pedigree 2**

3. Pedigree 1: Try out a pattern of autosomal dominant inheritance by assigning appropriate genotypes for an autosomal dominant pattern of inheritance to each person in this pedigree. Pedigree 2: Try out a pattern of X-linked dominant inheritance by assigning appropriate genotypes for this pattern of inheritance to each person in your pedigree. Which pattern is correct? _____

4. What is your key for this trait?

    Key: _____ normal eyelashes   _____ double row of eyelashes

5. Use correct genotypes to show a cross between Henry and Isabella and from experience with crosses, state the expected phenotypic ratio among the offspring:

    Cross   Henry   Isabella   **Phenotypic ratio:**

    _____ × _____   _____

6. What are the percentage chances that Henry and Isabella will have a child with double eyelashes? _____

Laboratory Review 21

_____   1. Mary's father does not have freckles, but Mary does. What genotypes could Mary's mother have?

_____   2. A cross results in a 3:1 phenotypic ratio. What are the genotypes of the parents?

_____   3. Parents who are $AAbb \times aaBB$ will have children with what genotype?

_____   4. The alleles of which parent, regardless of the phenotype, determine color blindness in a son?

_____   5. What is the genotype of a female with hemophilia?

_____   6. If a person has type AB blood, what is their genotype?

_____   7. Mary has blood type A and Don has blood type B; can they be the parents of a child with type O blood?

_____   8. What term refers to paired chromosomes arranged by size and shape?

_____   9. What pair of chromosomes is not homologous in a normal male karyotype?

_____   10. Name a common autosomal trisomy.

_____   11. What syndrome is inherited when an egg carrying two X chromosomes is fertilized by a sperm carrying one Y chromosome?

_____   12. What does a geneticist construct to show the inheritance pattern of a genetic disorder within a family?

_____   13. If the parents are not affected and a child is affected, what is the inheritance pattern?

_____   14. If only males are affected in a pedigree, what is the likely inheritance pattern for the trait?

Thought Questions

15. What inheritance pattern in a pedigree would allow you to decide that a characteristic is X-linked?

16. Why are X-linked disorders such as hemophilia generally more common in males?

17. Bob has attached earlobes and both his parents have unattached earlobes. Sally, Bob's wife, has unattached earlobes. Sally's mother has attached earlobes and her father has unattached earlobes. Sally's brother has attached earlobes.

  a. Identify the genotypes of all involved.

  b. What is the probability that Bob and Sally's first child will have unattached earlobes?

DNA Biology and Technology

Introduction

This laboratory pertains to molecular genetics and biotechnology. Molecular genetics is the study of the structure and function of **DNA (deoxyribonucleic acid),** the genetic material. **Biotechnology** is the manipulation of DNA for the benefit of human beings and other organisms.

First we will study the structure of DNA and see how that structure facilitates DNA replication in the nucleus of cells. DNA replicates prior to cell division; following cell division, each daughter cell has a complete copy of the genetic material. DNA replication is also needed to pass genetic material from one generation to the next. You may have an opportunity to use models to see how replication occurs.

Then we will study the structure of **RNA (ribonucleic acid)** and how it differs from that of DNA, before examining how DNA, with the help of RNA, specifies protein synthesis. The linear construction of DNA, in which nucleotide follows nucleotide, is paralleled by the linear construction of the primary structure of protein, in which amino acid follows amino acid. Essentially, we will see that the sequence of nucleotides in DNA codes for the sequence of amino acids in a protein. We will also review the role of three types of RNA in protein synthesis. DNA's code is passed to messenger RNA (mRNA), which moves to the ribosomes containing ribosomal RNA (rRNA). Transfer RNA (tRNA) brings the amino acids to the ribosomes, and they become sequenced in the order directed by mRNA.

We now understand that a mutated gene has an altered DNA base sequence, which can lead to a genetic disorder. You will have an opportunity to carry out a laboratory procedure that detects whether an individual is normal, has sickle cell disease, or is a carrier.

22.1 DNA Structure and Replication

The structure of DNA lends itself to **replication,** the process that makes a copy of a DNA molecule. DNA replication is a necessary part of chromosome duplication, which precedes cell division. It also makes possible the passage of DNA from one generation to the next.

DNA Structure

DNA is a polymer of nucleotide monomers (Fig. 22.1). Each nucleotide is composed of three molecules: deoxyribose (a 5-carbon sugar), a phosphate, and a nitrogen-containing base.

Figure 22.1   Overview of DNA structure.
Diagram of DNA double helix shows that the molecule resembles a twisted ladder. Sugar-phosphate backbones make up the sides of the ladder, and hydrogen-bonded bases make up the rungs of the ladder. Complementary base pairing dictates that A is bonded to T and G is bonded to C and vice versa.

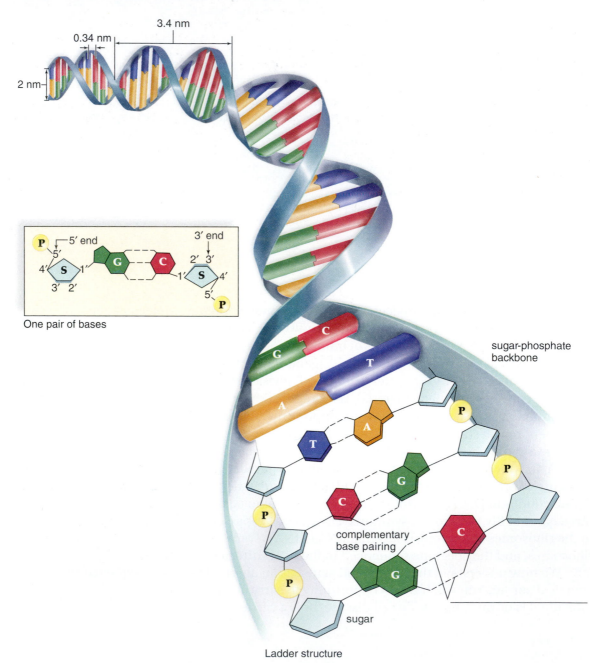

Ladder structure

1. A boxed nucleotide pair is shown in Figure 22.1. If you are working with a kit, draw a representation of one of your nucleotides here. *Label phosphate, base pair, and deoxyribose in your drawing.*

2. Notice the four types of bases: cytosine (C), thymine (T), adenine (A), and guanine (G). What is the color of each of the four types of bases in Figure 22.1? In your kit? Complete Table 22.1 by writing in the colors of the bases.

Table 22.1 Base Colors		
	In Figure 22.1	**In Your Kit**
Cytosine		
Thymine		
Adenine		
Guanine		

3. Using Figure 22.1 as a guide, join several nucleotides together. Observe the entire DNA molecule. What types of molecules make up the backbone (uprights of ladder) of DNA (Fig. 22.1)? _____ and _____ In the backbone, the phosphate of one nucleotide is bonded to a sugar of the next nucleotide.

4. Using Figure 22.1 as a guide, join the bases together with hydrogen bonds. *Label a hydrogen bond in Figure 22.1.* Dashes are used to represent hydrogen bonds in Figure 22.1 because hydrogen bonds are (strong or weak) _____.

5. Notice in Figure 22.1 and in your model that the base A is always paired with the base _____, and the base C is always paired with the base _____. This is called complementary base pairing.

6. In Figure 22.1, what molecules make up the rungs of the ladder? _____

7. Each half of the DNA molecule is a DNA strand. Why is DNA also called a double helix (Fig. 22.1)?

    _____

DNA Replication

During replication, the DNA molecule is duplicated so that there are two identical DNA molecules. We will see that complementary base pairing makes replication possible.

Observation: DNA Replication

1. Before replication begins, DNA is unzipped. Using Figure 22.2a as a guide, break apart your two DNA strands. What bonds are broken in order to unzip the DNA strands? _____

2. Using Figure 22.2b as a guide, attach new complementary nucleotides to each strand using complementary base pairing.

3. Show that you understand complementary base pairing by completing Table 22.2.

4. You now have two DNA molecules (Fig. 22.2c). Are your molecules identical?

   _____

5. Because of complementary base pairing, each new double helix is composed of an _____ strand and a _____ strand. *Write old or new in 1–10, Figure 22.2a, b, and c. Conservative means to save something from the past.* Why is DNA replication called semiconservative?

   _____

   _____

   _____

Figure 22.2   DNA replication.
Use of the ladder configuration better illustrates how replication takes place. **a.** The parental DNA molecule. **b.** The "old" strands of the parental DNA molecule have separated. New complementary nucleotides available in the cell are pairing with those of each old strand. **c.** Replication is complete.

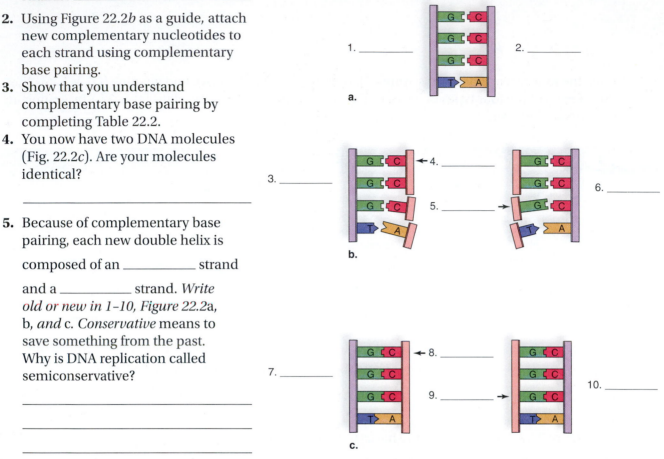

6. Genetic material has to be inherited from cell to cell and organism to organism. Consider that because of DNA replication, a chromosome is composed of two chromatids and each chromatid is a DNA double helix. The chromatids separate during cell division so that each daughter cell receives a copy of each chromosome. Does replication provide a means for passing DNA from cell to cell and organism to organism? _____ Explain. _____

_____

Table 22.2	DNA Replication
Old strand	G G G T T C C A T T A A A T T C C A G A A A T C A T A
New strand	

22.2 RNA Structure

Like DNA, RNA is a polymer of nucleotides (Fig. 22.3). In an RNA nucleotide, the sugar ribose is attached to a phosphate molecule and to a nitrogen-containing base, C, U, A, or G. In RNA, the base uracil replaces thymine as one of the bases. RNA is single stranded, whereas DNA is double stranded.

Figure 22.3 Overview of RNA structure.
RNA is a single strand of nucleotides. *Label the boxed nucleotide as directed in the next Observation.*

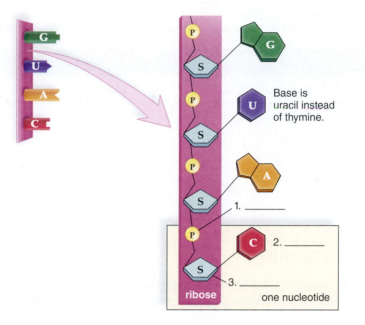

1. Describe the backbone of an RNA molecule. _____

2. Where are the bases located in an RNA molecule? _____

3. Complete Table 22.3 to show the complementary DNA bases for the RNA bases.

Table 22.3 DNA and RNA Bases			
RNA Bases C	U	A	G
DNA Bases			

1. If you are using a kit, draw a nucleotide for the construction of mRNA. *Label the ribose (the sugar in RNA), the phosphate, and the base in your drawing and in 1–3, Figure 22.3.*

2. Complete Table 22.4 by writing in the colors of the bases for Figure 22.3 and for your kit.

Table 22.4 Base Colors		
	In Figure 22.3	**In Your Kit**
Cytosine		
Uracil		
Adenine		
Guanine		

3. The base uracil substitutes for the base thymine in RNA. Complete Table 22.5 to show the several other ways RNA differs from DNA.

Table 22.5 DNA Structure Compared with RNA Structure		
	DNA	**RNA**
Sugar	Deoxyribose	
Bases	Adenine, guanine, thymine, cytosine	
Strands	Double stranded with base pairing	
Helix	Yes	

22.3 DNA and Protein Synthesis

Protein synthesis requires the processes of transcription and translation. During **transcription,** which takes place in the nucleus, an RNA molecule called **messenger RNA (mRNA)** is made complementary to one of the DNA strands. This mRNA leaves the nucleus and goes to the ribosomes in the cytoplasm. Ribosomes are composed of **ribosomal RNA (rRNA)** and proteins in two subunits.

During **translation,** RNA molecules called **transfer RNA (tRNA)** bring amino acids to the ribosome, and they join in the order prescribed by mRNA. This sequence of amino acids was originally specified by DNA. This is the information that DNA, the genetic material, stores.

What is the role of each of these participants in protein synthesis?

DNA _____

mRNA _____

tRNA _____

Transcription

During transcription, complementary RNA is made from a DNA template (Fig. 22.4). A portion of DNA unwinds and unzips at the point of attachment of the enzyme RNA polymerase. A strand of mRNA is produced when complementary nucleotides join in the order dictated by the sequence of bases in DNA. Transcription occurs in the nucleus, and the mRNA passes out of the nucleus to enter the cytoplasm.

Label Figure 22.4. For number 1, note the name of the enzyme that carries out mRNA synthesis. For number 2, note the name of this molecule.

Observation: Transcription

1. If you are using a kit, unzip your DNA model so that only one strand remains. This strand is the **sense strand,** the strand that is transcribed.
2. Using Figure 22.4 as a guide, construct a messenger RNA (mRNA) molecule by first lining up RNA nucleotides complementary to the sense strand of your DNA molecule. Join the RNA nucleotides together to form mRNA.
3. A portion of DNA has the sequence of bases shown in Table 22.6. *Complete Table 22.6 to show the sequence of bases in mRNA.*
4. If you are using a kit, unzip mRNA transcript from the DNA. Locate the end of the strand that will move to

the _____ in the cytoplasm.

Figure 22.4 Messenger RNA (mRNA).
Messenger RNA complementary to a section of DNA forms during transcription.

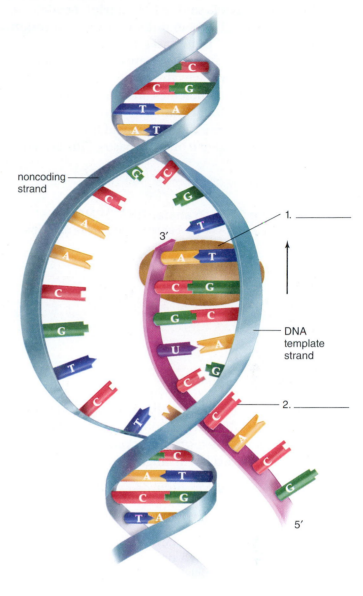

noncoding strand

1. _____

3′

DNA template strand

2. _____

5′

Table 22.6	Transcription
DNA	T A C A C G A G C A A C T A A C A T
mRNA	

Translation

DNA specifies the sequence of amino acids in a polypeptide because every three bases code for an amino acid. Therefore, DNA is said to have a **triplet code.** The bases in mRNA are complementary to the bases in DNA. Every three bases in mRNA are called a **codon.** One codon of mRNA represents one amino acid. Thus, the sequence of DNA bases serves as the blueprint for the sequence of amino acids assembled to make a protein. The correct sequence of amino acids in a polypeptide is the message that mRNA carries.

Messenger RNA leaves the nucleus and proceeds to the ribosomes, where protein synthesis occurs. Transfer RNA (tRNA) molecules are so named because they transfer amino acids to the ribosomes. Each RNA has a specific tRNA amino acid at one end and a matching **anticodon** at the other end (Fig. 22.5). *Label Figure 22.5,* where the amino acid is represented as a colored ball, the tRNA is green, and the anticodon is the sequence of three bases. (The anticodon is complementary to the mRNA codon.)

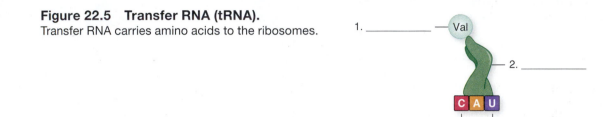

Figure 22.5 Transfer RNA (tRNA).
Transfer RNA carries amino acids to the ribosomes.

1. _____ — Val

2. _____

3. _____

Observation: Translation

1. Figure 22.6 shows seven tRNA–amino acid complexes. Every amino acid has a name; in the figure, only the first three letters of the name are inside the ball. *Using the mRNA sequence given in Table 22.7, number the tRNA–amino acid complexes in the order they will come to the ribosome.*

2. If you are using a kit, arrange your tRNA–amino acid complexes in the order consistent with

   Table 22.7. *Complete Table 22.7.* Why are the codons and anticodons in groups of three? _____

   _____

Figure 22.6 Transfer RNA diversity.
Each type of tRNA carries only one particular amino acid, designated here by the first three letters of its name.

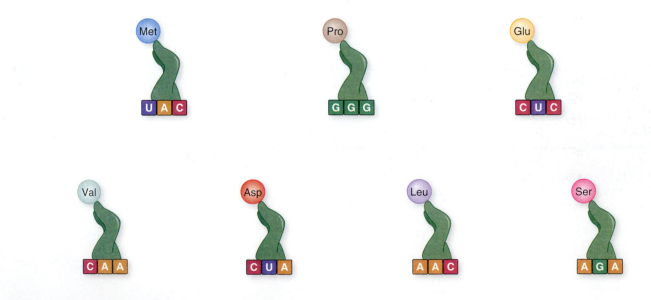

Table 22.7 Translation

mRNA codons	AUG	CCC	GAG	GUU	GAU	UUG	UCU
tRNA anticodons							
Amino acid*							

*Use three letters only. See Table 22.8 for the full names of these amino acids.

Table 22.8 Names of Amino Acids

Abbreviation	Name
Met	methionine
Pro	proline
Asp	aspartate
Val	valine
Glu	glutamate
Leu	leucine
Ser	serine

3. Figure 22.7 shows the manner in which the polypeptide grows. A ribosome has room for two tRNA complexes at a time. As the first tRNA leaves, it passes its amino acid or peptide to the next tRNA-amino acid complex. Then the ribosome moves forward, making room for the next tRNA–amino acid. This sequence of events occurs over and over until the entire polypeptide is borne by the last tRNA to come to the ribosome. Then a release factor releases the polypeptide chain from the ribosome. *In Figure 22.7, label the ribosome, the mRNA, and the peptide.*

Figure 22.7 Protein synthesis.

1. A ribosome has a binding site for two tRNA–amino acid complexes. **2.** Before a tRNA leaves, an RNA passes its attached peptide to a newly arrived tRNA–amino acid complex. **3.** The ribosome moves forward, and the next tRNA–amino acid complex arrives.

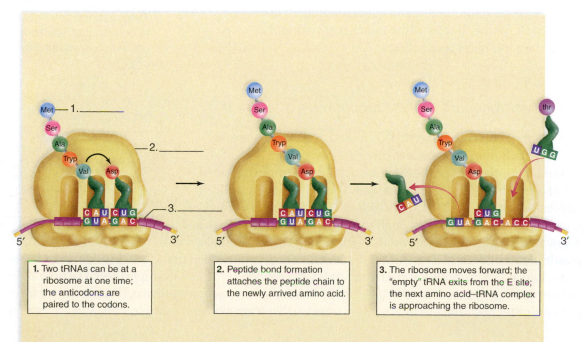

1. Two tRNAs can be at a ribosome at one time; the anticodons are paired to the codons.

2. Peptide bond formation attaches the peptide chain to the newly arrived amino acid.

3. The ribosome moves forward; the "empty" tRNA exits from the E site; the next amino acid–tRNA complex is approaching the ribosome.

22.4 Isolation of DNA and Biotechnology

In the first Experimental Procedure in this section, you will isolate DNA from the cells of an organism using a modified procedure like that used worldwide in biotechnology laboratories. In the second Experimental Procedure you will perform gel electrophoresis using the DNA as instructed by your instructor.

Experimental Procedure: Isolating DNA

You will extract DNA from an onion filtrate that contains DNA in solution. To prepare the filtrate, your instructor homogenized an onion with a detergent. The detergent emulsifies and forms complexes with the lipids and proteins of the plasma membrane, causing them to precipitate out of solution. Cell contents, including DNA, become suspended in solution. The cellular mixture is then filtered to produce the filtrate that contains DNA. The DNA molecule is easily degraded (broken down), so it is important to wear gloves and to follow all instructions closely. Handle glassware carefully to prevent nucleases in your skin from contaminating the glassware.

1. Obtain a large, clean test tube, and place it in an ice bath. Let stand for a few minutes to make sure the test tube is cold. Everything must be kept very cold.
2. Obtain approximately 4 mL of the onion filtrate, and add it to your test tube while keeping the tube in the ice bath.
3. Obtain and add 2 mL of cold meat tenderizer solution to the solution in the test tube, and mix the contents slightly with a stirring rod or Pasteur pipette. Let stand for 10 minutes.
4. Use a graduated cylinder or pipette to slowly add an equal volume (approximately 6 mL) of ice-cold 95% ethanol along the inside of the test tube. Keep the tube in the ice bath, and tilt it to a 45° angle. You should see a distinct layer of ethanol over the white filtrate. Let the tube sit for 2 to 3 minutes.
5. Insert a glass rod or a Pasteur pipette into the tube until it reaches the bottom of the tube. Gently swirl the glass rod or pipette, always in the same direction. (You are not trying to mix the two layers; you are trying to wind the DNA onto the glass rod like cotton candy.) This process is called "spooling" the DNA. The stringy, slightly gelatinous material that attaches to the pipette is DNA (Fig. 22.8). If the DNA has been damaged, it will still precipitate, but as white flakes that cannot be collected on the glass rod.

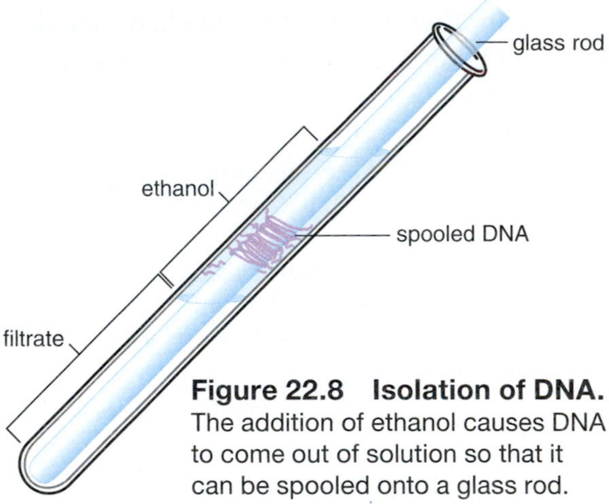

Figure 22.8   Isolation of DNA. The addition of ethanol causes DNA to come out of solution so that it can be spooled onto a glass rod.

Experimental Procedure: Gel Electrophoresis

During gel electrophoresis, charged DNA molecules migrate across a span of gel (gelatinous slab) because they are placed in a powerful electrical field. In the present experiment, each DNA sample is placed in a small depression in the gel called a well. The gel is placed in a powerful electrical field. The electricity causes DNA fragments, which are negatively charged, to move through the gel according to their size.

> ⚠️ **Gel electrophoresis** Students should wear personal protective equipment: safety goggles and smocks or aprons while loading gels and during electrophoresis and protective gloves while staining.

Almost all DNA gel electrophoresis is carried out using horizontal gel slabs (Fig. 22.9). First, the gel is poured onto a plastic plate, and the wells are formed. After the samples are added to the wells, the gel and the plastic plate are put into an electrophoresis chamber, and buffer is added. The DNA samples begin to migrate after the electrical current is turned on. With staining, the DNA fragments appear as a series of bands spread from one end of the gel to the other according to their size because smaller fragments move faster than larger fragments.

Figure 22.9 Equipment and procedure for gel electrophoresis.

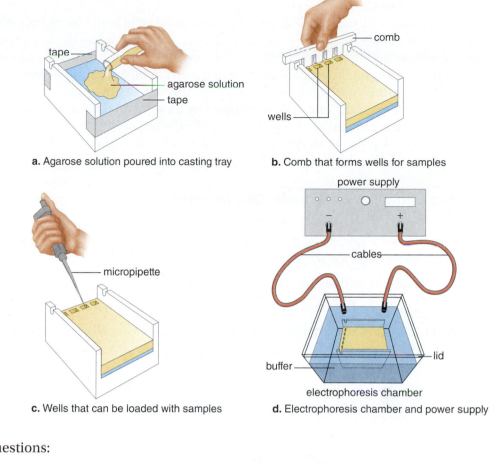

a. Agarose solution poured into casting tray

b. Comb that forms wells for samples

c. Wells that can be loaded with samples

d. Electrophoresis chamber and power supply

Answer the following questions:

a. What is biotechnology? (See page 303.) _____

_____

_____

b. Speculate how the ability to isolate DNA and run gel electrophoresis of DNA relates to biotechnology.

_____

_____

_____

c. Name a biotechnology product someone you know is now using or taking as a medicine. (Examples include insulin, vaccines, ingredients of ice cream, cheese, make-up, detergents, dye for blue jeans, etc.) _____

22.5 Detecting Genetic Disorders

The base sequence of DNA in all the chromosomes is an organism's genome. Now that the Human Genome Project is finished, we know the usual order of all the 3.6 billion nucleotide bases in the human genome. Someday it will be possible to sequence anyone's genome within a relatively short time, and thereby determine what particular base sequence alterations signify that he or she has a disorder or will have one in the future. In this laboratory, you will study the alteration in base sequence that causes a person to have sickle-cell disease.

In persons with sickle-cell disease, the red blood cells aren't biconcave disks like normal red blood cells—they are sickle-shaped. Sickle-shaped cells can't pass along narrow capillary passageways. They clog the vessels and break down, causing the person to suffer from poor circulation, anemia, and poor resistance to infection. Internal hemorrhaging leads to further complications, such as jaundice, episodic pain in the abdomen and joints, and damage to internal organs.

Sickle-shaped red blood cells are caused by an abnormal hemoglobin (Hb^S). Individuals with the Hb^AHb^A genotype are normal; those with the Hb^SHb^S genotype have sickle-cell disease, and those with the Hb^AHb^S have sickle-cell trait. Persons with sickle-cell trait do not usually have sickle-shaped cells unless they experience dehydration or mild oxygen deprivation.

Genetic Sequence for Sickle-Cell Disease

Examine Figure 22.10, *a* and *b,* which shows the DNA base sequence, the mRNA codons, and the amino acid sequence for a portion of the gene for Hb^A and the same portion for Hb^S. Today many genetic disorders can be detected by genomic sequencing.

1. In what one base does Hb^A differ from Hb^S? Hb^A _____ Hb^S _____

2. What are the codons that contain this base? Hb^A _____ Hb^S _____

3. What is the amino acid difference? Hb^A _____ Hb^S _____

Figure 22.10   Sickle cell disease.
a. When red blood cells are normal, the base sequence (in one location) for Hb^A alleles is CTC. **b.** In sickle cell disease at these locations, it is CAC.

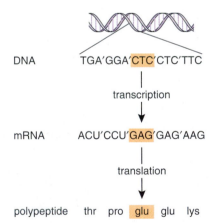
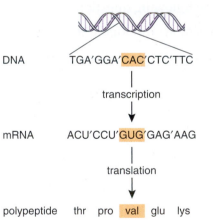

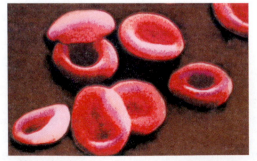
a. Normal red blood cells.

b. Sickle-shaped red blood cell.

This amino acid difference causes the polypeptide chain in sickle-cell hemoglobin to pile up as firm rods that push against the plasma membrane and deform the red blood cell into a sickle shape:

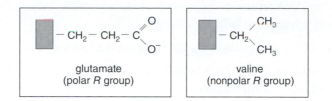

glutamate
(polar R group)

valine
(nonpolar R group)

Detection of Sickle-Cell Disease by Gel Electrophoresis

Three samples of hemoglobin have been subjected to protein gel electrophoresis. Protein gel electrophoresis is carried out in the in the same manner as DNA gel electrophoresis (see Fig. 22.9) except the gel has a different composition.

1. Sickle-cell hemoglobin (Hb^S) migrates slower toward the positive pole than normal hemoglobin (Hb^A) because the amino acid valine has no polar R groups, whereas the amino acid glutamate does have a polar R group.

2. In Figure 22.11, which lane contains only Hb^S, signifying that the individual is Hb^SHb^S?

   _____

3. Which lane contains only Hb^A, signifying that the individual is Hb^AHb^A? _____

4. Which lane contains both Hb^S and Hb^A, signifying that the individual is Hb^AHb^S?

   _____

Figure 22.11   Gel electrophoresis of hemoglobins.

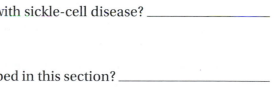

+Pole

−Pole

Lane 1   Lane 2   Lane 3

Detection by Genomic Sequencing

You are a genetic counselor. A young couple seeks your advice because sickle-cell disease occurs among the family members of each. You order DNA base sequencing to be done. The results come back that at one of the loci for normal hemoglobin, each has the abnormal sequence CAC instead of CTC. The other locus is normal. What are the chances that this couple will have a child with sickle-cell disease? _____

Conclusion: Detecting Genetic Disorders

- What two methods of detecting sickle-cell disease were described in this section? _____

  _____

- Which method is more direct and probably requires more expensive equipment to do? _____

  _____

- Which method probably preceded the other method as a means to detect sickle-cell disease? _____

  _____

Laboratory Review 22

_____ 1. The DNA structure resembles a twisted ladder. What molecules make up the uprights of the ladder?

_____ 2. In DNA, what makes up the rungs of the ladder?

_____ 3. Do the two DNA double helices following DNA replication have the same, or a different, composition?

_____ 4. If DNA has 20% of adenine bases, what would be the percentage of thymine?

_____ 5. If the codons are AUG, CGC, and UAC, what are the anticodons?

_____ 6. In what part of a cell does transcription occur?

_____ 7. During transcription, what type of RNA is formed?

_____ 8. In what part of the cell does translation occur?

_____ 9. During translation, what type of RNA carries amino acids to the ribosomes?

_____ 10. What type molecule has to be stripped from DNA before it will precipitate out of solution?

_____ 11. During gel electrophoresis, do DNA sequences with negative polar or nonpolar groups travel more quickly toward the positive pole?

_____ 12. What procedure allows investigations to determine the present or future genetic disorders of an individual?

_____ 13. When a person has sickle cell disease, what type molecule is not functioning in the usual manner?

Thought Questions

14. In general, describe the relationship between the DNA a person inherits and the proteins that are in the person's cells.

15. What role does mRNA play in transcription and translation?

16. Below is a sequence of bases associated with the template DNA strand:
TAC CCC GAG CTT

   a. Identify the sequence of bases in the mRNA resulting from the transcription of the above DNA sequence.

   b. Identify the sequence of bases in the tRNA anticodon that will bind with the first codon on the mRNA identified above.

Evidences of Evolution

Learning Outcomes

23.1 Evidence from the Fossil Record
- Use the geologic timescale to trace the evolution of life in broad outline.   318–20
- Describe several types of fossils and explain how fossils help establish the sequence in the evolution of life.   320–23
- Explain how scientists use fossils to establish that organisms are related by common descent.   324

23.2 Evidence from Comparative Anatomy
- Explain how comparative anatomy provides evidence that humans are related to other groups of vertebrates and also other groups of primates.   324–26
- Compare the human skeleton with the chimpanzee skeleton, and explain the differences on the basis of their different ways of life.   326–29
- Compare hominid skulls and hypothesize a possible evolutionary sequence.   329–31

23.3 Molecular Evidence
- Explain how molecular evidence aids the study of how humans are related to all other groups of organisms on Earth.   331–32
- Explain how molecular evidence also helps show the degree to which humans are related to other vertebrates and primates.   332–33

Introduction

Evolution is the process by which organisms are related by **common descent:** all organisms can trace their ancestry to the first cells. The process of evolution is amazingly simple: a group of organisms change over time because the members of a group most suited to the natural environment have more offspring than the others in the group. So, for example, among bacteria those which can withstand an antibiotic leave more offspring and with time the entire group of bacteria becomes resistant to the antibiotic. Reproduction and therefore evolution has been going on since the first cells appeared on Earth and through studying (1) the fossil record, (2) comparative anatomy, both anatomical and embryological, and (3) molecular evidence, science is able to show that all organisms are related to one another.

Students digging for fossils.

23.1 Evidence from the Fossil Record

The geologic timescale, which was developed by both geologists and paleontologists, depicts the history of life based on the fossil record (Table 23.1). A **fossil** is any evidence of the existence of an organism in ancient times as opposed to modern times. Paleontologists specialize in removing fossils from the Earth's crust (see page 317). In this section, we will study the geologic timescale and then examine some fossils.

Geologic Timescale

Divisions of the Timescale

Notice that the timescale divides the history of Earth into eras, then periods, and then epochs. The four eras span the greatest amounts of time, and the epochs are the shortest time frames. Notice that only the periods of the Cenozoic era are divided into epochs, meaning that more attention is given to the evolution of primates and flowering plants than to the earlier evolving organisms. List the four eras in the timescale starting with Precambrian time: _____

1. Using the geologic timescale, you can trace the history of life by beginning with Precambrian time at the bottom of the timescale. The timescale indicates that the first cells (the prokaryotes) arose some 3,500 MYA. The prokaryotes evolved before any other group. Why do you read the timescale starting at the bottom? _____

2. The Precambrian time was very long, lasting from the time the Earth first formed until 542 MYA. The fossil record during the Precambrian time is meager, but the fossil record from the Cambrian period onward is rich (for reasons still being determined). This helps explain why the timescale usually does not show any periods until the Cambrian period of the Paleozoic era. You can also use the timescale to check when certain groups evolved and/or flourished.
   Example: During the Ordovician period, the nonvascular plants appear on land, and the first jawless and jawed fishes appear in the seas.

   During the _____ era and the _____ period, the first flowering plants appear. How many million years ago was this? _____

3. On the timescale, note the Carboniferous period. During this period great swamp forests covered the land. These are also called coal-forming forests because, with time, they became the coal we burn today. How do you know that the plants in this forest were not flowering trees as most of our trees are today? _____

   What type of animal was diversifying at this time? _____

4. You should associate the Cenozoic era with the evolution of humans. Among mammals, humans are primates. During what period and epoch did primates appear? _____

   Among primates, humans are hominids. During what period and epoch did hominids appear? _____ _____ The scientific name for humans is *Homo sapiens*.

   What period and epoch is the age of *Homo sapiens*? _____

Dating within the Timescale

The timescale provides both relative dates and absolute dates. When you say, for example, "Flowering plants evolved during the Jurassic period," you are using relative time, because flowering plants evolved earlier or later than groups in other periods. If you use the dates that are given in millions of years (MYA), you are using absolute time. Absolute dates are usually obtained by measuring the amount of a radioactive isotope in the rocks surrounding the fossils. Why wouldn't you expect to find human fossils and dinosaur fossils together in rocks dated similarly? _____

Table 23.1   The Geologic Timescale: Major Divisions of Geological Time and Some of the Major Evolutionary Events That Occurred

Era	Period	Epoch	Millions of Years Ago	Plant Life	Animal Life
Cenozoic*	Quaternary	Holocene	0.01–present	Human influence on plant life	Age of *Homo sapiens*
				Significant Mammalian Extinction	
		Pleistocene	1.8–0.01	Herbaceous plants spread and diversify.	Presence of Ice Age mammals Modern humans appear.
	Tertiary	Pliocene	5.3–1.8	Herbaceous angiosperms flourish.	First hominids appear.
		Miocene	23–5.3	Grasslands spread as forests contract.	Apelike mammals and grazing mammals flourish; insects flourish.
		Oligocene	33.9–23	Many modern families of flowering plants evolve.	Browsing mammals and monkeylike primates appear.
		Eocene	55.8–33.9	Subtropical forests with heavy rainfall thrive.	All modern orders of mammals are represented.
		Paleocene	65.5–55.8	Flowering plants continue to diversify.	Primitive primates, herbivores, carnivores, and insectivores appear.
Mesozoic				**Mass Extinction: Dinosaurs and Most Reptiles**	
	Cretaceous		145.5–65.5	Flowering plants diversify; conifers persist.	Placental mammals appear; modern insect groups appear.
	Jurassic		199.6–145.5	Flowering plants appear.	Dinosaurs flourish; birds appear.
				Mass Extinction	
	Triassic		251–199.6	Forests of conifers and cycads dominate.	First mammals appear; first dinosaurs appear; corals and molluscs dominate seas.
Paleozoic				**Mass Extinction**	
	Permian		299–251	Gymnosperms diversify.	Reptiles diversify; amphibians decline.
	Carboniferous		359.2–299	Age of great coal-forming forests: Ferns, club mosses, and horsetails flourish.	Amphibians diversify; first reptiles appear; first great radiation of insects.
				Mass Extinction	
	Devonian		416–359.2	First seed plants appear. Seedless vascular plants diversify.	Jawed fishes diversify and dominate the seas; first insects and first amphibians appear.
	Silurian		443.7–416	Seedless vascular plants appear.	First jawed fishes appear.
				Mass Extinction	
	Ordovician		488.3–443.7	Nonvascular land plants appear. Marine algae flourish.	Invertebrates spread and diversify; jawless fishes (first vertebrates) appear.
	Cambrian		542–488.3	First plants appear on land. Marine algae flourish.	All invertebrate phyla present; first chordates appear.
Precambrian Time			600	Oldest soft-bodied invertebrate fossils	
			1,400–700	Protists evolve and diversify.	
			2,200	Oldest eukaryotic fossils	
			2,700	O_2 accumulates in atmosphere.	
			3,500	Oldest known fossils (prokaryotes)	
			4,600	Earth forms.	

*Many authorities divide the Cenozoic era into the Paleogene period (contains the Paleocene, Eocene, and Oligocene epochs) and the Neogene period (contains the Miocene, Pliocene, Pleistocene, and Holocene epochs).

Limitations of the Timescale

Because the timescale tells when various groups evolved and flourished, it might seem that evolution has been a series of events leading only from the first cells to humans. This is not the case; for example, prokaryotes (bacteria and archaea) never declined and are still the most abundant and successful organisms on Earth. Even today, they constitute up to 90% of the total weight of living things.

Then, too, the timescale lists mass extinctions, but it doesn't tell when specific groups became extinct. **Extinction** is the total disappearance of a species or a higher group; **mass extinction** occurs when a large number of species disappear in a few million years or less. For lack of space, the geologic timescale can't depict in detail what happened to the members of every group mentioned. Figure 23.1 does show how mass extinction affected a few groups of animals. Which of the animals shown in Figure 23.1 suffered the most during the **P-T extinction** (Permian-Triassic extinction)? _____

The **K-T extinction** occurred between the Cretaceous and the Tertiary periods. Which animals shown in Figure 23.1 became extinct during the K-T extinction? _____

Figure 23.1 shows only periods and no eras. *Fill in the eras on the lines provided in the figure.*

Figure 23.1 Mass extinctions.

Five significant mass extinctions and their effects on the abundance of certain forms of marine and terrestrial life. The width of the horizontal bars indicates the varying number of each life-form considered.

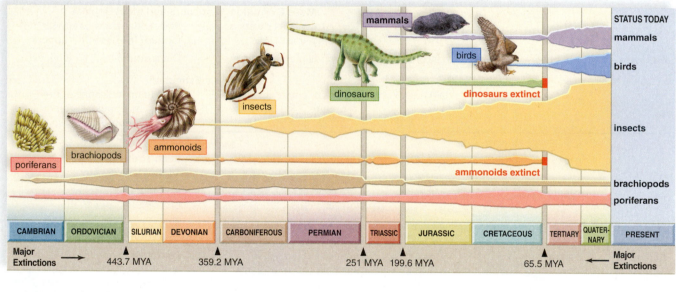

Era   _____   _____   _____

Fossils

The fossil record depends heavily on anatomical data to show the evolutionary changes that have occurred as one group gives rise to another group. Why would that be? _____

_____

Invertebrate Fossils

The **invertebrates** are animals without a backbone (see Fig. 23.2). Even the chordates, which is the phylum that contains the **vertebrates** (animals with a backbone), contain a couple of insignificant groups of invertebrates.

In order to familiarize yourself with possible invertebrate fossils, a few of the most common fossilized invertebrate groups are depicted in Figure 23.2.

Figure 23.2 Invertebrate fossils.

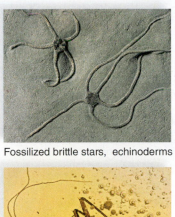
Fossilized brittle stars, echinoderms

Fossilized snails, mollusks

Fossilized trilobites, arthropods

Fossilized mantis in amber

ossilized ammonite, a mollusc

Fossilized pseudoscorpion, an arthropod

Observation: Invertebrate Fossils

1. Obtain a box of selected fossils. If the fossils are embedded in rocks, examine the rock until you have found the fossil. Fossils are embedded in rocks because the sediment that originally surrounded them hardened over time. Most fossils consist of hard parts such as shells, bones, or teeth because these parts are not consumed or decomposed over time. One possible reason the Cambrian might be rich in fossils is that organisms now had _____ whereas before they did not.

2. The kit you are using, or your instructor, will identify which of the fossils are invertebrate animals. Fill in the title of Table 23.2. List the names of these fossils in Table 23.2 and give a description of the fossil part that survived over time. Use the Geologic Timescale (Table 23.1) to tell the era and period they were most abundant; list the fossils from the latest (top) to earliest (bottom).

Table 23.2	Invertebrate Fossils	
Type of Fossil	**Era, Period**	**Description of Hard Part**

Observation: Vertebrate Fossils

The various groups of vertebrates are shown in Figure 23.3. Today it is generally agreed that birds are reptiles rather than being a separate group. That means that the major groups of vertebrates are (1) various types of fishes (jawless, jawed, cartilaginous and bony fishes), (2) amphibians such as frogs and salamanders, (3) reptiles such as lizards and crocodiles, and (4) mammals. There are many types of mammals from whales to mice to humans. The fossil record can rely on skeletal differences, such as limb structure, to tell if an animal is a mammal.

In order to familiarize yourself with possible vertebrate fossils, a few of the most common fossilized vertebrate groups are depicted in Figure 23.3. Which of the fossils available to you are vertebrates? _____

_____

Use Table 23.1 to associate each fossil with the particular era and period when this type animal was most abundant. Fill in Table 23.3 according to sequence of the time frames from the latest *(top)* to earliest *(bottom)*.

Fossilized bird, a reptile

Fossilized bony fish

Fossilized deerlike mammal

Fossilized frog, an amphibian

Fossilized duckbill dinosaur, a reptile

Fossilized snake, a reptile

Figure 23.3   Vertebrate fossils.

Table 23.3   Vertebrate Fossils

Type of Fossil	Era, Period	Description of Hard Part

Observation: Plant Fossils

See Figure 23.4, which shows the evolution of plants including the bryophytes, ferns and their allies, gymnosperms, and angiosperms. The fossil record for plants is not as good as that for invertebrates and vertebrates because plants have no hard parts that are easily fossilized. In order to familiarize yourself with possible plant fossils, a few of the most common fossilized plant groups are depicted in Figure 23.4.

ossilized ferns

Fossilized flower

Fossilized maple leaf

ossilized poplar leaf

Fossilized sassafras leaf

Fossilized early seed plant leaves

Figure 23.4   Plant fossils.

Plants that have no hard parts become fossils when their impressions are filled in by minerals. Use Table 23.1 to associate each fossilized plant in your kit with a particular era and period. Assume trees are flowering plants and associate them with the era and period when flowering plants were most abundant. Fill in Table 23.4 according in sequence from the latest *(top)* to earliest *(bottom)*.

Table 23.4 Plant Fossils		
Type of Fossil	**Era, Period**	**Description of Fossil**

Fossils and Molecular Evidence

Today we know that organisms evolve because genomes (all the genes an organism has) undergo mutations (permanent changes in their genes). If the change happens to be suited to the environment, then this member of the population has more offspring than other members. Using high-powered laboratory equipment, it is now possible to sequence the DNA bases (the sequence of A=T, G=C base pairs). Geneticists reason that the more closely related two groups of organisms, the fewer changes there will be in their DNA base sequences. Paleontologists, those who excavate fossils, and geneticists work closely to conclude which groups of organisms are more closely related to one another through evolution.

Summary

In this section, you studied the geologic timescale and various fossils represented in the record. The geologic timescale gives powerful evidence of evolution because

1. Fossils are _____.
2. Fossils can be arranged _____.
3. Younger fossils and not older fossils are more like _____.
4. In short, the fossil record shows that _____.

23.2   Evidence from Comparative Anatomy

In the study of evolutionary relationships, parts of organisms are said to be **homologous** if they exhibit similar basic structures and embryonic origins. If parts of organisms are similar in function only, they are said to be **analogous.** Only homologous structures indicate an evolutionary relationship and are used to classify organisms.

Comparison of Adult Vertebrate Forelimbs

The limbs of vertebrates are homologous structures (Fig. 23.5). Homologous structures share a basic pattern, although there may be specific differences. The similarity of homologous structures is explainable by descent from a common ancestor.

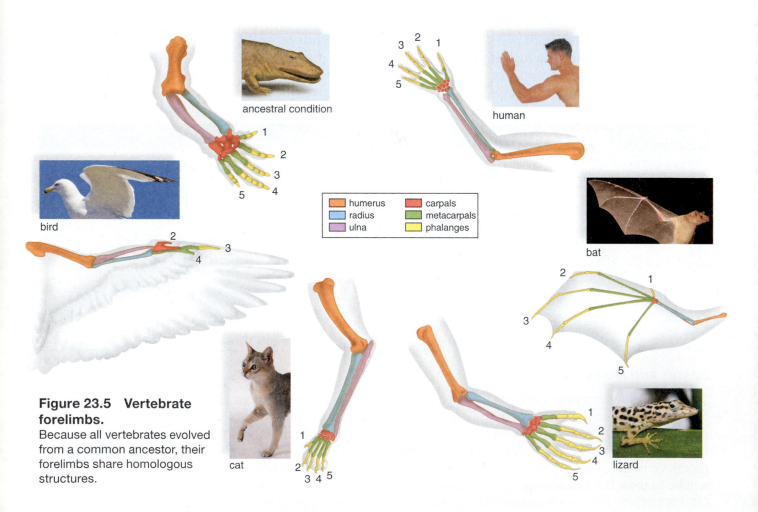

Figure 23.5   Vertebrate forelimbs.
Because all vertebrates evolved from a common ancestor, their forelimbs share homologous structures.

Observation: Vertebrate Forelimbs

1. Find the forelimb bones of the ancestral vertebrate in Figure 23.6. The basic components are the humerus (h), ulna (u), radius (r), carpals (c), metacarpals (m), and phalanges (p) in the five digits.
2. *Label the corresponding forelimb bones of the lizard, the bird, the bat, the cat, and the human in Figure 23.5.*
3. Fill in Table 23.5 to indicate which bones in each specimen appear to most resemble the ancestral condition and which most differ from the ancestral condition.
4. Adaptation to a way of life can explain the modifications that have occurred. Relate the change in bone structure to mode of locomotion in two examples.

   Example 1: _____

   Example 2: _____

Table 23.5 Comparison of Vertebrate Forelimbs		
Animal	Bones That Resemble Common Ancestor	Bones That Differ from Common Ancestor
Lizard		
Bird		
Bat		
Cat		
Human		

Conclusion: Vertebrate Forelimbs

- Vertebrates are descended from a _____, but they are adapted to _____.

Comparison of Vertebrate Embryos

The anatomy shared by vertebrates extends to their embryological development. During early developmental stages, all animal embryos resemble each other closely but as development proceeds the different types of vertebrates take on their own shape and form. Using new kinds of microscopes and modern techniques of studying genes, geneticists have discovered genes whose differential expression can bring about changes in body shapes. The same types of regulatory genes, called *Hox* genes, occur in all organisms despite millions of years of divergent evolution. *Hox* genes, recognized by a particular short sequence of bases, have diversified differently between vertebrates. For example, the *Hox* gene called *Tbx5* turns on different genes in birds and mammals accounting for why birds have wings and mammals do not.

In this observation, you will see that as embryos all vertebrates have a postanal tail, somites (segmented blocks of mesoderm lying on either side of the notochord), a dorsal tubular nerve cord, and paired pharyngeal pouches. In aquatic animals, these pouches become functional gills (Fig. 23.6). In humans, the first pair of pouches becomes the cavity of the middle ear and auditory tube, the second pair becomes the tonsils, and the third and fourth pairs become the thymus and parathyroid glands.

Figure 23.6 Vertebrate embryos.

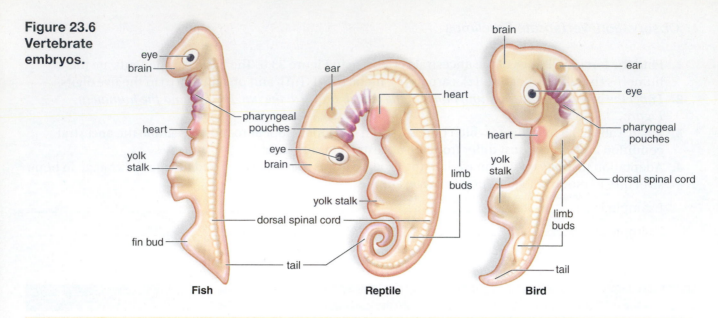

Fish   Reptile   Bird

1. Obtain prepared slides of vertebrate embryos at comparable stages of development. Observe each of the embryos using a stereomicroscope.
2. List five similarities of the embryos:

   a. _____

   b. _____

   c. _____

   d. _____

   e. _____

Comparison of Chimpanzee and Human Skeletons

Chimpanzees and humans are closely related, as is apparent from an examination of their skeletons. However, they are adapted to different ways of life. Chimpanzees are adapted to living in trees and are herbivores—they eat mainly plants. Humans are adapted to walking on the ground and are omnivores— they eat both plants and meat.

Observation: Chimpanzee and Human Skeletons

Posture

Chimpanzees are arboreal and climb in trees. While on the ground, they tend to knuckle-walk, with their hands bent. Humans are terrestrial and walk erect. In Table 23.6, compare:

1. **Head and torso:** Where are the head and trunk with relation to the hips and legs—thrust forward over the hips and legs or balanced over the hips and legs (See Fig. 23.7)?
2. **Spine:** Which animal has a long and curved lumbar region, and which has a short and stiff lumbar region?

   How does this contribute to an erect posture in humans? _____

   _____

3. **Pelvis:** Chimpanzees sway when they walk because lifting one leg throws them off balance. Which animal has a narrow and long pelvis, and which has a broad and short pelvis? Record your observations in Table 23.6.

Table 23.6 Comparison of Chimpanzee and Human Postures

Skeletal Part	Chimpanzee	Human
1. Head and torso		
2. Spine		
3. Pelvis		
4. Femur		
5. Knee joint		
6. Foot: Opposable toe		
Arch		

4. **Femur:** In humans, the femur better supports the trunk. In which animal is the femur angled between articulations with the pelvic girdle and the knee? In which animal is the femur straight with no angle? Record your observations in Table 23.6.

5. **Knee joint:** In humans, the knee joint is modified to support the body's weight. In which animal is the femur larger at the bottom and the tibia larger at the top? Record your observations in Table 23.6.

6. **Foot:** In humans, the foot is adapted for walking long distances and running with less chance of injury.

   In which animal is the big toe opposable? _____ How does an opposable toe assist chimpanzees?

   _____

   Which foot has an arch? _____ How does an arch assist humans?

   _____

   Record your observations in Table 23.6.

7. How does the difference in the position of the foramen magnum, a large opening in the base of the skull for the spinal cord, correlate with the posture and stance of the two organisms?

   _____

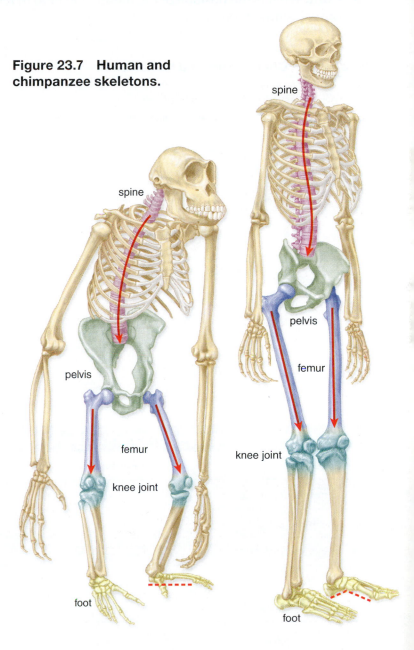

Figure 23.7 Human and chimpanzee skeletons.

spine

spine

pelvis

femur

knee joint

foot

pelvis

femur

knee joint

foot

Skull Features

Humans are omnivorous. A diet rich in meat does not require strong grinding teeth or well-developed facial muscles. Chimpanzees are herbivores, and a vegetarian diet requires strong teeth and strong facial muscles that attach to bony projections. Compare the skulls of the chimpanzee and the human in Figure 23.8 and answer the following questions:

1. **Supraorbital ridge:** For which skull is the supraorbital ridge (the region of frontal bone just above the eye socket) thicker? Record your observations in Table 23.7.
2. **Sagittal crest:** Which skull has a sagittal crest, a projection for muscle attachments that runs along the top of the skull? Record your observation in Table 23.7.
3. **Frontal bone:** Compare the slope of the frontal bones of the chimpanzee and human skulls. How are they different? Record your observations in Table 23.7.
4. **Teeth:** Examine the teeth of the adult chimpanzee and adult human skulls. Are the incisors (two front teeth) vertical or angled? Do the canines overlap the other teeth? Are the molars larger or moderate in size? Record your observations in Table 23.7
5. **Chin:** What is the position of the mouth and chin in relation to the profile for each skull? Record your observations in Table 23.7.

Figure 23.8   Chimpanzee and human skulls.

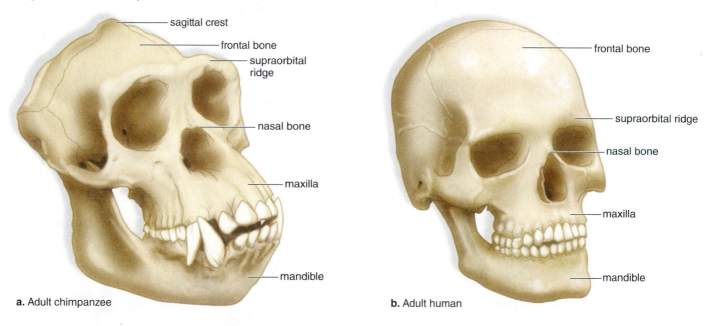

a. Adult chimpanzee

b. Adult human

Table 23.7 Skull Features of Chimpanzees and Humans		
Feature	**Chimpanzee**	**Human**
1. Supraorbital ridge		
2. Sagittal crest		
3. Slope of frontal bone		
4. Teeth		
5. Chin		

Conclusion: Chimpanzee and Human Skeletons

- Do your observations show that the skeletal differences between chimpanzees and humans can be related to posture? _____ Explain. _____

- Do your observations show that diet can be related to the facial features of chimpanzees and humans? _____ Explain. _____

Comparison of Hominid Skulls

The designation *hominid* includes humans and primates that are humanlike. Paleontologists have uncovered several fossils dated from 7.5 MYA (millions of years ago) to 30,000 years BP (before present), when humans called Cro-Magnons arose that are virtually identical to modern humans *(Homo sapiens)*. (Fig. 23.9).

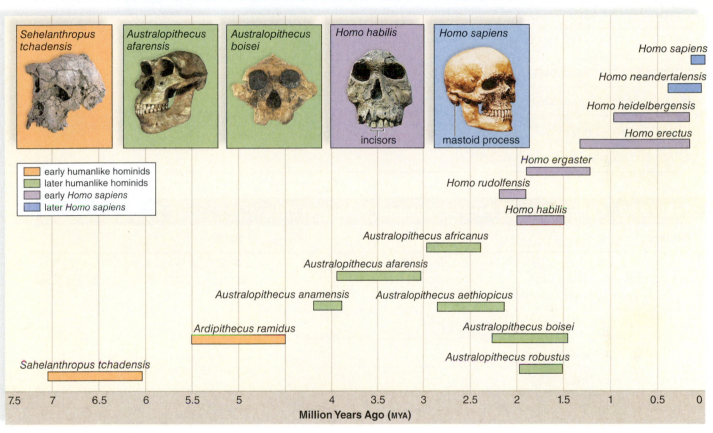

Figure 23.9 Human evolution.

Observation of Hominid Skulls

Several of the skulls noted in Figure 23.9 are on display. Use Tables 23.8, 23.9 and 23.10 to record data pertaining to the cranium (or braincase), the face, and the teeth. Compare the early skulls on display with a modern human skull. For example, is the forehead like or more flat than the human skull? List at least three skulls you examined for each feature.

Conclusions

- Do your data appear to be consistent with the evolutionary sequence of the hominids in Figure 23.9? Explain. _____

- Report here any data you collected that would indicate a particular hominid was closer in time to humans than indicated in Figure 23.9. _____

- Report here any data you collected that would indicate a particular hominid was more distant in time from humans than indicated in Figure 23.9. _____

Table 23.8 Other Hominid Craniums Compared to Human Cranium

Feature	Skulls		
	1.	2.	3.
a. Frontal bone (like or more flat?)			
b. Supraorbital ridge (divided or continuous?)			
c. Sagittal crest (present?)			
d. Mastoid process (flat or projecting?)			

Table 23.9 Other Hominid Faces Compared to Human Face

Feature	Skulls		
	1.	2.	3.
a. Nasal bones (raised or flat?)			
b. Nasal opening (larger?)			
c. Chin (projecting forward?)*			
d. Width of Face (wider?)**			

*If your instructor directs you to, measure from the edge of the foramen magnum to between the incisors.
**If your instructor directs you to, measure the width of the face from mid-zygomatic arch to the other mid-arch.

Figure 23.10 Other Hominid Dentition Compared to Human Dentition

Feature	Skulls		
	1.	2.	3.
a. Teeth rows (parallel or diverging from each other?)			
b. Incisors (vertical or angled?)			
c. Canine teeth (overlapping other teeth?)			
d. Molars (more massive?)			

Summary of Evidence from Comparative Anatomy

1. The similarity between the bones in all vertebrate forelimbs and in a common ancestor shows that today's vertebrates are (descended, not descended) from the common ancestor. However, the various vertebrates are adapted to _____ ways of life.

2. The similarity in the appearance of vertebrate embryos also shows that today's vertebrates are (related, not related) _____.

3. A study of the chimpanzee and the human skeleton shows that their differences are simply due to _____ posture of humans. A contrast in skulls shows that the facial differences are due to a difference in _____.

4. Differences in fossil skull features suggest the sequence in which humans _____.

23.3   Molecular Evidence

Molecular data substantiates (1) the comparative and developmental data that biologists have accumulated over the years. The activity of *Hox* genes differ and this can account, for example, for why vertebrates have a dorsally placed nerve cord while it is ventrally placed in invertebrates. (2) Sequencing DNA data shows which organisms are closely related. Molecular data among primates is of extreme interest because it can help determine which of the primates we are most closely related to. Chromosomal and genetic data allows us to conclude that we are more closely related to chimpanzees than to other types of apes.

In this section, we note that scientists can compare the amino acid sequence in proteins to determine the degree to which any two groups of organisms are related. The sequence of amino acids in cytochrome *c,* a carrier of electrons in the electron transport chain found in mitochondria, has been determined per a variety of organisms. On the basis of the number of amino acid *differences* reported in Figure 23.10, it is concluded that the evolutionary relationship between humans and these organisms decreases in the order stated: monkeys, pigs, ducks, turtles, fishes, moths, and yeast. This conclusion agrees with the sequence of dates these organisms are found in the fossil record. Why can comparing amino acid data lead to the same conclusions as comparing DNA data?

_____

_____

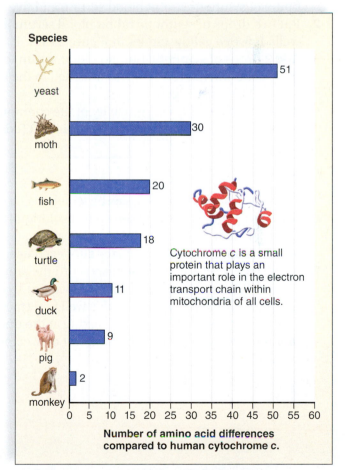

Cytochrome *c* is a small protein that plays an important role in the electron transport chain within mitochondria of all cells.

Number of amino acid differences compared to human cytochrome c.

Figure 23.10   Amino acid differences in cytochrome c.
The few differences found in cytochrome *c* between monkeys and humans show that of these organisms, humans are most closely related to monkeys.

Protein Similarities

The immune system makes **antibodies** (proteins) that react with foreign proteins, termed **antigens.** Antigen-antibody reactions are specific. An antibody will react only with a particular antigen. In today's procedure, it is assumed rabbit antibodies to human antigens are in rabbit serum (Fig. 23.11). When these antibodies are allowed to react against the antigens of other animals, the stronger the antibody-antigen reaction (determined by the amount of precipitate), the more closely related the animal is to humans.

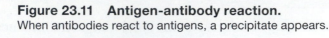

Figure 23.11   Antigen-antibody reaction.
When antibodies react to antigens, a precipitate appears.

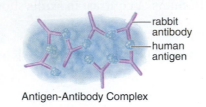

rabbit antibody

human antigen

Antigen-Antibody Complex

Experimental Procedure: Protein Similarities

1. Obtain a chemplate (a clear glass tray with wells), one bottle of synthetic human blood serum, one bottle of synthetic rabbit blood serum, and five bottles (I–V) of blood serum test solution.
2. Put two drops of synthetic rabbit blood serum in each of the six wells in the chemplate. Label the wells 1–6. See yellow circles in Figure 23.12.
3. Add 2 drops of synthetic human blood serum to each well. See red circles in Figure 23.12. Stir with the plastic stirring rod that was attached to the chemplate. The rabbit serum now contains antibodies against human antigens.
4. Rinse the stirrer. (The large cavity of the chemplate may be filled with water to facilitate rinsing.)
5. Add 4 drops of blood serum test solution III (contains human antigens) to well 6. Describe what

   you see. _____

   _____

   This well will serve as the basis by which to compare all the other samples of test blood serum.
6. Now add 4 drops of blood serum test solution I to well 1. Stir and observe. Rinse the stirrer. Do the same for each of the remaining blood serum test solutions (II–V)—adding II to well 2, III to well 3, and so on. Be sure to rinse the stirrer after each use.
7. At the end of 10 and 20 minutes, record the amount of precipitate in each of the six wells in Figure 23.12. Well 6 is recorded as having ++++ amount of precipitate after both 10 and 20 minutes. Compare the other wells with this well (+ = trace amount; 0 = none). Holding the plate slightly above your head at arm's length and looking at the underside toward an overhead light source will allow you to more clearly determine the amount of precipitate.

Figure 23.12 Molecular evidence of evolution.

The greater the amount of precipitate, the more closely related an animal is to humans.

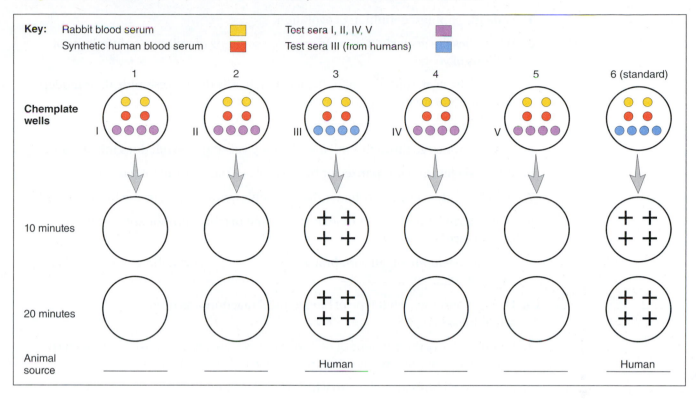

Conclusions: Protein Similarities

- The last row in Figure 23.12 tells you that the test serum in well 3 is from a human. How do your test results confirm this? _____

- Aside from humans, the test sera (supposedly) came from a pig, a monkey, an orangutan, and a chimpanzee. Which is most closely related to humans—the pig or the chimpanzee? _____

- Judging by the amount of precipitate, complete the last row in Figure 23.12 by indicating which serum you believe came from which animal. On what do you base your conclusions?

  _____

- In Section 23.2, a comparison of bones in vertebrate forelimbs showed that vertebrates share a common ancestor. Molecular evidence shows us that of the vertebrates studied, _____ and _____ are most closely related. In evolution, closely related organisms share a recent common ancestor. Humans share a more recent common ancestor with chimpanzees than they do with pigs.

_____ **1.** State the simplest definition of evolution.

_____ **2.** What does the geologic timescale encompass—tens of millions or hundreds of millions of years?

_____ **3.** What is the best explanation for the fact that fossils do not resemble their modern-day representatives?

_____ **4.** Fossils are the _____ of past life.

_____ **5.** All vertebrates go through similar embryological stages. What does this suggest?

_____ **6.** Which skeleton—chimpanzee or human—has a narrow and long pelvis?

_____ **7.** Which has thicker supraorbital ridges, the chimpanzee skull or the human skull?

_____ **8.** Which term—homologous or analogous—means that components are similar in structure?

_____ **9.** During development, all vertebrates have _____, even though only fish have gill slits as adults.

_____ **10.** In this laboratory, what type of biochemical reaction was used to determine relatedness?

_____ **11.** If antibodies to the antigens of one species react strongly against the antigens of another species, the two species are (closely or distantly) related.

_____ **12.** All apes and humans belong to what group of mammals?

_____ **13.** All vertebrates have what type of supporting rod during development?

Thought Questions

14. If a characteristic is found in bacteria, fungi, pine trees, snakes, and humans, in which type organism did it most likely evolve? Explain.

15. What do mutations have to do with amino acid changes in a protein? How do such changes help determine relatedness of organisms?

16. Homologous structures arise from common ancestry; analogous structures do not. Given this information, how might you explain the existence of analogous structures?

24
Microbiology

Introduction

The history of life began with the evolution of the prokaryotic cell. Although the prokaryotic cell contains genetic material, it is not located in a nucleus, and the cell also lacks any other type of membranous organelle. At one time prokaryotes were believed to be a unified group, but based on molecular data, they are now divided into two major groups—**domain Bacteria** and **domain Archaea.** The eukaryotic cell is more closely related to the archaea than the bacteria. Eukaryotes in **domain Eukarya** have a membrane-bounded nucleus and membranous organelles. Prokaryotes resemble each other structurally but are metabolically diverse. Eukaryotes, on the other hand, are structurally diverse and exist as protists, fungi, plants, and animals (Fig. 24.1).

Figure 24.1   The world of living things.
Prokaryotes are represented in this illustration by the bacteria. The protists, fungi, plants, and animals are all eukaryotes.

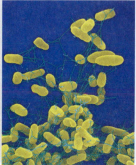

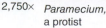

Bacteria   2,750×   *Paramecium,*   Morel, a fungus   Sunflower, a plant   Snow goose, an animal
                           a protist

1,000×

24.1 Bacteria

In this laboratory you will first relate the general structure of a bacterium to its ability to cause disease. The specific shape, growth habit, and staining characteristics of bacteria are often used to identify them. Therefore, you will observe a variety of bacteria using the microscope. Aside from their medical importance, bacteria are essential in ecosystems because, along with fungi, they are decomposers that break down dead organic remains, and thereby return inorganic nutrients to plants.

Pathogenic Bacteria

Pathogenic bacteria are infectious agents that cause disease. Infectious bacteria are able to invade and multiply within a host. Some also produce a toxin. Antibiotic therapy is often an effective treatment against a bacterial infection.

We will explore how it is possible to relate the structure of a bacterium to its ability to be invasive and avoid destruction by the immune system. We will also consider what morphophysiological attributes allow bacteria to be resistant to antibiotics and to pass the necessary genes on to other bacteria.

Observation: Structure of a Bacterium

1. Study the generalized structure of a bacterium in Figure 24.2 and, if available, examine a model or view a CD-ROM of a bacterium.
2. Identify the following:

   capsule, a gel-like coating outside the cell wall. Capsules often allow bacteria to stick to surfaces such as teeth. They also prevent phagocytic white blood cells from taking them up and destroying them.

   fimbriae, hairlike bristles that allow adhesion to surfaces. This can be how a bacterium clings to and gains access to the body prior to an infection.

   conjugation pilus, an elongated, hollow appendage used to transfer DNA to other cells. Genes that allow bacteria to be resistant to antibiotics can be passed in this manner.

   flagellum, a rotating filament that pushes the cell forward.

   cell wall, a structure that provides support and shapes the cell. Antibiotics that prevent the formation of a cell wall are most effective against Gram-positive rather than Gram-negative bacteria.

   plasma membrane, a sheet that surrounds the cytoplasm and regulates entrance and exit of molecules. Resistance to antibiotics can be due to plasma membrane alterations that do not allow the drug to bind to the membrane or cross the membrane, or to a plasma membrane that increases the elimination of the drug from the bacteria.

   ribosomes, site of protein synthesis. Some bacteria possess antibiotic-inactivating enzymes that make them resistant to antibiotics.

   nucleoid, the location of the bacterial chromosome.

Figure 24.2 Generalized structure of a bacterium.

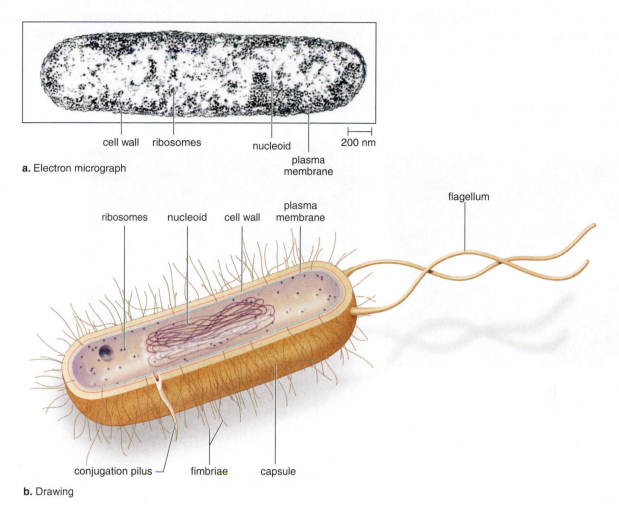

cell wall   ribosomes   nucleoid   200 nm

plasma membrane

a. Electron micrograph

flagellum

ribosomes   nucleoid   cell wall   plasma membrane

conjugation pilus   fimbriae   capsule

b. Drawing

Also, some bacteria contain plasmids, small rings of DNA that replicate independently of the chromosomes and can be passed to other bacteria. Genes that allow bacteria to be resistant to antibiotics are often located in a plasmid.

Conclusions: Structure of a Bacterium

- Which portions of a bacterial cell aid the ability of a bacterium to cause infections?

  _____

- Which portions of a bacterial cell aid the ability of a bacterium to be resistant to antibiotics?

  _____

Colony Morphology

On a nutrient material called agar, bacteria grow as colonies. A **colony** contains cells descended from one original cell. Sometimes, it is possible to identify the type of bacterium by the appearance of the colony (Fig. 24.3).

Figure 24.3 Colony morphology.
Colonies of bacteria on agar plates.

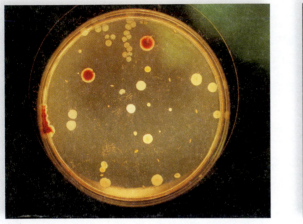

a.

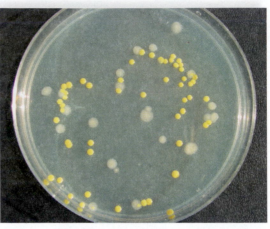

b.

Observation: Colony Morphology

1. View agar plates that have been inoculated with bacteria and then incubated. Notice the "colonies" of bacteria growing on the plates.
2. Compare the colonies' color, surface, and margin, and note your observations in Table 24.1. It is not necessary to identify the type of bacteria.

Table 24.1 Agar Plates	
Plate Number	**Description of Colonies**

3. If available, obtain a sterile agar plate, and inoculate the plate with your thumbprint, or use a swab and inoculate the plate with material from around your teeth or inside your nose. Put your name on the plate, and place it where directed by your instructor. Remember to view the plate next laboratory period. Describe your plate:

4. If available, obtain a sterile agar plate, and expose it briefly (at most for 10 minutes) anywhere you choose, such as in the library, your room, or your car. No matter where the plate is exposed, it subsequently will show bacterial colonies. Describe your plate:

Shape of Bacterial Cell

Most bacteria are found in three basic shapes: **spirillum** (spiral or helical), **bacillus** (rod), and **coccus** (round or spherical) (Fig. 24.4). Bacilli may form long filaments, and cocci may form clusters or chains. Some bacteria form endospores. An **endospore** contains a copy of the genetic material encased by heavy protective spore coats. Spores survive unfavorable conditions and germinate to form vegetative cells when conditions improve.

Observation: Shape of Bacterial Cell

1. View the microscope slides of bacteria on display. What magnification is required to view bacteria?

   _____

2. Using Figure 24.4 as a guide, identify the three different shapes of bacteria. _____

3. Do any of the slides on display show bacterial cells with endospores? _____

   What is an endospore, and why does it have survival value? _____

   _____

Figure 24.4   Diversity of bacteria.
a. Spirillum, a spiral-shaped bacterium. **b.** Bacilli, rod-shaped bacteria. **c.** Cocci, round bacteria.

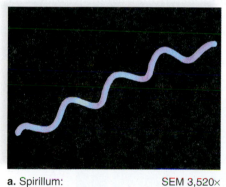

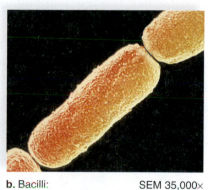

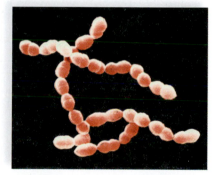

a. Spirillum: SEM 3,520×
Spirillum volutans

b. Bacilli: SEM 35,000×
Bacillus anthracis

c. Cocci: SEM 6,250×
Streptococcus thermophilus

Cyanobacteria

Cyanobacteria were formerly called blue-green algae because their general growth habit and appearance through a compound light microscope are similar to green algae. Electron microscopic study of cyanobacteria, however, revealed that they are structurally similar to other bacteria, particularly other photosynthetic bacteria. Although cyanobacteria do not have chloroplasts, they do have thylakoid membranes, where photosynthesis occurs.

Gloeocapsa

1. Prepare a wet mount of a *Gloeocapsa* culture, if available, or examine a prepared slide, using high power (45×) or oil immersion (if available). The single cells adhere together because each is surrounded by a sticky, gelatinous sheath (Fig. 24.5).

2. What is the estimated size of a single cell? _____

a. Micrograph at low magnification.   150×

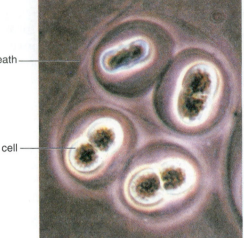

gelatinous sheath —

cell —

b. Micrograph at high magnification.   500×

Figure 24.5   *Gloeocapsa*.

Oscillatoria

1. Prepare a wet mount of an *Oscillatoria* culture, if available, or examine a prepared slide, using high power (45×) or oil immersion (if available). This is a filamentous cyanobacterium with individual cells that resemble a stack of pennies (Fig. 24.6).

2. *Oscillatoria* takes its name from the characteristic oscillations that you may be able to see if your sample is alive. If you have a living culture, are oscillations visible? _____

Figure 24.6   *Oscillatoria*.

250×

Anabaena

1. Prepare a wet mount of an *Anabaena* culture, if available, or examine a prepared slide, using high power (45×) or oil immersion (if available). This is also a filamentous cyanobacterium, although its individual cells are barrel-shaped (Fig. 24.7).

2. Note the thin nature of this strand. If you have a living culture, what is its color? _____

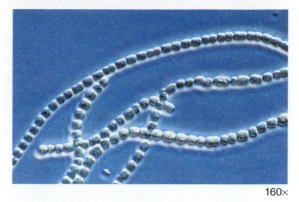

160×

Figure 24.7   *Anabaena*.

24.2   Protists

Protists were the first eukaryotes to evolve. Their diversity and complexity make it difficult to categorize them. However, the diagram in Figure 24.8 may be helpful.

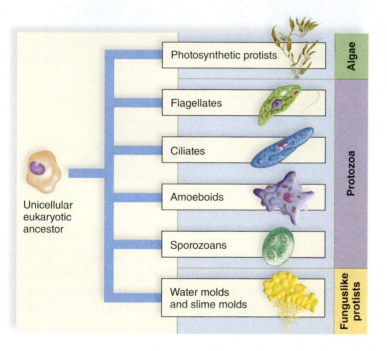

Figure 24.8   Major groups of protists.
Because the precise evolutionary relationships between these groups are not yet known, they are grouped here by major shared characteristics.

Algae

Algae is a term that has been used for aquatic organisms that photosynthesize in the same manner as land plants. All photosynthetic protists contain green chlorophyll, but they also may contain other pigments that mask the chlorophyll color, and this accounts for their common names—the green algae, red algae, brown algae, and golden-brown algae.

If available, view a film loop showing the many forms of green algae. Notice that green algae can be single cells, filaments, colonies, or multicellular sheets. You will examine a filamentous form *(Spirogyra)* and a colonial form *(Volvox).* A **colony** is a loose association of cells.

1. *Spirogyra* is a filamentous alga, lives in fresh water, and often is seen as a green scum on the surface of ponds and lakes. The most prominent feature of the cells is the spiral, ribbonlike chloroplast (Fig. 24.9). How do you think *Spirogyra* got its name? _____

_____

Figure 24.9 *Spirogyra.*
a. *Spirogyra* is a filamentous green alga, in which each cell has a ribbonlike chloroplast. **b.** During conjugation, the cell contents of one filament enter the cells of another filament. Zygote formation follows.

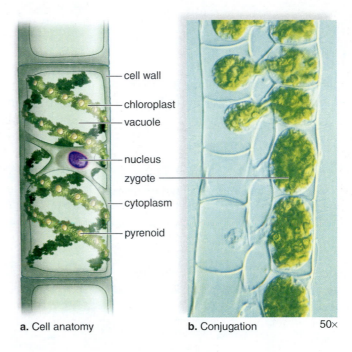

cell wall
chloroplast
vacuole
nucleus
zygote
cytoplasm
pyrenoid

a. Cell anatomy   b. Conjugation   50×

Spirogyra's chloroplast contains a number of circular bodies, the **pyrenoids,** centers of starch polymerization. The nucleus is in the center of the cell, anchored by cytoplasmic strands. Your slide may show **conjugation,** a sexual means of reproduction illustrated in Figure 24.9*b*. If it does not, obtain a slide that does show this process. Conjugation tubes form between two adjacent filaments, and the contents of one set of cells enter the other set. As the nuclei fuse, a zygote is formed. The zygote overwinters, and in the spring, meiosis and, subsequently, germination occur. The resulting adult protist is therefore haploid.

Make a wet mount of live *Spirogyra,* or observe a prepared slide.

Volvox is a green algal colony. It is motile (capable of locomotion) because the thousands of cells that make up the colony have flagella. These cells are connected by delicate cytoplasmic extensions (Fig. 24.10).

Volvox is capable of both asexual and sexual reproduction. Certain cells of the adult colony can divide to produce **daughter colonies** (Fig. 24.10) that reside for a time within the parental colony. A daughter colony escapes the parental colony by releasing an enzyme that dissolves away a portion of the matrix of the parental colony. During sexual reproduction, some colonies of *Volvox* have cells that produce sperm, and others have cells that produce eggs. The resulting zygote undergoes meiosis and the adult *Volvox* is haploid.

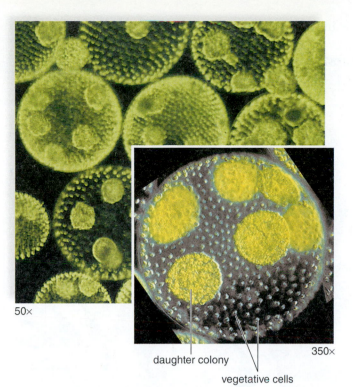

Figure 24.10 Volvox.
Volvox is a colonial green alga. The adult Volvox colony often contains daughter colonies, asexually produced by special cells.

50×

daughter colony

vegetative cells

350×

Using a depression slide, make a wet mount of live *Volvox,* or study a prepared slide.

In Table 24.2, list the genus names of each of the green algae specimens available, and give a brief description.

Table 24.2	Green Algae Diversity	
Specimen	**Description**	

Observation: Brown Algae

Brown algae are commonly called *seaweed,* along with the multicellular green and red algae. Brown algae contain brown pigments that mask chlorophyll's green color. These algae are large and have specialized parts.

Fucus is called rockweed because it is seen attached to rocks at the seashore when the tide is out (Fig. 24.11). If available, view a preserved specimen. Note the dichotomously branched body plan, so called because the **stipe** repeatedly divides into two branches (Fig. 24.11). Note also the **holdfast** by which the alga anchors itself to the rock; the **air vesicles,** or bladders, that help hold the thallus erect in the water; and the **receptacles,** or swollen tips. The receptacles are covered by small raised areas, each with a hole in the center. These areas are cavities in which the sex organs are located, with the gametes escaping to the outside through the holes. *Fucus* is unique among algae in that as an adult it is diploid (2n) and always reproduces sexually.

If available, study preserved specimens of other brown algae (Fig. 24.11). *Laminaria* algae are called **kelps.**

Figure 24.11 Brown algae.
Laminaria and *Fucus* are seaweeds known as kelps. They live along rocky coasts of the north temperate zone. The other brown algae featured, *Macrocystis* and *Nereocystis,* form spectacular underwater "forests" at sea.

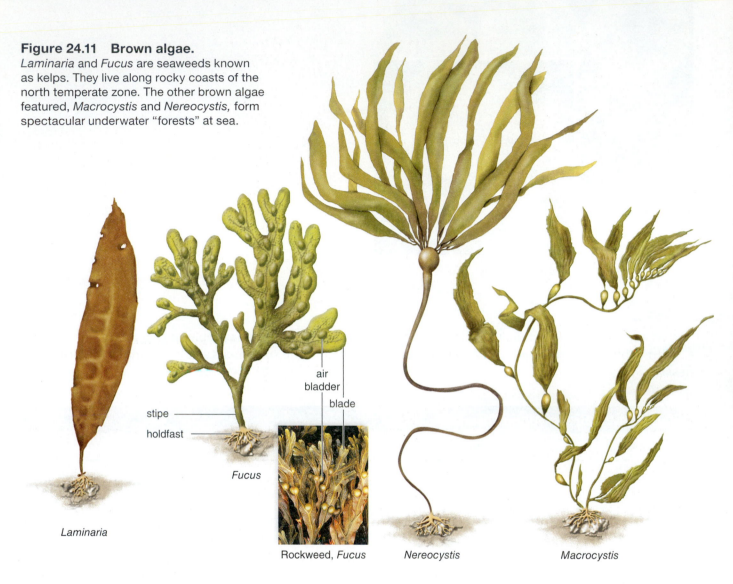

air bladder

blade

stipe

holdfast

Fucus

Laminaria

Rockweed, *Fucus*

Nereocystis

Macrocystis

Observation: Red Algae

Like most brown algae, the red algae (Fig. 24.12) are multicellular, but they occur chiefly in warmer seawater, growing both in shallow waters and as deep as light penetrates. Some forms of red algae are filamentous, but more often, they are complexly branched with a feathery, flat, and expanded or ribbonlike appearance. Coralline algae are red algae that have cell walls impregnated with calcium carbonate ($CaCO_3$).

In Table 24.3, list the genus names of each of the brown and red algae specimens available, and give a brief description.

Figure 24.12 Red algae.

Generally, red algae are smaller and more delicate than brown algae. **a.** *Sebdenia* has a pronounced filamentous structure. **b.** Calcium carbonate is deposited in the walls of the red alga, *Corallina*.

a.

b.

Table 24.3	Brown and Red Algae	
Specimen	Genus	Description
1		
2		
3		
4		

Observation: Diatoms

Diatoms (golden-brown algae) have a yellow-brown pigment that, in addition to chlorophyll, gives them their color.

The cell wall of **diatoms** is in two sections, with the larger one fitting over the smaller as a lid fits over a box. Since the cell wall is impregnated with silica, diatoms are said to "live in glass houses." The glass cell walls of diatoms do not decompose, so they accumulate in thick layers subsequently mined as diatomaceous earth and used in filters and as a natural insecticide. Diatoms, being photosynthetic and extremely abundant, are important food sources for the small heterotrophs (organisms that must acquire food from external sources) in both marine and freshwater environments.

Make a wet mount of live diatoms, or view a prepared slide (Fig. 24.13). Describe what you see:

350×

Figure 24.13 Diatoms.
Diatoms, photosynthetic protists of the oceans.

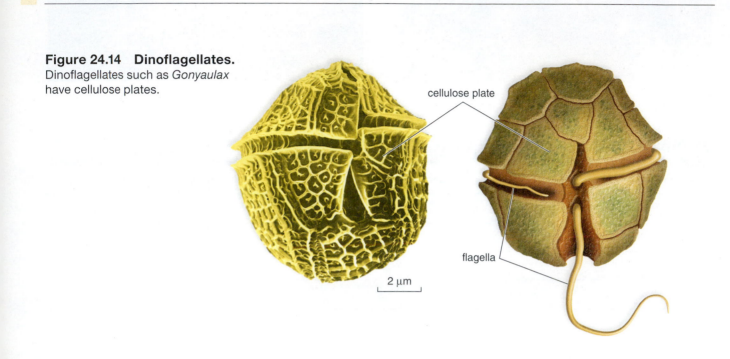

Dinoflagellates are photosynthetic, but they have two flagella; one is free, but the other is located in a transverse groove that encircles the animal. The beating of these flagella causes the organism to spin like a top. The cell wall, when present, is frequently divided into closely joined polygonal plates of cellulose. At times there are so many of these organisms in the ocean that they cause a condition called "red tide." The toxins given off in these red tides cause widespread fish kills and can cause paralysis in humans who eat shellfishes that have fed on the dinoflagellates.

Make a wet mount of live dinoflagellates or view a prepared slide (Fig. 24.14). Describe what you see:

_____

_____

Figure 24.14   Dinoflagellates.
Dinoflagellates such as *Gonyaulax* have cellulose plates.

cellulose plate

flagella

2 µm

Protozoans

The term *protozoan* refers to unicellular eukaryotes and is often restricted to heterotrophic organisms that ingest food by forming **food vacuoles.** Other vacuoles, such as **contractile vacuoles** that rid the cell of excess water, are also typical. Usually protozoans have some form of locomotion; some use **pseudopodia,** some move by **cilia,** and some use **flagella** (Fig. 24.15).

Plasmodium vivax, a common cause of malaria, is an apicomplexan, a protozoan that contains a special organelle called an apicoplast. The apicoplast assists the parasite in penetrating host cells. This type of protozoan is also often called a **sporozoan** because it goes through an asexual phase in which it exists as particulate spores (Fig. 24.15*d*). During its asexual phase, *Plasmodium vivax* lives inside red blood cells and the chills and fever of malaria occur when the red blood cells burst to release the spores. Sporozoans have no obvious means of locomotion as do the other types of protozoans illustrated in Figure 24.15. How do sporozoans differ from other types of protozoans? _____

_____

_____

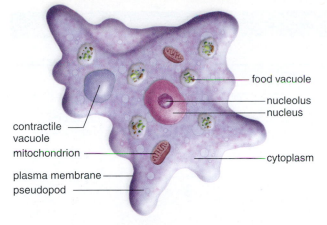

contractile vacuole
mitochondrion
plasma membrane
pseudopod

food vacuole
nucleolus
nucleus
cytoplasm

a. *Amoeba* moves by pseudopods.

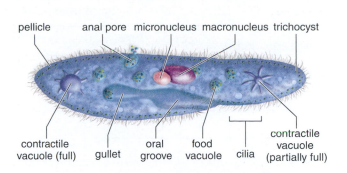

pellicle   anal pore   micronucleus   macronucleus   trichocyst

contractile vacuole (full)   gullet   oral groove   food vacuole   cilia   contractile vacuole (partially full)

b. *Paramecium* moves by cilia.

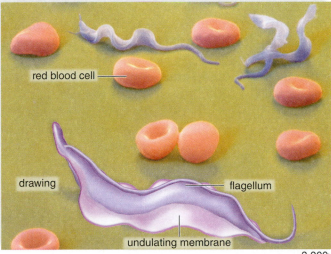

red blood cell

drawing

flagellum

undulating membrane

3,000×

c. *Trypanosoma*, which lives in bloodstream of host, moves by flagella.

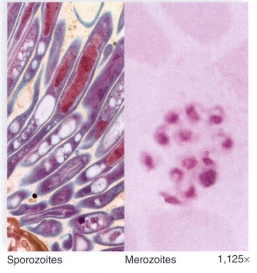

Sporozoites   Merozoites   1,125×

d. *Plasmodium* exists as nonmotile spores.

Figure 24.15   Protozoan diversity.
Protozoans are motile by the means illustrated, except for sporozoans such as *Plasmodium*. **a.** *Amoeba;* **b.** *Paramecium;* **c.** *Trypanosoma* in host; **d.** *Plasmodium* exists in several stages; merozoites invade red blood cells and divide to produce more spores.

Observation: Protozoans

Individual Protozoans

You may already have had the opportunity to observe a protozoan such as *Paramecium* or *Euglena* in Laboratory 2. However, your instructor may want you to observe these organisms again. *Euglena* have flagella but many also have chloroplasts and therefore they do not match our definition of a protozoan. Watch a video if available, and note the various forms of protozoans. Prepare wet mounts or examine prepared slides of protozoans as directed by your instructor. Complete Table 24.4, listing the structures for locomotion in the types of protozoans you have observed.

Table 24.4 Heterotrophic Protists

Name	Structures for Locomotion	Observations

Observation: Euglena

Euglena (Fig. 24.16) typifies the problem of classifying protists. One third of all *Euglena* genera have chloroplasts; the rest do not. This discrepancy can be explained: the chloroplasts are probably green algae taken up by phagocytosis (engulfing them). A pyrenoid is a region of the chloroplast where a special type of carbohydrate is formed.

Euglena has a long flagellum that projects out of a vase-like indentation and a much shorter one that does not project out. It moves very quickly and you will be advised to add Protoslo to your wet mount to slow it down. Like some of the protozoans discussed previously, *Euglena* is bounded by a flexible pellicle made of protein. This means it can also assume all sorts of shapes. *Euglena* lives in freshwater and contains a contractile vacuole that collects water and then contracts, ridding the body of excess water.

Make a wet mount of *Euglena* by using a drop of a *Euglena* culture and adding a drop of Protoslo (methyl cellulose solution) onto the slide to slow it down. Describe what you see: _____

_____

_____

Figure 24.16 *Euglena.*
Euglena is a unicellular, flagellated protist.

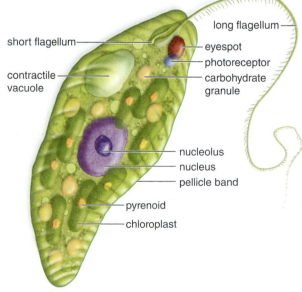

a. Drawing

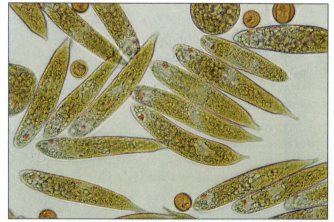

b. Photomicrograph   150×

Observation: Pond Water

1. Make a wet mount of pond water by taking a drop from the bottom of a container of pond water.
2. Scan the slide for organisms: Start at the upper left-hand corner, and move the slide forward and back as you work across the slide from left to right.
3. Experiment by using all available objective lenses, by focusing up and down with the fine-adjustment knob, and by adjusting the light so that it is not too bright.
4. Identify the organisms you see by consulting Figure 24.17, and use any pictorial guides provided by your instructor.

Figure 24.17   Microorganisms found in pond water. Drawings are not actual sizes of the organisms.

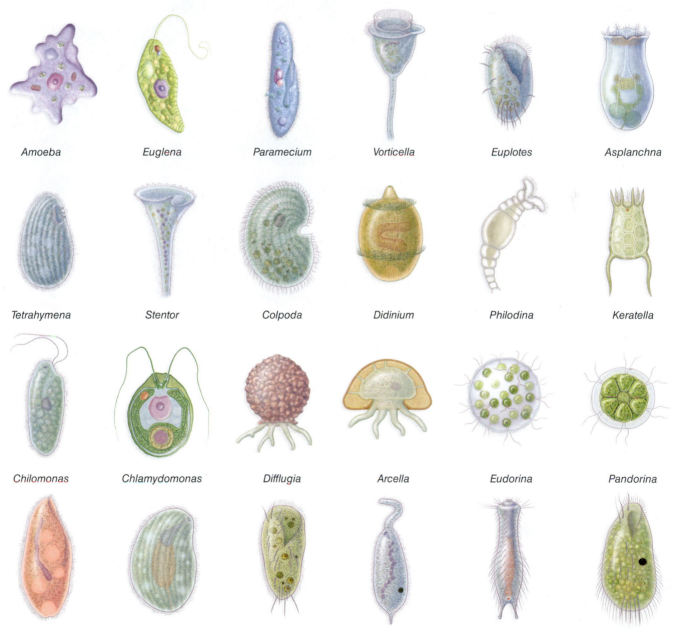

| Amoeba | Euglena | Paramecium | Vorticella | Euplotes | Asplanchna |

| Tetrahymena | Stentor | Colpoda | Didinium | Philodina | Keratella |

| Chilomonas | Chlamydomonas | Difflugia | Arcella | Eudorina | Pandorina |

| Blepharisma | Chilodonella | Stylonychia | Dileptus | Chaetonotus | Oxytricha |

24.3 Fungi

Fungi (kingdom Fungi) (Fig. 24.18) are saprotrophic in the same manner as bacteria. Both fungi and bacteria are often referred to as "organisms of decay" because they break down dead organic matter and release inorganic nutrients for plants. A fungal body, called a **mycelium,** is composed of many strands, called **hyphae** (Fig. 24.19). Sometimes, the nuclei within a hypha are separated by walls called septa.

Fungi produce windblown **spores** (small, haploid bodies with a protective covering) when they reproduce sexually or asexually.

Figure 24.18   Diversity of fungi.
a. Scarlet hood, an inedible mushroom. **b.** Spores exploding from a puffball. **c.** Common bread mold. **d.** Morel, an edible fungus.

d.

Figure 24.19   Body of a fungus.
a. The body of a fungus is called a mycelium. **b.** A mycelium contains many individual chains of cells, and each chain is called a hypha.

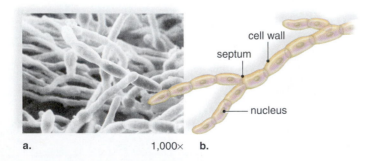

a.   1,000×   b.

Black Bread Mold

In keeping with its name, black bread mold grows on bread and any other type of bakery goods. Notice in Figure 24.20 the sporangia at the tips of aerial hyphae that produce spores in both the asexual and sexual life cycles. A **zygospore** is diploid (2n); otherwise, all structures in the asexual and sexual life cycles of bread mold are haploid (n).

Figure 24.20 Black bread mold.
The mycelium of this mold (1) uses sporangia to produce windblown spores. (2) During sexual reproducing, the ends of plus and minus hyphae fuse as (3) fertilization occurs. (4) The resulting zygospore undergoes meiosis to produce spores that germinate on the bread to produce a new mycelium (5).

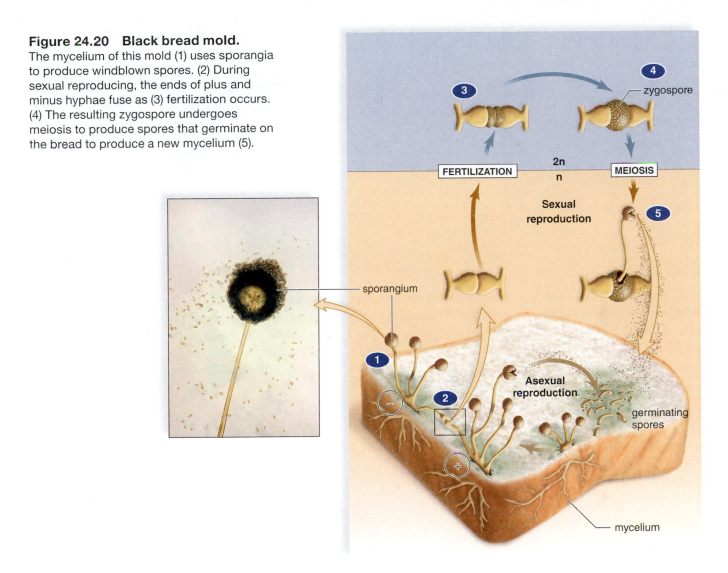

Observation: Black Bread Mold

1. If available, examine bread that has become moldy. Do you recognize black bread mold on the bread?
2. Obtain a petri dish that contains living black bread mold. Observe with a stereomicroscope. *Label the mycelium and a sporangium in Figure 24.21a.*
3. View a prepared slide of *Rhizopus*, using both a stereomicroscope and the low-power setting of a light microscope. The absence of cross walls in the hyphae is an identifying feature of zygospore fungi. *Label the mycelium and zygospore in Figure 24.21b.*

Figure 24.21 **Microscope slides of black bread mold.**
a. Asexual life cycle structures. **b.** Sexual life cycle structures.

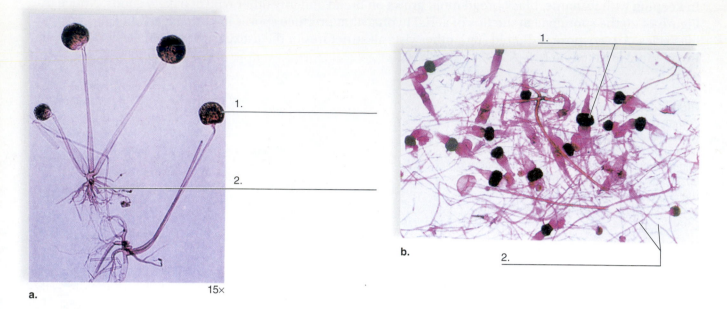

1.

2.

1.

b.

2.

a.   15×

Club Fungi

Club fungi are just as familiar as black bread mold to most laypeople because they include the mushrooms. A gill mushroom consists of a stalk and a terminal cap with gills on the underside (Fig. 24.22). The cap, called a **basidiocarp,** is a fruiting body that arises following the union of + and – hyphae. The gills bear basidia, club-shaped structures where nuclei fuse, and meiosis occurs during spore production. The spores are called **basidiospores.**

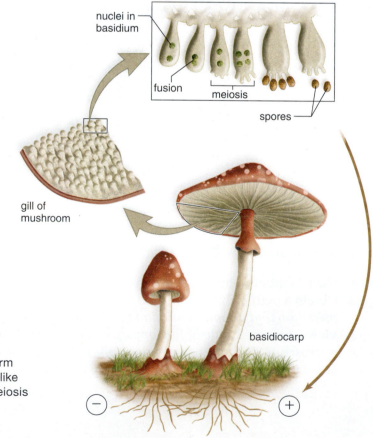

nuclei in basidium

fusion

meiosis

spores

gill of mushroom

basidiocarp

Figure 24.22 **Sexual reproduction produces mushrooms.**
Fusion of + and – hyphae tips results in hyphae that form the mushroom (a fruiting body). The nuclei fuse in clublike structures attached to the gills of a mushroom, and meiosis produces spores.

–

+

1. Obtain an edible mushroom—for example, *Agaricus*—and identify as many of the following structures as possible:
    a. **Stalk:** The upright portion that supports the cap.
    b. **Annulus:** A membrane surrounding the stalk where the immature (button-shaped) mushroom was attached.
    c. **Cap:** The umbrella-shaped basidiocarp of the mushroom.
    d. **Gills:** On the underside of the cap, radiating lamellae on which the basidia are located.
    e. **Basidia:** On the gills, club-shaped structures where basidiospores are produced.
    f. **Basidiospores:** Spores produced by basidia.
2. View a prepared slide of a cross section of *Coprinus*. Using all three microscope objectives, look for the gills, basidia, and basidiospores.
3. Can you see individual hyphae in the gills?
4. Are the basidiospores inside or outside of the basidia?

5. What type of nuclear division does the zygote undergo to produce the basidiospores? _____

6. Can you suggest a reason for some of the basidia having fewer than four basidiospores? _____

7. What happens to the basidiospores after they are released? _____

Fungi and Human Diseases

Fungi cause a number of human diseases. Oral thrush is a yeast infection of the mouth common in newborns and AIDS patients (Fig. 24.23*a*). Ringworm is a group of related diseases caused by the fungus *Tinea*. The fungal colony grows outward, forming a ring of inflammation (Fig. 24.23*b*). Athlete's foot is a form of *Tinea* that affects the foot, mainly causing itching and peeling of the skin between the toes (Fig. 24.23*c*).

Figure 24.23 Human fungal diseases.
a. Thrush, or oral candidiasis, is characterized by the formation of white patches on the tongue.
b. Ringworm and **c.** athlete's foot are caused by *Tinea* spp.

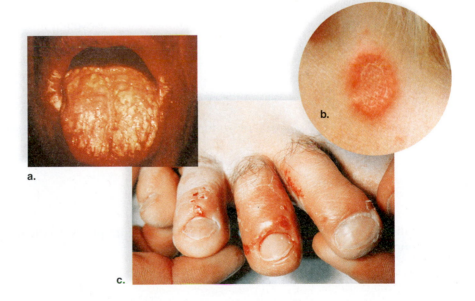

_____ 1. What role do bacteria and fungi play in ecosystems?

_____ 2. What type of semisolid medium is used to grow bacteria?

_____ 3. What is the scientific name for spherical bacteria?

_____ 4. It is sometimes said that diatoms live in what kind of "houses"?

_____ 5. What type of nutrition do algae have?

_____ 6. Name a colonial alga studied today.

_____ 7. What part of a bacterium can protect it from phagocytosis by white blood cells?

_____ 8. Where would you find the bacterial chromosome?

_____ 9. Once called the blue-green algae, cyanobacteria are now classified as what?

_____ 10. What do you call the projection that allows amoeboids to move and feed?

_____ 11. What type nutrition do protozoans have?

_____ 12. A mushroom consists of a stalk and _____.

_____ 13. What type of nutrition do fungi have?

_____ 14. What do fungi produce during both asexual and sexual reproduction?

_____ 15. Where do you find the basidia of a mushroom?

_____ 16. Name three morphological shapes associated with bacterial cells.

Thought Questions

17. Are all the organisms studied today in the same domain? Explain.

18. In general, how does sexual reproduction differ from asexual reproduction among fungi?

19. What is the benefit of sexual reproduction compared to asexual reproduction?

20. Penicillin prevents a bacterium from producing a unique molecule in its cell wall. Why would you expect pencillin to kill bacteria but not human cells?

25
Seedless Plants

Introduction

Plants are multicellar photosynthetic eukaryotes; their evolution is marked by adaptations to a land existence. Among aquatic green algae (phylum Chlorophyta), the **charophytes** are most closely related to the plants that now live on land. The charophytes have several features that would have promoted the evolution of land plants, including retention of and care of the zygote. The most successful land plants are those that protect all phases of reproduction (sperm, egg, zygote, and embryo) from drying out and that have an efficient means of dispersing offspring on land.

In this laboratory, you will have the opportunity to examine the various adaptations of plants to living on land. Although this lab will concentrate on land plant reproduction, you will also see that much of a land plant's body is covered by a **waxy cuticle** that prevents water loss, and that land plants require structural support to oppose the force of gravity and to lift their leaves up toward the sun. **Vascular tissue** offers this support and transports water to, and nutrients from, the leaves.

Also, you will examine the adaptations of nonvascular plants and the seedless vascular plants to a land existence, and you will see that reproduction in these plants requires an outside source of water. In Laboratory 26, we will note that reproduction in seed plants is not dependent on water at all.

25.1 The Evolution and Diversity of Land Plants

Figure 25.1 shows the evolution of land plants. What evolutionary events led to adaptation of plants to a land existence? _____

_____

Figure 25.1 Evolutionary relationships among the plants.

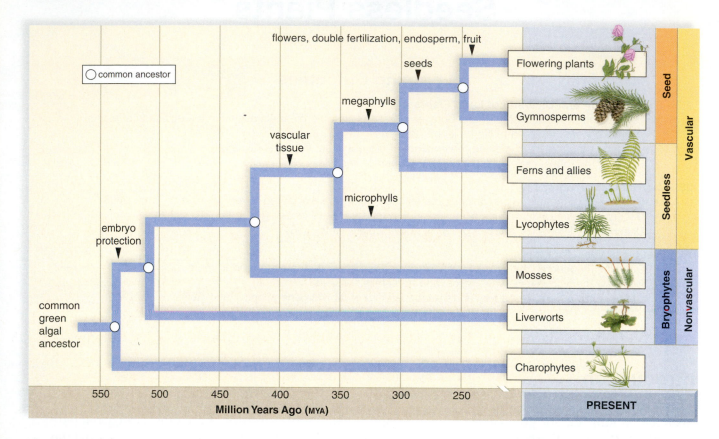

Algal Ancestor of Land Plants

In the evolutionary tree, the charophytes are green algae that share a common ancestor with land plants. This common ancestor may have resembled a Charales, such as *Chara,* which can live in warm, shallow ponds that occasionally dry up. Adaptation to such periodic desiccation may have facilitated the ability of certain members of the common ancestor population to invade land, and with time, become the first land plants.

Observation: Chara

Examine a living *Chara* (Fig. 25.2). How does it superficially resemble a land plant? _____

_____

   Chara is a filamentous green alga that consists of a primary branch and many side branches (Fig. 25.2). Each branch has a series of very long cells. Note where one cell begins and the other begins. Measure the length of one cell. _____ Gently pick up and handle *Chara* while you are examining it. What does it feel like? _____ Its cell walls are covered with calcium carbonate deposit.

Figure 25.2 *Chara*.

Chara is an example of a stonewort, the type of green alga believed to be most closely related to the land plants.

Chara, several individuals   One individual

branch

main axis

node

Conclusion: **Chara**

- What characteristics cause *Chara* to resemble land plants? _____

  _____

- Why are *Chara* called stoneworts? _____

Alternation of Generations

Land plants have a two-generation life cycle called **alternation of generations.**

1. The **sporophyte** (diploid) **generation** produces haploid spores by meiosis. Spores develop into a haploid generation, the gametophyte.

2. The **gametophyte** (haploid) **generation** produces **gametes** (eggs and sperm) by mitosis. The gametes then unite to form a diploid zygote.

   Figure 25.3 contrasts the plant life cycle (alternation of generations) with the animal life cycle **(diploid).**

1. In the plant life cycle, meiosis occurs during the production of _____.

2. In the human life cycle, meiosis occurs during the production of _____.

3. In the plant life cycle, the generation that produces gametes is (n or 2n) _____.

4. In the human life cycle, the individual that produces gametes is (n or 2n) _____.

Figure 25.3 Plant and animal life cycles.

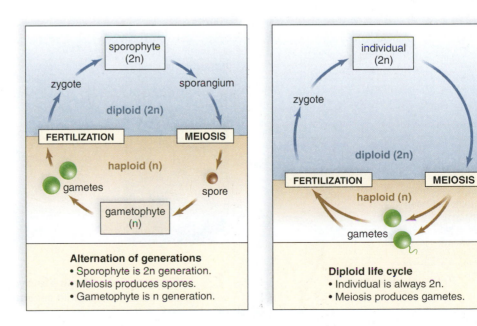

25.2 Nonvascular Plants

The nonvascular plants include the **mosses** (phylum Bryophyta) and **liverworts** (phylum Hepaticophyta). The gametophyte is dominant in all nonvascular plants. The gametophyte produces eggs within archegonia and flagellated sperm in **antheridia**. The sperm swim to the egg, and the embryo develops within **archegonia**. The nonvascular sporophyte grows out of the archegonium. The sporophyte is dependent on the female gametophyte (Fig. 25.4).

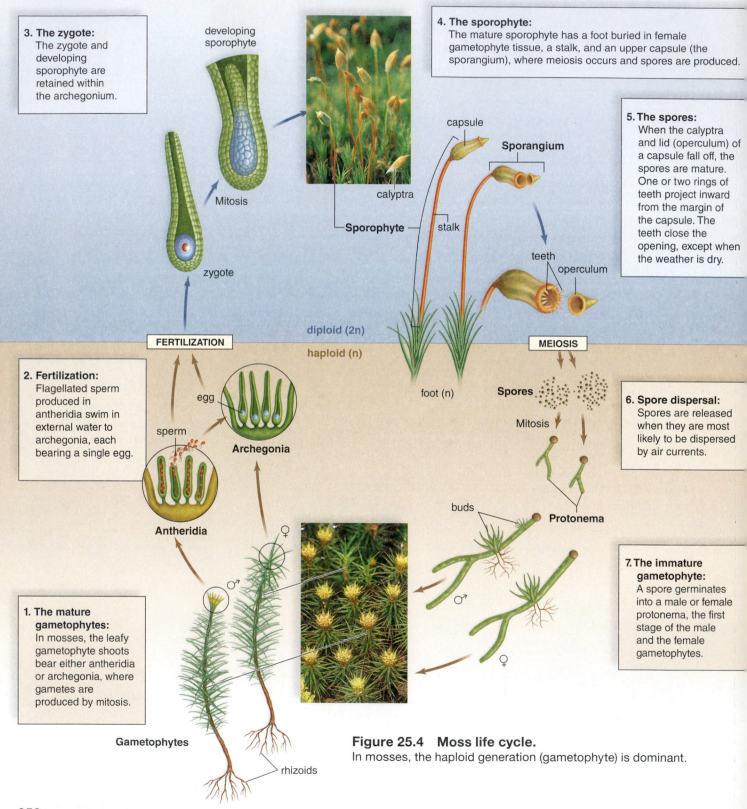

3. The zygote:
The zygote and developing sporophyte are retained within the archegonium.

developing sporophyte

Mitosis

zygote

4. The sporophyte:
The mature sporophyte has a foot buried in female gametophyte tissue, a stalk, and an upper capsule (the sporangium), where meiosis occurs and spores are produced.

capsule

Sporangium

calyptra

Sporophyte — stalk

teeth

operculum

5. The spores:
When the calyptra and lid (operculum) of a capsule fall off, the spores are mature. One or two rings of teeth project inward from the margin of the capsule. The teeth close the opening, except when the weather is dry.

FERTILIZATION

diploid (2n)

haploid (n)

MEIOSIS

foot (n)

Spores

Mitosis

2. Fertilization:
Flagellated sperm produced in antheridia swim in external water to archegonia, each bearing a single egg.

egg

sperm

Archegonia

Antheridia

♂ ♀

6. Spore dispersal:
Spores are released when they are most likely to be dispersed by air currents.

buds

Protonema

7. The immature gametophyte:
A spore germinates into a male or female protonema, the first stage of the male and the female gametophytes.

1. The mature gametophytes:
In mosses, the leafy gametophyte shoots bear either antheridia or archegonia, where gametes are produced by mitosis.

Gametophytes

rhizoids

Figure 25.4 Moss life cycle.
In mosses, the haploid generation (gametophyte) is dominant.

1. Put a check mark beside the phrases that describe nonvascular plants:

        I   II

_____ No vascular tissue to transport water   _____ Vascular tissue to transport water

_____ Flagellated sperm that swim to egg   _____ Sperm protected from drying out

_____ Dominant gametophyte   _____ Dominant sporophyte

2. Which listing of features (**I** or **II**) would you expect to find in a plant fully adapted to a land environment? _____ Explain. _____

3. In nonvascular plants, windblown spores are dispersal agents, and some species forcefully expel their spores. How are windblown spores an adaptation to reproduction on land? _____

_____

Observation: Moss Gametophyte

Living or Plastomount

Obtain a living moss gametophyte or a plastomount of this generation. Describe its appearance. _____

_____

    The leafy green shoots of a moss are said to lack true roots, stems, and leaves because, by definition, roots, stems, and leaves are structures that contain vascular tissue.

Microscope Slide

1. Study a slide of the top of a male moss shoot that contains antheridia, the reproductive structures where sperm are produced (Fig. 25.5). What is the chromosome number (choose 2n or n) of the sperm (see Fig. 25.4)? _____ Are the surrounding cells haploid or diploid? _____

2. Study a slide of the top of a female moss shoot that contains archegonia, the reproductive structures where eggs are produced (Fig. 25.6). What is the chromosome number of the egg? _____ Are the surrounding cells haploid or diploid? _____ When sperm swim from the antheridia to the archegonia, a zygote results. The zygote develops into the sporophyte. Is the sporophyte haploid or diploid? _____

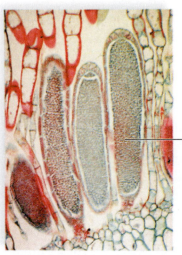

— sperm

Figure 25.5   Moss antheridia.
Flagellated sperm are produced in antheridia.

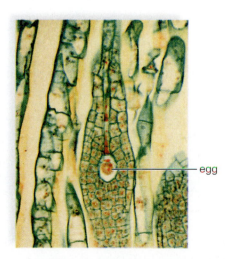

— egg

Figure 25.6   Moss archegonia.
Eggs are produced in archegonia.

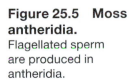

Living Sporophyte

1. Examine the living sporophyte of a moss in a minimarsh, or obtain a plastomount of a female shoot with the sporophyte attached. Identify the capsule **(sporangium)** where spores are produced and released through a lid (the **operculum**).

2. *Bracket and label the gametophyte and sporophyte in Figure 25.7a. Place an* n *beside the gametophyte and a* 2n *beside the sporophyte.*

Figure 25.7   Moss sporophyte.
a. The moss sporophyte is dependent on the female gametophyte. **b.** The sporophyte produces spores by meiosis.

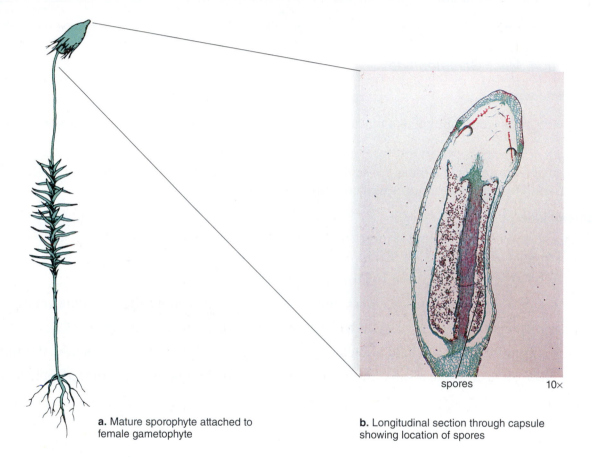

spores   10×

a. Mature sporophyte attached to female gametophyte

b. Longitudinal section through capsule showing location of spores

Microscope Slide

Examine a slide of a longitudinal section through a moss sporophyte (Fig. 25.7*b*). Identify the stalk and the sporangium, where spores are being produced. By what process are the spores being produced?

_____

When spores germinate, what generation begins to develop? _____

Why is it proper to say that spores are dispersal agents? _____

_____

Obtain a living sample of the liverwort *Marchantia* (Fig. 25.8) thallus, body of the plant. Examine it under the stereomicroscope. What generation is this sample? _____ Identify the following:

1. The gametophyte consisting of lobes. Each lobe is about a centimeter or so in length; the upper surface is smooth, and the lower surface bears numerous **rhizoids** (rootlike hairs).
2. **Gemma cups** on the upper surface of the thallus. These contain groups of cells called **gemmae** that can asexually start a new plant.
3. Disk-headed stalks, which bear antheridia, where flagellated sperm are produced.
4. Umbrella-headed stalks, which bear archegonia, where eggs are produced. Following fertilization, tiny sporophytes arise from the archegonia.

Figure 25.8   Liverwort, *Marchantia*.
a. Gemmae can detach and start a new plant. **b.** Antheridia are present in disk-shaped structures. **c.** Archegonia are present in umbrella-shaped structures.

a. Gemma cup

Thallus with gemmae cups

b. Male gametophytes bear antheridia.

c. Female gametophytes bear archegonia.

25.3   Seedless Vascular Plants

Seedless vascular plants include the **lycophytes** (club mosses) and **pteridophytes**—ferns and their allies, the whisk ferns and horsetails. These plants were prevalent and quite large during the swampy forest Carboniferous period. At that time, the coal deposits still used today were formed.

The sporophyte is dominant in the seedless vascular plants. The dominant sporophyte has adaptations for living on land; it has vascular tissue and produces windblown spores. The spores develop into a separate gametophyte generation that is very small (less than 1 centimeter). The gametophyte generation lacks vascular tissue and produces flagellated sperm.

1. Place a check mark beside the phrases that describe seedless vascular plants:

I | II

_____ Independent gametophyte   _____ Gametophyte dependent on sporophyte, which has vascular tissue

_____ Flagellated sperm   _____ Sperm protected from drying out

2. Which listing (**I** or **II**) would you expect to find in a plant fully adapted to a land environment? _____ Explain. _____

_____

Are seedless vascular plants fully adapted to living on land? _____

Observation: Lycophytes

Lycophytes (phylum Lycophyta) are commonly called **club mosses.** Lycophytes are representative of the first vascular plants. They have an aerial stem and a horizontal root (rhizome with attached rhizoids) both of which have vascular tissue. The leaves are called **microphylls** because they have only one strand of vascular tissue.

Ground Pines

1. Examine a living or preserved specimen of *Lycopodium* (Fig. 25.9).
2. Note the shape and the size of the microphylls and the branches of the stems.
3. Note the terminal clusters of leaves, called **strobili,** that are club-shaped and bear sporangia.
4. *Label strobili, leaves, stem, and rhizoids in Figure 25.9.*
5. Examine a prepared slide of a *Lycopodium* that shows the sporangia with spores inside. The spore develops into a tiny microscopic gametophyte that remains in the soil.

Figure 25.9 *Lycopodium.*

In the club moss *Lycopodium*, green photosynthetic stems are covered by scalelike leaves, and spore-bearing leaves are clustered in strobili.

Observation: Pteridophytes

Molecular (DNA) studies tell us that the whisk ferns, horsetails, and ferns are closely related.

Whisk Ferns

Psilotum is representative of **whisk ferns,** named for their resemblance to whisk brooms.

1. Examine a preserved specimen of *Psilotum,* and note that it has no leaves. The underground stem, called a **rhizome,** gives off upright, aerial stems with a dichotomous branching pattern, where bulbous sporangia are located (Fig. 25.10).

   What generation are you examining? _____

2. *Label a sporangium, the stem, and the rhizome in Figure 25.10b.*

Figure 25.10 Whisk fern, *Psilotum.*

This whisk fern has no roots or leaves—the branches carry on photosynthesis. The sporangia are yellow.

a.

b.

1. _____

scale _____

2. _____

rhizoid _____

3. _____

Horsetails

In **horsetails** a rhizome produces aerial stems that stand about 1.3 meters.

1. Examine *Equisetum,* a horsetail, and note the minute, scalelike leaves (Fig. 25.11).
2. Feel the stem. *Equisetum* contains a large amount of silica in its stem. For this reason, these plants are sometimes called scouring rushes and may be used by campers for scouring pots.
3. Strobili appear at the tips of the stems, or else special buff-colored stems bear the strobili. Sporangia are in the strobili.
4. *Label the horsetail in Figure 25.11.*

Figure 25.11 Horsetail.

In the horsetail *Equisetum,* whorls of branches or tiny leaves appear in the joints of the stem. The sporangia are borne in strobili.

1. _____
2. _____
3. _____
4. _____
5. _____
6. _____

Ferns

Ferns are quite diverse and range in size from those that are low growing and resemble mosses to those that are as tall as trees. The rhizome grows horizontally, which allows ferns to spread without sexual reproduction (Fig. 25.12). The gametophyte (called a **prothallus**) is small (about 0.5 cm) and usually heart-shaped. The prothallus contains both archegonia and antheridia. Ferns are largely restricted to moist, shady habitats because sexual reproduction requires adequate moisture. Why? _____

_____

Figure 25.12   Fern life cycle.
The sporophyte is the frond, and the gametophyte is the heart-shaped prothallus.

Fern spores are produced by meiosis in structures called sporangia, which in many species occur on the underside of large leaves called **fronds.** These leaves are called **megaphylls** because they are broad leaves with several strands of vascular tissue. The spores are released when special cells of the **annulus** (a line or ring of thickened cells on the outside of the sporangium) dry out, and the sporangium opens.

How do ferns disperse offspring? _____

Observe the ferns on display, and then complete Table 25.1.

Table 25.1 Fern Diversity	
Type of Fern	**Description of Frond**

Observation: Fern Sporophyte

Study the life cycle of the fern (Fig. 25.12), and find the sporophyte generation. This large, complexly divided leaf is known as a frond. Fronds arise from an underground stem called a rhizome.

Living or Preserved Frond

Examine a living or preserved specimen of a frond, and on the underside, notice a brownish clump called a **sorus** (pl., sori), each a cluster of many sporangia (Fig. 25.13).

What is being produced in the sporangia? _____

Given that this is the generation we call the fern, what generation is dominant in ferns? _____

Figure 25.13   Underside of frond leaflets.
Sori occur on the underside of frond leaflets.

sorus

Microscope Slide of Sorus

1. Examine a prepared slide of a cross section of a frond leaflet. Using Figure 25.14 as a guide, locate the fern leaf above and the sorus below.

Figure 25.14   Micrograph of cross section of a frond leaflet.
Micrograph of the internal anatomy of a sorus depicts many sporangia, where spores are produced.

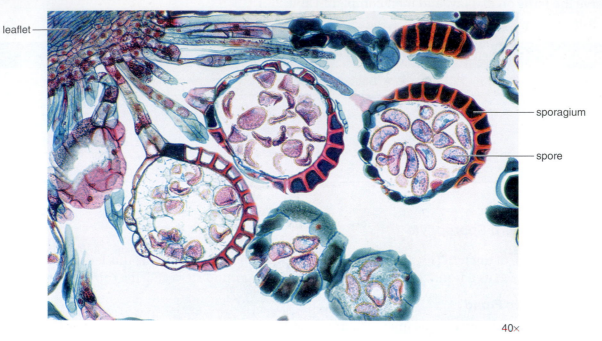

leaflet

sporagium

spore

40×

2. Within the sorus, find the sporangia and spores. Look for an **indusium** (not present in all species), a shelflike structure that protects the sporangia until they are mature. Does this fern have

an indusium? _____

Observation: Fern Gametophyte

Plastomount

1. Examine a plastomount showing the fern life cycle.
2. Notice the prothallus, a small, heart-shaped structure. Can you find this structure in your fern

   minimarsh (if available)? _____ The prothallus is the gametophyte generation of the fern. Most

   persons do not realize that this structure exists as a part of the fern life cycle. What is the function of

   this structure? _____

   _____

Microscope Slide

1. Examine whole-mount slides of fern prothallium-archegonia and fern prothallium-antheridia (Fig. 25.15).
2. If you focus up and down very carefully on an archegonium, you may be able to see an egg inside.
   What is being produced inside an antheridium? _____ When sperm produced in an antheridium
   swim to the archegonium in a film of water, what results? _____
   This structure develops into what generation? _____

Conclusions: Ferns

- How are ferns dispersed from one area to another? _____
- Is either generation in the fern dependent for any length of time on the other generation? _____
  Explain. _____

Figure 25.15 Fern prothallus.
The underside of the heart-shaped fern prothallus contains archegonia, where eggs are produced, and antheridia, where flagellated sperm are produced.

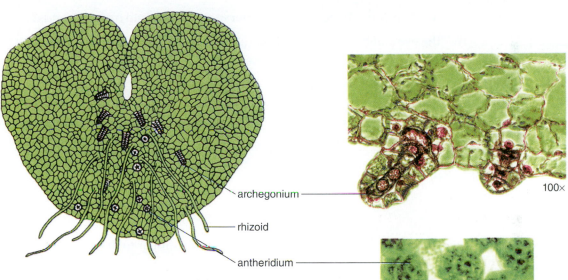

archegonium

rhizoid

antheridium

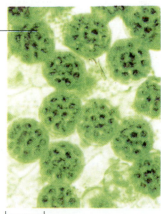

100×

50 μm

Laboratory Review 25

_____ 1. What type of life cycle do plants have?

_____ 2. Which generation in the plant life cycle produces gametes?

_____ 3. In the plant life cycle, meiosis produces _____.

_____ 4. What generation is dominant in mosses?

_____ 5. Leafy green shoot describes which generation of a moss?

_____ 6. What reproductive structure in mosses produces eggs?

_____ 7. What type of structure in mosses disperses the offspring?

_____ 8. What type of habitat do mosses need for sexual reproduction to occur?

_____ 9. What do ferns have that mosses lack?

_____ 10. What type of plant studied today has stems impregnated with silica?

_____ 11. Which generation is dominant in seedless vascular plants?

_____ 12. In whisk ferns, which part of the plant carries out photosynthesis?

_____ 13. A small, heart-shaped structure describes which generation in ferns?

_____ 14. Are sperm protected from drying out in the fern life cycle?

_____ 15. What do you call the clusters of sporangia often found on fern fronds?

_____ 16. The evolution of what type tissue allows vascular plants to typically grow taller than nonvascular plants?

Thought Questions

17. Contrast the life cycle of plants to that of animals (e.g., human beings).

18. Aside from appearance, how is the gametophyte generation in ferns similar to that of mosses? How is it different?

19. During the alternation of generations life cycle typical of plants, spores are produced via meiosis within the sporangium. Why must these spores develop into a gametophyte by mitosis rather than meiosis?

26

Seed Plants

Introduction

Among plants, **gymnosperms** and **angiosperms** are seed plants. Review the plant evolutionary tree (see Fig. 25.1) and note that the gymnosperms and angiosperms share a common ancestor, which produced **seeds,** a structure that contains the next sporophyte generation. We shall see that the use of seeds to disperse the next generation requires a major overhaul of the alternation of generations life cycle. Nonseed plants disperse the gametophyte, the n generation, by production of spores. Seed plants disperse the sporophyte, the 2n generation, by production of seeds. Which generation—gametophyte or sporophyte—is

better adapted to a land environment because it contains vascular tissue? _____

Nonseed plants utilize an archegonium to protect the egg; and following fertilization, the sporophyte develops immediately within the archegonium. Seed plants protect the entire female gametophyte within an **ovule,** which following fertilization develops into a sporophyte-containing seed. You will see that the formation of the ovule and also the **pollen grains** (contain male gametophytes) are radical innovations in the life cycle of seed plants. In gymnosperms, pollen grains are windblown, but many flowering plants have a mutualistic relationship with animals, particularly flying insects, which disperse pollen within the species. Animals also disperse the seeds of many flowering plants. Relationships with animals help explain why flowering plants are so diverse and widespread today.

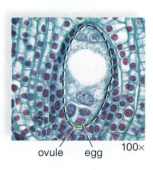

ovule   egg   100×

26.1 Life Cycle of Seed Plants

Figure 26.1 shows alternation of generations as it occurs in nonseed plants, and Figure 26.2 shows alternation of generations as it occurs in seed plants.

1. In which life cycle, nonseed or seed, do you note pollen sacs (microsporangia) and ovules (megasporangia)? _____ In which life cycle, nonseed or seed, do you note two types of spores, microspores and megaspores? _____ The formation of **heterospores** (unlike spores) is an innovation that leads to the production of pollen grains and the formation of ovules in seed plants. *Label heterospores where appropriate in Figure 26.2.* In which life cycle do you note male gametophyte (pollen grain) and female gametophyte (embryo sac in ovule)? _____

2. In nonseed plants, flagellated sperm must swim in external water to the egg. In seed plants, **pollination** (the transport of male gametophytes by wind or pollen carrier to the vicinity of female gametophytes) does not require external water. The production of pollen grains is an innovation in seed plants. *Label pollination where appropriate in Figure 26.2.*

3. In which life cycle does a seed appear between the zygote and the sporophyte? _____ What generation is present in a seed? _____ Formation of a seed from an embryo sac in the ovule is an innovation in seeds. In the life cycle of seed plants, note which structures are n and which are 2n. _____

Figure 26.2   Alternation of generations in flowering plants which produce seeds.

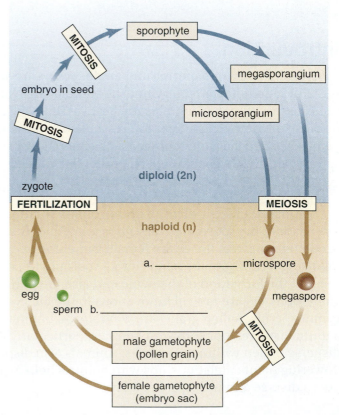

Figure 26.1   Alternation of generations in nonseed plants.

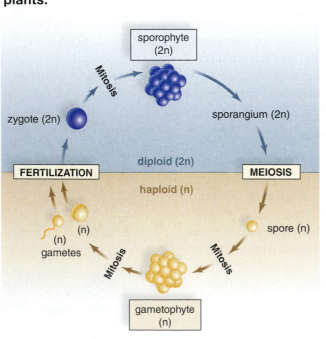

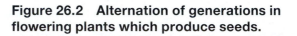

26.2 Gymnosperms

The term *gymnosperm* means "naked seed." The ovules and seeds of gymnosperm are exposed on a cone scale or some other comparable structure. Most gymnosperms do produce cones. Three groups of gymnosperms are especially familiar: cycads, ginkgoes, and conifers. Representatives of these plants may be available for you to examine. Compare them with Figure 26.3.

Conifers

The **conifers** (phylum Pinophyta) are by far the largest group of gymnosperms. Pines, hemlocks, and spruces are evergreen conifers because their leaves remain on the tree through all seasons. The cypress tree and larch (tamarack) are examples of conifers that are not evergreen.

Figure 26.3 Gymnosperms.
Three divisions of gymnosperms are well known: **a.** cycads, **b.** ginkgoes, and **c.** conifers.

a. Cycad, *Encephalartos humlis.*

b. Ginkgo, *Ginkgo biloba.* Female maidenhair tree with seeds.

c. Conifer, *Picea.* Spruce with pollen cones and seed cones.

Figure 26.4 Pine life cycle.

The sporophyte is the tree. The male gametophytes are windblown pollen grains shed by pollen cones. The female gametophytes are retained within ovules on seed cones. The ovules develop into windblown seeds.

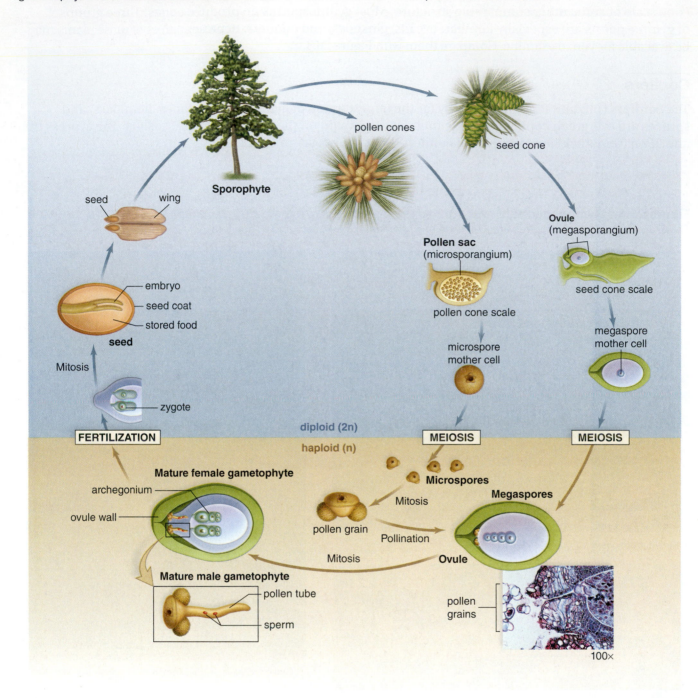

Pine Trees

The pine tree is the dominant sporophyte generation (Fig. 26.4). Vascular tissue extends from the roots, through the stem, to the leaves. Pine tree leaves are needlelike, leathery, and covered with a waxy, resinous cuticle. The **stomata**, openings in the leaves for gas exchange, are sunken. The structure of the leaf and the leaf's internal anatomy are adaptive to a drier climate.

In pine trees, the pollen sacs and ovules are located in cones. Pollen sacs on pollen cone scales contain cells which undergo meiosis to produce **microspores.** Microspores (n) become sperm-bearing **male gametophytes** (pollen grains). Seed cone scales bear ovules where cells undergo meiosis to produce **megaspores.** A megaspore (n) becomes an egg-bearing **female gametophyte.**

Pollination occurs when pollen grains are windblown to the seed cones. After **fertilization,** the egg becomes a sporophyte (2n) embryo enclosed within the ovule, which develops a seed coat. The seeds are winged and are dispersed by the wind.

1. Which part of the pine life cycle is the sporophyte? _____

2. Which part of the pine life cycle is the male gametophyte and which is the female gametophyte?

   _____

   _____

3. Where does fertilization and seed production occur? _____

   _____

Observation: Pine Leaf

Obtain a cluster of pine leaves (needles). A very short woody stem is at its base. Each type of pine has a typical number of leaves in a cluster (Fig. 26.5). How many leaves are in the cluster you are examining?

_____ What is the common name of your specimen? _____

b. Pitch pine

a. White pine

c. Red pine

Figure 26.5   Pine leaves (needles).
a. The needles of white pines are in clusters of five. **b.** The needles of pitch pines are in clusters of three. **c.** The needles of red pines are in clusters of two.

Preserved Cones

1. Compare a pine pollen cone to a pine seed cone. _____

_____

_____

Note the size and texture of the pollen cones relative to the larger, woody, seed cones.

2. Pollen cones. Remove a single scale (sporophyll) from the pollen cone and examine with a stereomicroscope. Note the two pollen sacs on the lower surface of each scale (Fig. 26.6*a*).

What do the pollen sacs produce? _____

_____

_____

3. Seed cones.
   a. Seed cones of three distinct ages are present. First-year cones are about 1 cm wide. Second-year cones are about 10 cm long. They are green and the scales are tightly closed together. In third-year cones the scales have opened, revealing the mature seeds at the base of the cone scale.
   b. If available, examine a first-year cone cut length-wise. The ovules are visible as small, milky-white domes at the base of the cone scale. Ovules will hold the female gametophyte generation, and then they become seeds.
   c. Examine mature seed cones. See if any seeds are present (Fig. 26.6*b*). Where are they located?

_____ Note the hardness of the seed coat.

What is the function of the seed coat? _____

If instructed to do so, use tweezers to pull out a seed and note the wing. What is the wing for?

_____

Replace the seed in the cone when you are finished.

4. Pine seeds. If available, examine a pine seed (called a pine nut) that has the seed coat removed. These seeds can be used for cooking foods such as pesto. Carefully cut the seed lengthwise and examine.

Can you find an embryo inside? _____

Figure 26.6 Pine cones.
a. The scales of pollen cones bear pollen sacs where microspores become pollen grains. **b.** The scales of seed cones bear ovules that develop into winged seeds.

pollen sac

scale

a. Pollen cones

wing

seed

scale

b. Seed cone

Conclusions: Pine Cones

- Are the pine seeds covered by tissue donated by the original sporophyte? _____

What does gymnosperm mean? _____

Explain. _____

Microscope Slides

1. Examine a prepared slide of a longitudinal section through a mature pine pollen cone. *Label a pollen sac in Figure 26.7a and a pollen grain in Figure 26.7b.* A pollen grain has a central body and two attached hollow bladders. How do these help in the dispersal of pine pollen? _____

_____

One pollen grain cell will divide to become two nonflagellated sperm, one of which fertilizes the egg after pollination. The other cell forms the **pollen tube** through which a sperm travels to the egg.

Figure 26.7 Pine pollen cone.
a. Pollen cones bear pollen sacs (microsporangia) in which microspores develop into pollen grains. **b.** Enlargement of pollen grains.

1. _____

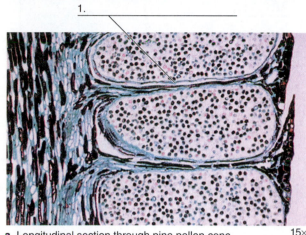

a. Longitudinal section through pine pollen cone, showing pollen grains within microsporangia. 15×

2. _____

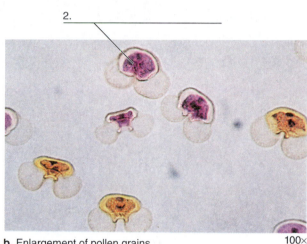

b. Enlargement of pollen grains. 100×

2. Examine a prepared slide of a longitudinal section through an immature pine seed cone. Seed cone scales bear ovules. The ovule contains a megaspore mother cell, which undergoes meiosis to produce four megaspores, three of which disintegrate. This megaspore (n) becomes an egg-bearing female gametophyte. *Label the ovule and the megaspore mother cell in Figure 26.8. Also, label the pollen grains that you can see just outside the ovule.*

Figure 26.8 Seed cone.
Seed cones bear ovules, each of which will contain a female gametophyte. Note pollen grains near the entrance.

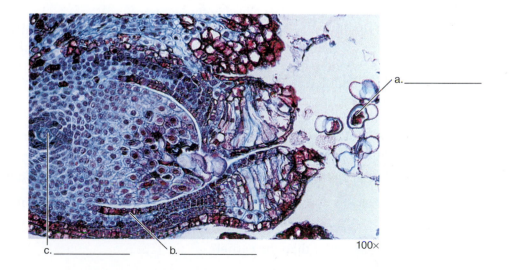

a. _____

c. _____ b. _____ 100×

26.3 Angiosperms

Flowering plants (phylum Magnoliophyta) are the dominant plants today. They occur as trees, shrubs, vines, and garden plants. At some point in their life cycle, all angiosperms bear flowers as also discussed in Laboratory 10.

Figure 26.9 Generalized flower.
A flower has four main kinds of parts: sepals, petals, stamens, and a carpel. A stamen has an anther and a filament. A carpel has a stigma, a style, and an ovary. An ovary contains ovules.

Observation: A Flower

1. With the help of Figure 26.9, identify the following structures on a model of a flower:
   a. **Receptacle:** The portion of a stalk to which the flower parts are attached.
   b. **Sepals:** An outermost whorl of modified leaves, collectively called the **calyx.** Sepals are green in most flowers. They protect a bud before it opens.
   c. **Petals:** Usually colored leaves that collectively constitute the **corolla.**
   d. **Stamen:** A swollen terminal **anther** and the slender supporting **filament.** The anther contains two pollen sacs, where microspores develop into microgametophytes (pollen grains).
   e. **Carpel:** A modified sporophyll consisting of a swollen basal ovary; a long, slender **style** (stalk); and a terminal **stigma** (sticky knob).
   f. **Ovary:** The enlarged part of the carpel that develops into a fruit.
   g. **Ovule:** The structure within the ovary where a megaspore develops into a female gametophyte (embryo sac). The ovule becomes a seed.

2. Carefully inspect a fresh flower. What is the common name of your flower? _____

3. Remove the sepals and petals by breaking them off at the base. How many sepals and petals are there?

   _____

4. Are the stamens taller than the carpel? _____

Flower parts in threes and multiples of three

Table 26.1 Monocots and Eudicots	
Monocots	**Eudicots**
One cotyledon	Two cotyledons
Flower parts in threes or multiples of three	Flower parts in fours or fives or multiples of four or five
Usually herbaceous	Woody or herbaceous
Usually parallel venation	Usually net venation
Scattered bundles in stem	Vascular bundles in a ring
Never woody	Can be woody

Flower parts in fours or fives and their multiples

5. Remove a stamen, and touch the anther to a drop of water on a slide. If nothing comes off in the water, crush the anther a little to release some of its contents. Place a coverslip on the drop, and observe with low- and high-power magnification. What are you observing? _____

6. Remove the carpel by cutting it free just below the base. Make a series of thin cross sections through the ovary. The ovary is hollow, and you can see nearly spherical bodies inside. What are these bodies? _____

_____

7. Flowering plants are divided into two classes, called **monocots** (class Monocotyledones) and **eudicots** (class Eudicotyledones). Table 26.1 lists significant differences between the two classes of plants. Is your flower a monocot or eudicot? _____

Life Cycle of Flowering Plants

The life cycle of a flowering plant is like that of the pine tree except for these innovations:

- The often brightly colored flower contains the pollen sacs and ovules. Locate these structures in Figure 26.10, and trace the life cycle of flowering plants from the sporophyte generation (the tree) through the various stages to the sporophyte generation once again.
- Pollination in flowering plants is sometimes accomplished by wind but more likely by the assistance of an animal pollinator. The pollinator acquires nutrients (e.g., nectar) from the flower and inadvertently collects pollen, which it takes to the next flower.
- Notice that flowering plants practice **double fertilization.** A mature pollen grain contains two sperm; one fertilizes the egg, and the other joins with the two polar nuclei to form **endosperm** (3n), which serves as food for the developing embryo.
- Also, flowering plants have seeds enclosed within fruits. Fruits protect the seeds and aid in seed dispersal. Sometimes, animals eat the fruits, and after the digestion process, the seeds are deposited far away from the parent plant. The term *angiosperm* means "covered seeds." The seeds of angiosperm are found in fruits, which develop from parts of the flower.

Figure 26.10 Flowering plant life cycle.

The parts of the flower involved in reproduction are the anthers of stamens and the ovules in the ovary of a carpel.

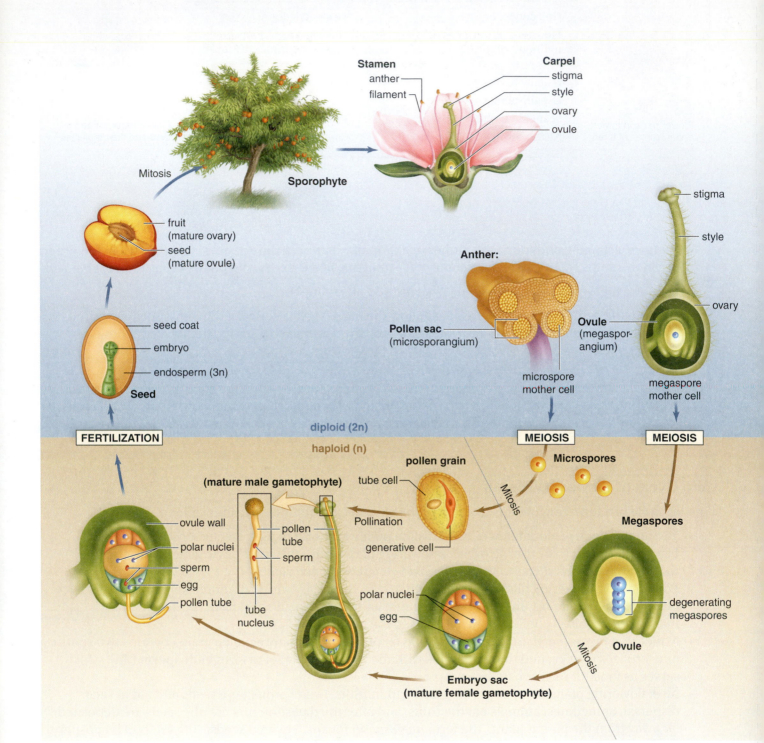

The Male Gametophyte

Notice that in the life cycle of flowering plants (Fig. 26.10), meiosis in pollen sacs produces four microspores, each of which will become a two-celled pollen grain. In flowering plants, **pollination** is the transfer of the pollen grain from the anther to the stigma, where the pollen grain germinates, and becomes the mature male gametophyte. How is pollination accomplished in the flowering plant

life cycle? _____

Observation: Pollen Grain Slide

1. A pollen grain released from the anther has two cells. The larger of the two cells is the **tube cell,** and the smaller is the **generative cell.** Examine a prepared slide of pollen grains. Identify the tube cell and the generative cell. Sketch your observation here.

2. The pollen grain's tube cell gives rise to the pollen tube. As it grows, the pollen tube passes through the stigma and grows through the style and into the ovary. Two sperm cells produced by division of the pollen grain's generative cell migrate through the pollen tube into the embryo sac. Examine a prepared slide of germinated pollen grains with pollen tubes. You should be able to see the tube cell nucleus and two sperm cells. What signifies that the mature pollen grain is the male gametophyte?

_____

*Experimental Procedure: Pollen Grains

Inoculate an agar-coated microscope slide with pollen grains. Invert the slide, and place it onto a pair of wooden supports in a covered petri dish. Leave the inverted slide in the covered petri dish for 1 hour. Then remove the slide, and examine it with the compound microscope. Have any of the pollen grains

germinated? _____

If so, describe. _____

_____

*Experimental Procedure from Richard Carter and Wayne R. Faircloth, *General Laboratory Studies,* Lab 17.2, 1991. Used with permission of Kendall/Hunt Publishing Company.

The Female Gametophyte

In the ovule, a megaspore undergoes three mitotic divisions to produce a seven-celled (eight-nuclei) female gametophyte called an **embryo sac** (Fig. 26.11).

Figure 26.11 Embryo sac in a lily ovule.
An embryo sac is the female gametophyte of flowering plants. It contains seven cells, one of which is the egg. The fate of each cell is noted.

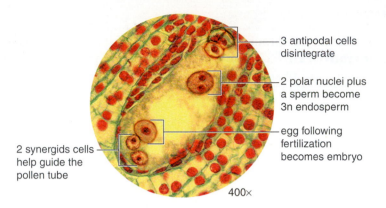

3 antipodal cells disintegrate

2 polar nuclei plus a sperm become 3n endosperm

egg following fertilization becomes embryo

2 synergids cells help guide the pollen tube

400×

Observation: Embryo Sac Slide

Examine the demonstration slide of the mature embryo sac of *Lilium*. Identify the egg labeled in Figure 26.11.

Answer these questions.

1. What signifies that the embryo sac is the female gametophyte? _____

_____

2. When the pollen tube delivers sperm to the embryo sac, double fertilization occurs. Due to double fertilization, what happens to the egg? _____

_____

What happens to the polar nuclei? _____

_____

3. Following fertilization, the ovules develop into seeds, and the ovary develops into the fruit.

What are the three parts of a seed (as shown in Figure 26.10)? _____

Label the flower remnants, the fruit, and the seeds in the drawings of a pea pod and apple provided. The flesh of an apple comes from the enlarged receptacle that grows up and around the ovary, while the ovary largely consists of the core.

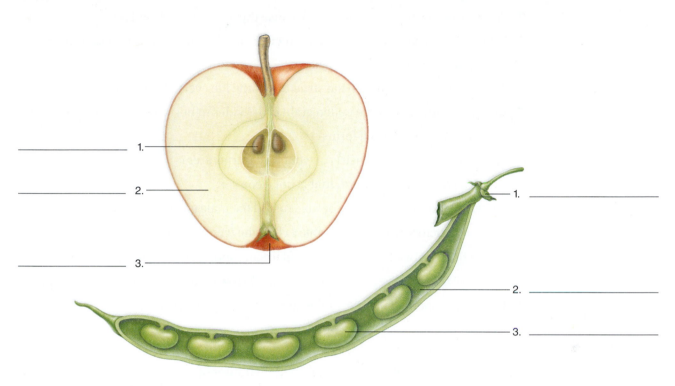

26.4   Comparison of Gymnosperms and Angiosperms

1. Complete Table 26.2 using "yes" or "no" to compare adaptations of gymnosperms and angiosperms.

Table 26.2 Comparison of Gymnosperms and Angiosperms					
	Heterospores	**Pollen grains/ Ovule**	**Cones**	**Flower**	**Fruit**
Gymnosperms					
Angiosperms					

2. What structure in gymnosperms and angiosperms delivers sperm to the vicinity of the egg?
   _____ Does delivery require external water? _____

3. What structure in gymnosperms and angiosperms becomes a seed? _____

4. The embryo of what generation is in a seed? _____

5. What innovation in angiosperms led to the production of seeds covered by fruit? _____

Laboratory Review 26

_____ **1.** What structure transports sperm to the ovule in seed plants?

_____ **2.** Which group of plants practices double fertilization?

_____ **3.** Gymnosperms have naked seeds because they are not enclosed by _____.

_____ **4.** What type of cell division produces microspores and megaspores in seed plants?

_____ **5.** In a conifer, is fruit present or absent?

_____ **6.** The pollen grain replaces what structure in the life cycle of seedless plants?

_____ **7.** In angiosperms, one pine pollen grain cell divides to become two nonflagellated sperm, and the other cell forms the _____.

_____ **8.** The carpel of a flower consists of what three structures?

_____ **9.** Name a type of gymnosperm that is not a conifer.

_____ **10.** On what structure would you be able to find the male gametophyte of a pine tree?

_____ **11.** In what structure would you be able to find the female gametophyte of a pine tree?

_____ **12.** Name the part of a flower that has a filament topped by the anther.

_____ **13.** What are the two classes of flowering plants? How do they differ from one another?

_____ **14.** How many sperm are in the pollen grain of a flowering plant?

Thought Questions

15. A pine tree, unlike a fern, is able to reproduce sexually in a dry environment. Explain.

16. What is the difference between pollination and fertilization?

17. Why are most pollen cones located at the tips of branches and not near the center of the tree?

18. How does the angiosperm life cycle differ from the gymnosperm life cycle? In what ways are these two life cycles similar to one another?

27

Introduction to Invertebrates

Introduction

In our survey of the animal kingdom, we will see that animals are very diverse in structure. Even so, all animals are multicellular and **heterotrophic,** which means their food consists of organic molecules made by other organisms. Consistent with the need to acquire food, animals have some means of locomotion by use of muscle fibers. Animals are always diploid, and during sexual reproduction, the embryo undergoes specific developmental stages.

> ⏱ **Planning Ahead** To see hydra and planarians feed, have students observe the animals at the start of lab, add food, and then check frequently until food engulfment occurs.

While we tend to think of animals in terms of **vertebrates** (e.g., dogs, fishes, squirrels), which have a backbone, most animal species are those that lack a backbone, commonly known as **invertebrates.** In this laboratory, we will examine those invertebrates that lack a true body cavity, called a **coelom.** A survey of the rest of the animal kingdom follows in Laboratory 28 and Laboratory 29.

27.1 Evolution of Animals

Today, molecular data is used to trace the evolutionary history of animals. These data tell us, as shown in the phylogenetic (evolutionary) tree (Fig. 27.1), that all animals share a common ancestor. This common ancestor was most likely a choanoflagellate consisting of a colony of flagellated cells. All but one of the phyla depicted in the tree consists of only invertebrates—the chordates contain a few invertebrates and also the vertebrates.

Figure 27.1 Evolutionary tree of animals.

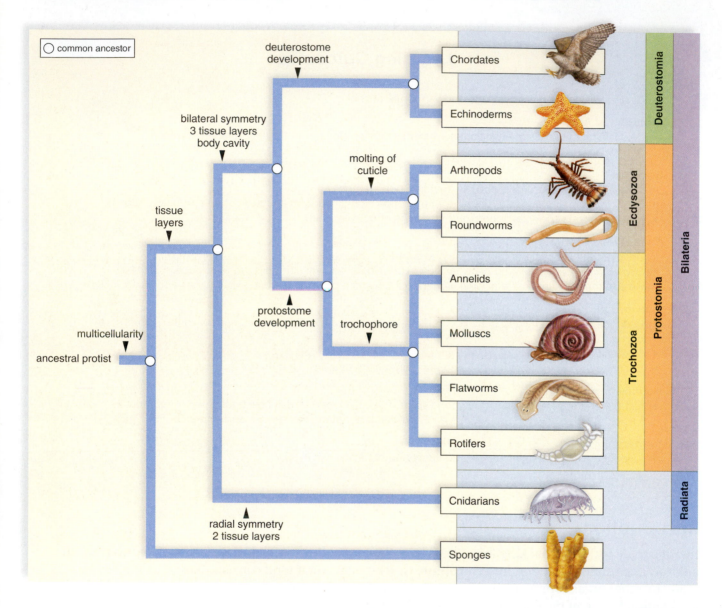

Certain anatomical features of animals are used in the tree. The first feature of interest is formation of tissue layers. Sponges have no true tissue layers. Which phyla in the tree have only two tissue layers? _____ The other phyla have three tissue layers.

Another feature of interest is **symmetry. Asymmetry** means the animal has no particular symmetry. Which phyla have **radial symmetry,** in which, as in a wheel, two identical halves are obtained no matter how the animal is longitudinally sliced? _____ The other phyla have **bilateral symmetry,** which means the adult animal has a definite right half and left half.

Finally, complex animals are either **protostomes** (first opening during development is the mouth) or **deuterostomes** (second opening during development is the mouth (the first one is the anus). Which pattern of development do the flatworms, rotifers, and roundworms—animals included in this laboratory—have? _____

27.2   Sponges

Sponges (phylum Porifera) live in water, mostly marine, attached to rocks, shells, and other solid objects. An individual sponge is typically shaped like a tube, cup, or barrel. Sponges grow singly or in colonies whose overall appearances vary widely. A single sponge can become a colony by asexual budding.

Anatomy of Sponges

Sponges consist of loosely organized cells and have no well-defined tissues. They are asymmetrical or radially symmetrical and **sessile** (immotile). They can reproduce asexually by budding or fragmentation, but they also reproduce sexually by producing eggs and sperm.

Sponges have a few types of specialized cells. Most notably they have flagellated **collar cells** (**choanocytes**). The movement of their flagella keep water moving through the pores into the central cavity and out the osculum of a sponge (Fig. 27.2). Collar cells also take in suspended food particles from the water and digest them for the benefit of all the other cells in a sponge.

Observation: Anatomy of Sponges

Preserved Sponge

1. Examine a preserved sponge (Fig. 27.2a). Note the main excurrent opening (**osculum**) and the multiple incurrent pores. Water is constantly flowing in through the pores and out the osculum. *Label the arrows in Figure 27.2a to indicate the flow of water.* Use the labels *water out* and *water in* through pores.
2. Examine a sponge specimen cut in half. Note the central cavity and the sponge wall. The wall is convoluted in some sponges, and the pores line small canals. Does this particular sponge have pore-lined canals? _____
3. You may be able to see **spicules,** fine projections over the body and especially encircling the osculum. Does this sponge have spicules? _____

Figure 27.2 Sponge anatomy.

a. Movement of water through pores into the central cavity and out the osculum is noted. **b.** Collar cells line the central cavity of a sponge and the movement of their flagella keeps the water moving through the sponge. **c.** Draw an enlargement of spicules here.

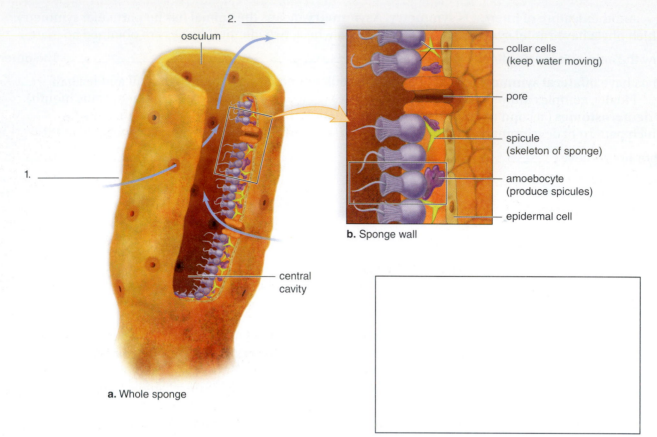

2. _____

osculum

1. _____

central cavity

a. Whole sponge

collar cells (keep water moving)

pore

spicule (skeleton of sponge)

amoebocyte (produce spicules)

epidermal cell

b. Sponge wall

c. Drawing of spicules by student

Prepared Slides

1. Examine a prepared slide of *Grantia*.

   a. Find the collar cells that line the interior (Fig. 27.2*b*). A sponge is a **sessile filter feeder.** Collar cells phagocytize (engulf) tiny bits of food that come through the pores along with the water flowing through the sponge. They then digest the food in food vacuoles. Explain the expression *sessile filter feeder.* _____

   _____

   b. Do you see any spicules? Do they project from the wall of a sponge? _____

   c. Depending on the sponge, spicules are made of either calcium carbonate, silica (glass), or protein. Calcium carbonate and silica produce hard, sharp spicules. Name two possible advantages of spicules to a sponge. _____

   _____

2. Examine a prepared slide of sponge spicules. What do you see? _____

   Draw a sketch of four spicules, each having a different appearance, in the space provided in Figure 27.2c.

Diversity of Sponges

Sponges are very diverse and come in many shapes and sizes. Some sponges live in fresh water although most live in the sea and are a prominent part of coral reefs, areas of abundant sea life discussed in the next section. Zoologists have described over 5,000 species of sponges, which are grouped according to the type spicule (Fig. 27.3).

a. Calcareous sponge, *Clathrina canariensis*

b. Bath sponge, *Xestospongia testudinaria*

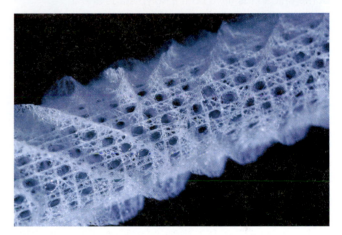

c. Glass sponge, *Euplectella aspergillum*

Figure 27.3 Diversity of sponges.
a. Calcareous (chalk) sponges have spicules of calcium carbonate. **b.** Bath sponges have a skeleton of spongin. **c.** Glass sponges have glassy spicules.

Conclusions: Anatomy of Sponges

- The anatomy and behavior of a sponge aid its survival and its ability to reproduce. How does a sponge
  a. Protect itself from predators? _____
  b. Acquire and digest food? _____
  c. Reproduce asexually and sexually? _____

27.3   Cnidarians

Cnidaria (phylum Cnidaria) are tubular or bell-shaped animals that live in shallow coastal waters, except for the oceanic jellyfishes. Two basic body forms are seen among cnidaria. The mouth of a **polyp** is directed upward, while the mouth of a jellyfish, or **medusa,** is directed downward. At one time, both body forms may have been a part of the life cycle of all cnidaria. When both are present, the sessile polyp stage produces medusae, and this motile stage produces egg and sperm (Fig. 27.4). Today in some cnidaria, one stage is dominant and the other is reduced; in other species, one form is absent altogether. Regardless, all cnidaria are radially symmetrical. How can radial symmetry be a benefit to an animal?

_____

How can a life cycle that involves two forms, called **polymorphism,** be of benefit to an animal, especially if one stage is sessile (stationary)? _____

_____

Figure 27.4   The life cycle of a cnidarian.
Some cnidarians have both a polyp stage and a medusa stage; in others, one stage may be dominant or absent altogether.

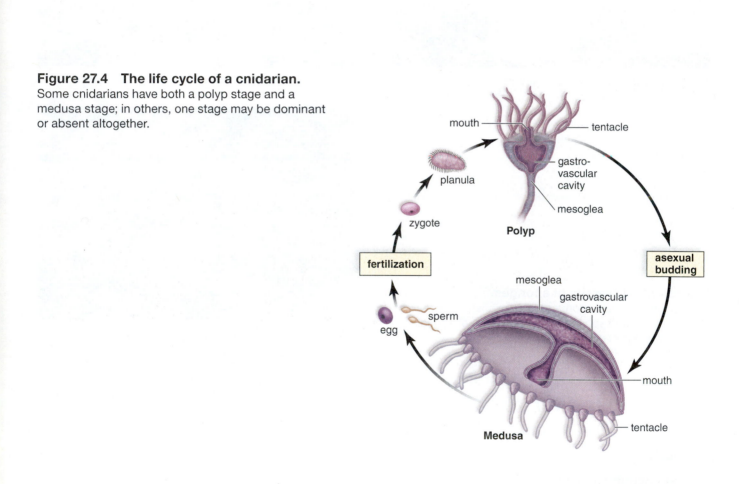

Anatomy of *Hydra*

Figure 27.5 shows the anatomy of a hydra, which will be studied as a typical cnidarian. Hydras exist only as sessile polyps; there is no alternate stage. Note the **tentacles** that surround the **mouth,** the large **gastrovascular cavity,** and the basal disk. A gastrovascular cavity has a single opening that is used as both an entrance for food and an exit for wastes.

Figure 27.5   Anatomy of *Hydra*.
Hydra typifies the anatomy of a cnidarian.

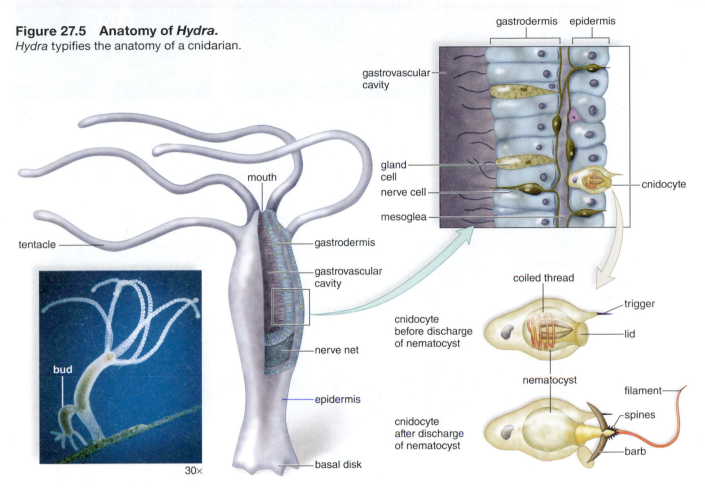

1. **Preserved *Hydra*.** With the aid of a hand lens, examine preserved specimens of *Hydra*. Hydras typically reproduce asexually by budding. Do you see any evidence of buds that are developing directly into small hydras? _____ The body wall can also produce ovaries and testes that produce eggs and sperm. The testes are generally located near the attachment of the tentacles; the ovaries appear farther down on the trunk, toward the basal disk.
2. **Slide of *Hydra*.** Examine prepared slides of cross and longitudinal sections of *Hydra*. With the help of Figure 27.5, note the epidermis, the mesoglea (a gelatinous material between the two tissue layers), and the gastrodermis, which lines the gastrovascular cavity. Switch to high power. Do you find any cells? _____ Describe them. _____

_____

Figure 27.6 Hydra feeding.

3. Living specimen. The tentacles of a hydra capture food, which is stuffed into the gastrovascular cavity (Fig. 27.6). Observe a living *Hydra* in a small petri dish for a few minutes. What is the current behavior of your hydra? _____

_____

Most often a hydra is attached to a hard surface by its basal disk. A hydra can move, however, by turning somersaults:

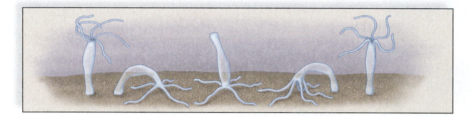

After a few minutes, tap the edge of the petri dish. What is the reaction of your hydra?

_____

_____

4. Mount a living *Hydra* on a depression glass slide with a coverslip and examine a tentacle. Unique to cnidarians are specialized stinging cells, called **cnidocytes**, which give the phylum its name. Each cnidocyte has a fluid-filled capsule called a **nematocyst** (see Fig. 27.5, far right), which contains a long, spirally coiled hollow thread. The threads trap and/or sting prey. Note the cnidocytes as swellings on the tentacles. Add a drop of vinegar (5% acetic acid) and note what happens to the cnidocytes. Did your hydra discard any nematocysts? _____ Describe. _____

_____

_____

Of what benefit is it to *Hydra* to have cnidocysts? _____

_____

Conclusions: Anatomy of Cnidarians

- The anatomy and behavior of a hydra aid its survival and its ability to reproduce. How does a hydra

  a. Acquire and digest food? _____

  _____

  b. Protect itself from predators? _____

  _____

  c. Reproduce asexually and sexually? _____

  _____

Diversity of Cnidarians

Cnidarians consist of a large number of mainly marine animals (Fig. 27.7). Sea anemones, sometimes called the flowers of the sea, are solitary polyps often found in coral reefs, areas of biological abundance in shallow tropical seas. Stony corals have a calcium carbonate skeleton that contributes greatly to the building of coral reefs. Portuguese man-of-war is a colony of modified polyps and medusae. Jellyfishes are a part of the zooplankton, suspended animals that serve as food for larger animals in the ocean.

Figure 27.7 Cnidarian diversity.

a. Sea anemone, *Corynactis*

b. Cup coral, *Tubastrea*

c. Portuguese man-of-war, *Physalia*

d. Jellyfish, *Aurelia*

27.4 Flatworms

Flatworms (phylum Platyhelminthes) are bilaterally symmetrical animals that can be either free-living or parasitic. Free-living flatworms, called planarians, are more complex than the cnidarians. In addition to the germ layers ectoderm and endoderm, mesoderm is also present. Flatworms are usually **hermaphroditic**, which means they possess both male and female sex organs (Fig. 27.8c). Why is it advantageous for an

animal to be hermaphroditic? _____

_____

Planarians practice cross-fertilization when the penis of one is inserted into the genital pore of the other. The fertilized eggs are enclosed in a cocoon and hatch as tiny worms in two or three weeks.

Planarians

Planarians such as *Dugesia* live in lakes, ponds, and streams, where they feed on small, living or dead organisms. In planarians, the three germ layers give rise to various organs aside from the reproductive organs (Fig. 27.8). The three-part **gastrovascular cavity** ramifies throughout the body; the excretory organs called **flame cells** (because their cilia reminded early investigators of a flickering flame of a candle) collect fluids from inside the body and send via a tube to an excretory pore; and the nervous system contains a brain and lateral nerve cords connected by transverse nerves. Therefore it is called **ladder-like**. *Complete the labels in both 27.8b and d. Label excretory canal, brain, and nerve cord.* Why would you expect an

animal that lives in fresh water to have a well-developed excretory system? _____

_____

Figure 27.8 Planarian anatomy.
a. When a planarian extends the pharynx, food is sucked up into a gastrovascular cavity that branches throughout the body. **b.** The excretory system has flame cells. **c.** The reproductive system has both male (blue) and female (pink) organs. **d.** The nervous system looks like a ladder.

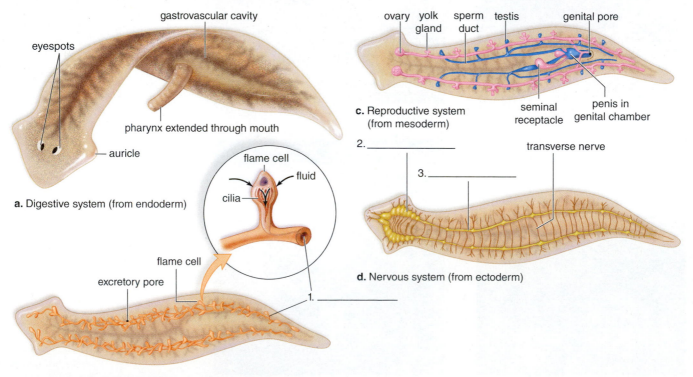

a. Digestive system (from endoderm)

b. Excretory system (from mesoderm)

c. Reproductive system (from mesoderm)

2. _____

3. _____

1. _____

d. Nervous system (from ectoderm)

Preserved Specimen

Examine a whole mount of a planarian that shows the branching gastrovascular cavity (Fig. 27.8*a*). What is the advantage of a gastrovascular cavity that ramifies through the body? _____

_____

Prepared Slide

Examine a cross section of a planarian under the microscope. Can you locate the structures shown in Figure 27.9? Does a planarian have a body cavity for its internal organs?

_____ Explain. _____

_____

Figure 27.9   Planarian micrograph.
Cross section of a planarian at the pharynx.

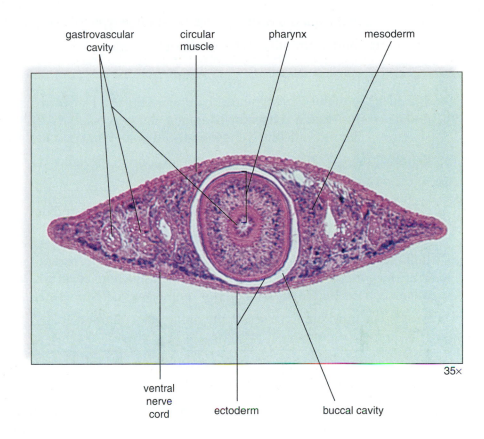

gastrovascular cavity   circular muscle   pharynx   mesoderm

ventral nerve cord   ectoderm   buccal cavity

35×

Living Specimen

1. Examine the behavior of a living planarian (Fig. 27.10) in a petri dish. Describe the behavior of the animal. _____

    _____

2. Does the animal move in a definite direction?_____ Why is it advantageous for a predator such as a planarian to have bilateral symmetry and a definite head region? _____

    _____

Figure 27.10 Micrograph of planarian feeding and the gastrovascular cavity.

a. Planarian feeding   10×   b. Gastrovascular cavity

3. Gently touch the animal with a probe. What three types of cells must be present for flatworms to be able to respond to stimuli and move about? _____

_____

The auricles on the side of the head are sense organs. Flatworms have well-developed muscles and a nervous system consisting of a brain and nerves.

4. If a strong light is available, shine it on the animal. What part of the animal would be able to detect light? _____ How does the animal respond to the light? _____

_____

5. Offer the worm some food, such as a small piece of liver, and describe its manner of eating.

_____

Roll the animal away from its food and note the pharynx extending from the body (Fig. 27.10).

6. Transfer your worm to a concave depression slide and cover with a coverslip. Examine with a microscope and note the cilia on the ventral surface. Numerous gland cells secrete a mucous material that assists movement. Describe the mode of locomotion. _____

_____

Conclusions: Planarians

- Planarians, with three germ layers, are more complex than cnidarians. Contrast a hydra with a planarian by stating in Table 27.1 significant organ differences between them.
- Planarians have no respiratory or circulatory system. As with cnidarians, each individual _____ takes care of its own needs for these two life functions.

Table 27.1 Contrasts Between a Hydra and a Planarian			
	Digestive System	Excretory System	Nervous Organization
Hydra			
Planarian			

Tapeworms

Tapeworms are parasitic flatworms known as cestodes. They live in the intestines of vertebrate animals, including humans (Fig. 27.11). The worms consist of a **scolex** (head), usually with suckers and hooks, and **proglottids** (segments of the body). Ripe proglottids detach and pass out with the host's feces, scattering fertilized eggs on the ground. If pigs or cattle happen to ingest these, larvae called bladder worms develop and eventually become encysted in muscle, which humans may then eat in poorly cooked or raw meat. A bladder worm that escapes from a cyst develops into a mature tapeworm attached to the intestinal wall.

1. How do humans get infected with the pig tapeworm? _____

2. What is the function of a tapeworm's hooks and suckers? _____

3. Why would you expect a tapeworm to have a reduced digestive system? _____

4. Proglottids mature into "bags of eggs." Given the life cycle of the tapeworm, why might a tapeworm have an expanded reproductive system compared to a planarian? _____

Figure 27.11   Life cycle of the tapeworm *Taenia*.
The pig host is the means by which the worm is dispersed to the human host.

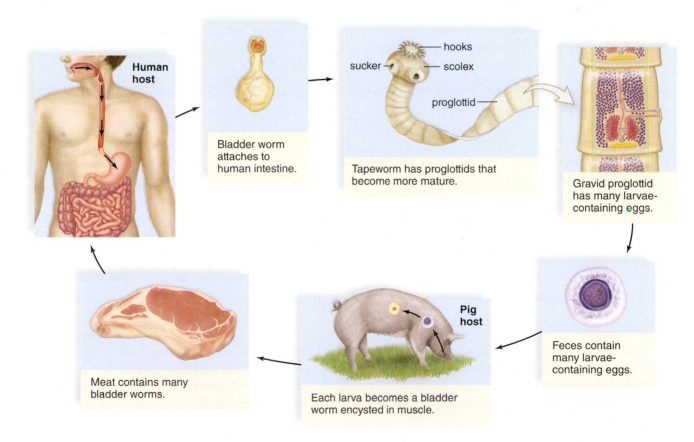

1. Examine a preserved specimen and/or slide of *Taenia pisiformis*, a tapeworm.
2. With the help of Figure 27.12, identify the scolex, with hooks and suckers, and the proglottids.

Figure 27.12 Anatomy of *Taenia*.
The adult worm is modified for its parasitic way of life. It consists of a scolex and many proglottids, which become bags of eggs.

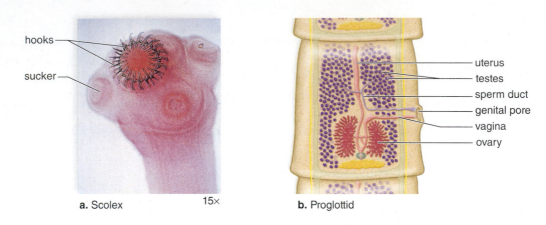

a. Scolex   15×   **b.** Proglottid

27.5 Roundworms

Like planarians, **roundworms** (phylum Nematoda) have three germ layers, bilateral symmetry, and various organs, including a well-developed nervous system. Both planarians and roundworms are nonsegmented; the body has no repeating units. In addition, roundworms have the following features:

1. **Complete digestive tract:** The digestive tract has both a mouth and an anus. What is the advantage of this? _____

_____

2. **Pseudocoelom:** A body cavity, which allows space for the organs, is incompletely lined with mesoderm.

How would the presence of these two features lead to complexity? For example, how would a complete digestive system lead to a greater number of specialized organs such as both a small intestine that assists digestion and a large intestine that assists elimination? _____

_____

How would a spacious body cavity promote the evolution of a greater number of diverse internal organs such as a pancreas and a liver? _____

_____

Roundworms are found in all aquatic habitats and in damp soil. Some even survive in hot springs, deserts, and cider vinegar. They parasitize (take nourishment from) both plants and animals. They are significant crop pests and also cause disease in humans. Both pinworms and hookworms are roundworms that cause intestinal difficulties; trichinosis and elephantiasis are also caused by roundworms.

Ascaris

Ascaris, a large, primarily tropical intestinal parasite, is often studied as an example of this phylum.

Observation: Ascaris

Examine preserved specimens of *Ascaris,* both male and female (Fig. 27.13). In roundworms, the sexes are separate. The male is smaller and has a curved posterior end. Be sure to examine specimens of each sex.

Figure 27.13   Roundworm anatomy.
a. Photograph of male *Ascaris.* **b.** Male reproductive system. **c.** Photograph of female *Ascaris.* **d.** Female reproductive system.

a. Male *Ascaris*

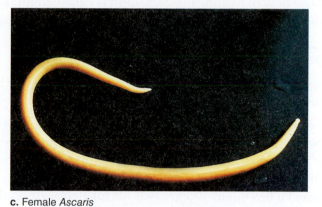

c. Female *Ascaris*

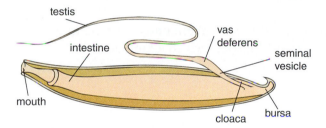

b. Male reproductive system

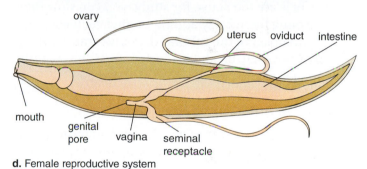

d. Female reproductive system

Trichinella

Trichinella is a parasitic roundworm that causes the disease **trichinosis.** When pigs or humans eat raw or undercooked pork infected with *Trichinella* cysts, juvenile worms are released in the digestive tract where they penetrate the wall of the small intestine and mature sexually. After male and female worms mate, females produce juvenile worms that migrate and form cysts in various muscles (Fig. 27.14). A human with trichinosis has muscular aches and pains that can lead to death if the respiratory muscles fail.

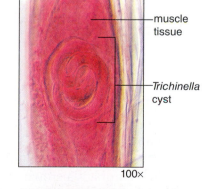

Figure 27.14   Larva of the roundworm *Trichinella* embedded in a muscle.
A larva coils in a spiral and is surrounded by a sheath derived from a muscle fiber.

Observation: Trichinella

1. Examine preserved, infected muscle or a slide of infected muscle, and locate the *Trichinella* cysts, which contain the juvenile worms.

2. How can trichinosis be prevented in humans? _____

_____

3. How can pig farmers help to stamp out trichinosis so that humans are not threatened by the disease?

_____

Filarial Worm

A roundworm called a **filarial worm** infects lymphatic vessels and blocks the flow of lymph. The condition is called **elephantiasis** because when a leg is affected, it becomes massively swollen.

Vinegar Eels

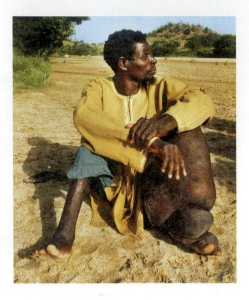

Vinegar eels are tiny, free-living nematodes that can live in unpasteurized vinegar.

1. Examine live vinegar eels, and observe their active, whiplike swimming movements. This thrashing motion may be a result of nematodes having longitudinal muscles only; they lack circular muscles.
2. Select a few larger vinegar eels for further study, and place them in a small drop of vinegar on a clean microscope slide. If the eels are too active for study, you can slow them by briefly warming them or by adding methyl cellulose.
3. Try to observe the tubular digestive tract, which begins with the mouth and ends with the anus. Also, you may be able to see some of the reproductive organs, particularly in a large female vinegar eel.

Conclusion: Anatomy of Roundworms

- Nematodes are extremely plentiful, in terms of both their variety and their overall number. From your knowledge of adaptive radiation, explain why there might be so many different types of

   nematodes. _____

   _____

   _____

27.6   Rotifers

Rotifers (phylum Rotifera) are common and abundant freshwater animals. They are important constituents of the plankton of lakes, ponds, and streams, and are a significant food source for many species of fish and other animals.

Like roundworms, rotifers have a pseudocoelom and a complete digestive tract with a mouth and anus. The corona is a crown of cilia around the mouth. To some, the movement of the cilia resembles that of a rapid wheel; and in Latin, *rotifer* means "wheel-bearer." The cilia draw water into the mouth and from there food is ground up by trophi (jaws) before entering the stomach from which nondigested remains pass through the cloaca and anus (Fig. 27.15).

Figure 27.15 Live *Philodina*, a common rotifer.
a. Micrograph. **b.** Line art.

a.   250×

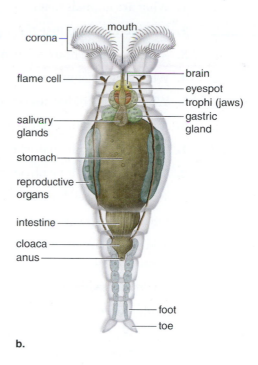

b.

corona
mouth
flame cell
brain
eyespot
trophi (jaws)
salivary glands
gastric gland
stomach
reproductive organs
intestine
cloaca
anus
foot
toe

Observation: Rotifers

Living Specimen

1. Use a pipette to obtain a living rotifer specimen near a clump of vegetation. Place the liquid and rotifer on a concave depression slide. Do not add a cover slip. Study the animal's behavior and appearance.

2. Describe the rotifer's behavior. _____

    _____

3. Observe the elongated, cylindrical body that can be divided into three general regions: the head, the trunk, and a posterior foot. The rotifers have no true segmentation, but the cuticle covering the body can be divided into a number of superficial segments.

4. Observe the telescoping of the segments when the animal retracts its head.

5. Note also the large, ciliated **corona** at the anterior end. Most rotifers have a conspicuous corona that serves both for locomotion and for feeding. It creates a current that brings smaller microorganisms (e.g., algae, protozoans, bacteria) close enough to be swallowed.

_____ 1. What are animals that lack a backbone called?

_____ 2. Water enters a sponge through _____ in the body wall.

_____ 3. Animals with a body cavity have a _____.

_____ 4. What type of gametes (sex cells) does a hermaphroditic animal produce?

_____ 5. What do the cnidocytes (stinging cells) of cnidarians contain?

_____ 6. A nerve net is characteristic of what group of animals?

_____ 7. Which of the animal phyla studied today contain three tissue layers?

_____ 8. Which of the animal phyla studied today contains animals with radial symmetry?

_____ 9. What type of cell lines the interior cavity of a sponge?

_____ 10. On what basis can you tell one type of sponge from another?

_____ 11. The cnidarian life cycle often includes two phases. One phase is absent in *Hydra*. Which phase is present?

_____ 12. *Hydra* uses which two anatomical structures to obtain food?

_____ 13. Name two characteristics of a planarian that may be associated with its free-living lifestyle.

_____ 14. What type of excretory system do planarians have?

_____ 15. What type of coelom does *Ascaris* have?

_____ 16. Upon examining a roundworm, how would you know it has a complete digestive tract?

_____ 17. What type of meat must be cooked thoroughly to prevent trichinosis?

_____ 18. What type of nervous system does a planarian have?

Thought Questions

19. Explain the difference between radial and bilateral symmetry, and associate these with the lifestyle of one of the animals studied.

20. Relate the number of germ layers to the complexity of the animal.

21. Why are planarians considered more complex than cnidarians?

28

Invertebrate Coelomates

Introduction

Animals are distinguished by their body cavity. The **acoelomates** have no body cavity, the **pseudocoelomates** have a body cavity incompletely lined with mesoderm, and the **coelomates** have a body cavity completely lined with mesoderm. (**Mesoderm,*** you'll recall is the last germ layer to develop.) A coelom offers many advantages:

- The digestive system and body wall can move independently.
- Internal organs can become more complex.
- Coelomic fluid can assist respiration, circulation, and excretion.
- The coelom also serves as a hydrostatic skeleton because muscles can work against a fluid-filled cavity.

All the phyla considered in today's laboratory are coelomates. The molluscs, annelids, and arthropods are protostomes, animals in which the first *(protos)* embryonic opening becomes the mouth *(stoma)*, while the echinoderms studied in this lab and the vertebrates studied in Laboratory 29 are deuterostomes. In the deuterostomes, the first opening becomes the anus, and the second *(deutero)* opening becomes the mouth.

* See Laboratory 19.

28.1   Molluscs

Most **molluscs** (phylum Mollusca) are marine, but there are also some freshwater and terrestrial molluscs (Fig. 28.1). Among molluscs, the grazing marine herbivores, known as **chitons**, have a body flattened dorsoventrally covered by a shell consisting of eight plates (Fig. 28.1a). The **bivalves** contain marine and freshwater sessile filter feeders, such as clams and scallops, with a body enclosed by a shell consisting of two valves (Fig. 28.1b). The **gastropods** contain marine, freshwater, and terrestrial species. In snails, the shell, if present, is coiled (Fig. 28.1c). The **cephalopods** contain marine active predators, such as squids and nautiluses. Tentacles are about the head (Fig. 28.1d).

All molluscs have a three-part body consisting of (1) a muscular **foot** specialized for various means of locomotion; (2) **visceral mass** that includes the internal organs; and (3) a **mantle,** a thin tissue that encloses the visceral mass and may secrete a shell. **Cephalization** is the development of a head region. On the lines provided in Figure 28.1, write *cephalization* or *no cephalization* as appropriate for this mollusc.

a. Chitons, *Tonicella*

Figure 28.1   Molluscan diversity.
a. You can see the exoskeleton of this chiton but not its dorsally flattened foot. **b.** A scallop doesn't have a foot but it does have strong adductor muscles to close the shell. In this specimen, the edge of the mantel bears tentacles and many blue eyes. **c.** A gastropod, such as a snail, is named for the location of its large foot beneath the visceral mass. **d.** In a cephalopod, such as this nautilus, a funnel (its foot) opens in the area of the tentacles and allows it to move by jet propulsion.

b. Scallop, *Argopecten*, is a bivalve.

c. Snail, *Helix*, is a gastropod.

d. Nautilus, *Nautilus*, is a cephalopod.

Anatomy of a Clam

Clams are bivalved because they have right and left shells secreted by the mantle. Clams have no head, and they burrow in sand by extending a **muscular foot** between the valves. Clams are **filter feeders** and feed on debris that enters the mantle cavity. In the visceral mass, the blood leaves the heart and enters sinuses (cavities) by way of anterior and posterior aortas. There are many different types of clams. The one examined here is the freshwater clam *Venus*.

External Anatomy

1. Examine the external shell (Fig. 28.2) of a preserved clam *(Venus)*. The shell is an **exoskeleton.**
2. Find the posterior and anterior ends. The more pointed end of the **valves** (the halves of the shell) is the posterior end.
3. Determine the clam's dorsal and ventral regions. The valves are hinged together dorsally.
4. What is the function of a heavy shell? _____

Figure 28.2   External view of the clam shell.

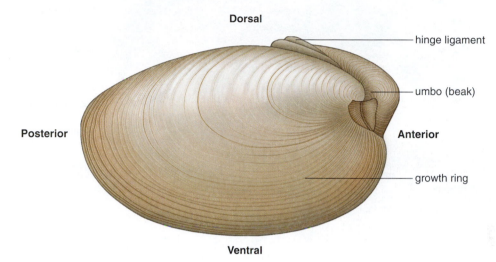

Internal Anatomy

1. Place the clam in the dissecting pan, with the **hinge ligament** and **umbo** (blunt dorsal protrusion) down. Carefully separate the **mantle** from the right valve by inserting a scalpel into the slight opening of the valves. What is a mantle? _____

   _____

2. Insert the scalpel between the mantle and the valve you just loosened.
3. The **adductor muscles** hold the valves together. Cut the adductor muscles at the anterior and posterior ends by pressing the scalpel toward the dissecting pan. After these muscles are cut, the valve can be carefully lifted away. What is the advantage of powerful adductor muscles? _____

   _____

4. Examine the inside of the valve you removed. Note the concentric lines of growth on the outside, the hinge teeth that interlock with the other valve, the adductor muscle scars, and the mantle line. The inner layer of the shell is mother-of-pearl.
5. Examine the rest of the clam (Fig. 28.3) attached to the other valve. Notice the adductor muscles and the mantle, which lies over the visceral mass and foot.
6. Bring the two halves of the mantle together. Explain the term *mantle cavity.* _____

   _____

7. Identify the **incurrent** (more ventral) and **excurrent siphons** at the posterior end (Fig. 28.3). Explain how water enters and exits the mantle cavity. _____

   _____

Figure 28.3 Anatomy of a bivalve.

The mantle has been removed to reveal the internal organs. **a.** Drawing. **b.** Dissected specimen.

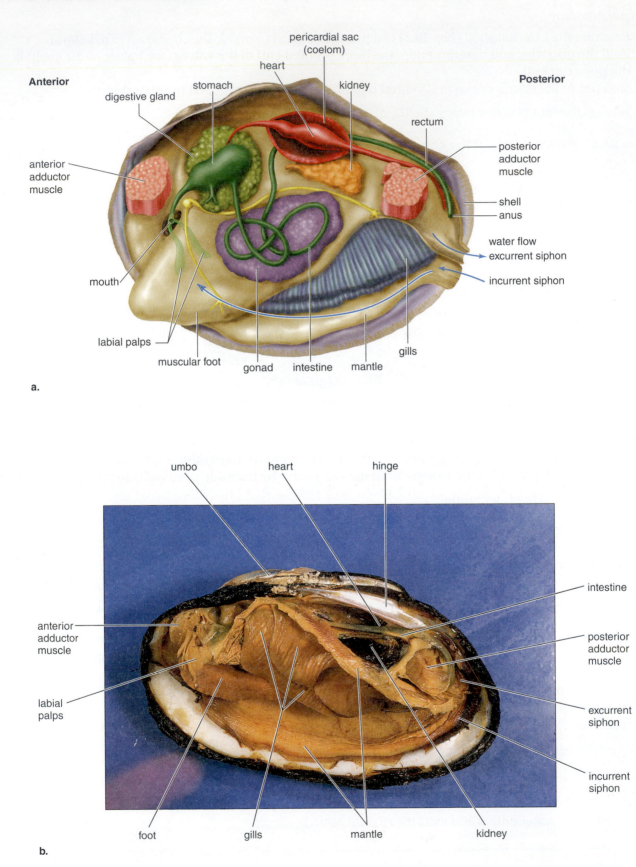

a.

b.

8. Cut away the free-hanging portion of the mantle to expose the **gills.** Does the clam have a respiratory organ? _____

    If so, what type of respiratory organ? _____

9. A mucous layer on the gills entraps food particles brought into the mantle cavity, and the cilia on the gills convey these food particles to the mouth. Why is the clam called a filter feeder?

    _____

    _____

10. The nervous system is composed of three pairs of ganglia (located anteriorly, posteriorly, and in the foot), all connected by nerves. The clam does not have a brain. A ganglion contains a limited number of neurons, whereas a brain is a large collection of neurons in a definite head region.

11. Identify the **foot,** a tough, muscular organ for locomotion, and the **visceral mass,** which lies above the foot and is soft and plump. The visceral mass contains the digestive and reproductive organs.

12. Identify the **labial palps** that channel food into the open mouth.

13. Identify the **anus,** which discharges into the excurrent siphon.

14. Find the **intestine** by its dark contents. Trace the intestine forward until it passes into a sac, the clam's only evidence of a coelom.

15. Locate the **pericardial sac (pericardium)** that contains the heart. The intestine passes through the heart. The heart pumps blood into the aortas, which deliver it to blood sinuses in the tissues.

    A clam has an **open circulatory system.** Explain. _____

    _____

16. Cut the visceral mass and the foot into exact left and right halves, and examine the cut surfaces. Identify the digestive glands, greenish-brown; the stomach, embedded in the digestive glands; and the intestine, which winds about in the visceral mass. Reproductive organs are also present.

Anatomy of a Squid

Squids are cephalopods because they have a well-defined head; the foot became the funnel surrounded by two arms and the many tentacles about the head. The head contains a brain and bears sense organs. The squid moves quickly by jet propulsion of water, which enters the mantle cavity by way of a space that encircles the head. When the cavity is closed off, water exits by means of the funnel. Then the squid moves rapidly in the opposite direction.

The squid seizes fish with its tentacles; the mouth has a pair of powerful, beaklike jaws and a **radula,** a beltlike organ containing rows of teeth. The squid has a **closed circulatory system** composed of vessels and three hearts, one of which pumps blood to all the internal organs, while the other two pump blood to the gills located in the mantle cavity.

Observation: Anatomy of a Squid

1. Examine a preserved squid.

2. Refer to Figure 28.4 for help in identifying the mouth (defined by beaklike jaws and containing a radula) and the tentacles and arms, which encircle the mouth.

3. Locate the head with its sense organs, notably the large, well-developed eye.

4. Find the funnel, where water exits from the mantle cavity, causing the squid to move backward.

5. If the squid has been dissected, note the heart, gills, and blood vessels.

Figure 28.4 Anatomy of a squid.

The squid is an active predator and lacks the external shell of a clam. It captures fish with its tentacles and bites off pieces with its jaws. A strong contraction of the mantle forces water out the funnel, resulting in "jet propulsion." **a.** Drawing. **b.** Dissected specimen.

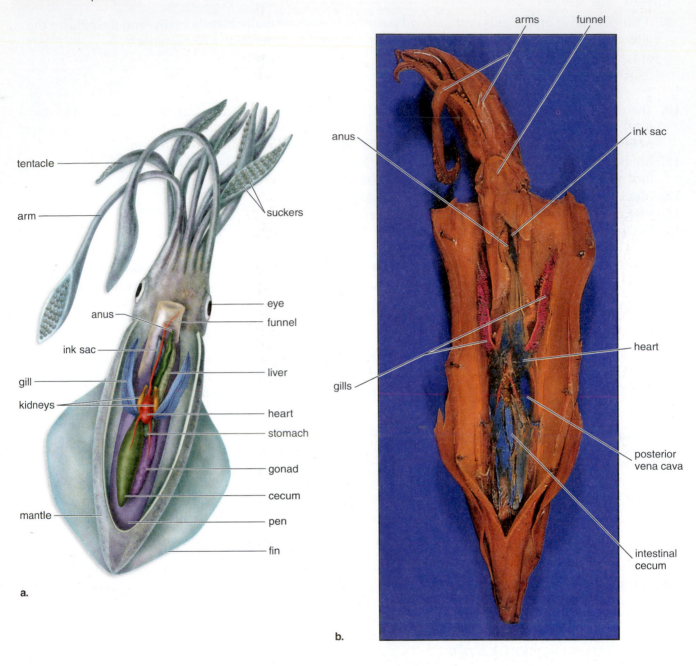

a.

b.

Conclusion: Comparison of Clam to Squid

- Compare clam anatomy to squid anatomy by completing Table 28.1.
- Explain how both clams and squids are adapted to their way of life.

_____

_____

_____

Table 28.1 Comparison of Clam to Squid

	Clam	Squid
Feeding mode		
Skeleton		
Circulation		
Cephalization		
Locomotion		
Nervous system	Three separate ganglia	

28.2 Annelids

Annelids (phylum Annelida) are the segmented worms, so called because the body is divided into a number of segments and has a ringed appearance. The circular and longitudinal muscles work against the fluid-filled coelom to produce changes in width and length (Fig. 28.5). Therefore, annelids are said to have a **hydrostatic skeleton.**

Figure 28.5 Locomotion in the earthworm.
Contraction of first circular muscles and then longitudinal muscles allow the earthworm to move forward.

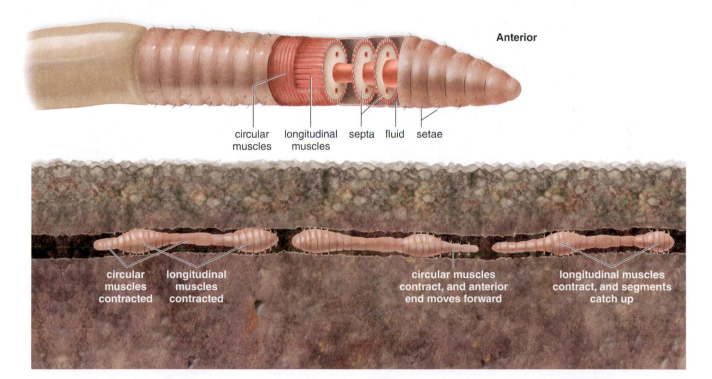

circular muscles   longitudinal muscles   septa   fluid   setae   Anterior

circular muscles contracted   longitudinal muscles contracted   circular muscles contract, and anterior end moves forward   longitudinal muscles contract, and segments catch up

Among annelids (Fig. 28.6), **polychaetes** have many slender bristles called **setae.** The polychaetes, almost all marine, are plentiful from the intertidal zone to the ocean depths. They are quite diverse, ranging from jawed forms that are carnivorous to fanworms that live in tubes and extend feathery filaments when filter feeding. Earthworms are called **oligochaetes** because they have few setae. Earthworms, which have a worldwide distribution in almost any soil, occur in large number, and reach a length of as much as 3 meters. **Leeches,** another type of annelid, include the medicinal leech, which has been used in the practice of bloodletting for centuries. Most people simply called them bloodsuckers.

Show that the annelids are the segmented worms by *labeling a segment in 28.6*a, b, d. In which group would you expect the animals to be predators based on the type of head region? _____

Figure 28.6 Annelid diversity.

The annelids include (**a**) the earworm, an oligochaete, (**b**) the marine clam worm and (**c**) the giant fanworm, which are polychaetes, and (**d**) the leech, well known for being a blood sucker.

anterior region

a. Earthworms, *Lumbricus*, mating

head
(sense organs
and jaws)

b. Clam worm, *Nereis*

mouth
appendages
(feathery arms)

c. Christmas tree worm, *Spirobranchus*

anterior region
(anterior sucker)

posterior
sucker

d. Leech, *Hirudo*

Anatomy of the Earthworm

Earthworms are segmented in that the body has a series of ringlike segments. Earthworms have no head, and burrow in the soil by alternately expanding and contracting segments along the length of the body.

Earthworms are scavengers that feed on decaying organic matter in the soil. They have a well-developed coelomic cavity, providing room for a well-developed digestive tract and both sets of reproductive organs. Earthworms are **hermaphroditic.**

Observation: Anatomy of the Earthworm

External Anatomy

1. Examine a live or preserved specimen of an earthworm. Locate the small projection that sticks out over the mouth. Has cephalization occurred? _____ Explain. _____

   _____

2. Count the total number of segments, beginning at the anterior end. The sperm duct openings are on segment 15 (somite XV) (Fig. 28.7). The enlarged section around a short length of the body is the **clitellum.** The clitellum secretes mucus that holds the worms together during mating. It also functions as a cocoon, in which fertilized eggs hatch and young worms develop. The anus is located on the worm's terminal segment.

3. Lightly pass your fingers over the earthworm's ventral and lateral sides. Do you feel the setae? _____

   _____

   Earthworms insert these slender bristles into the soil. Setae, along with circular and longitudinal muscles, enable the worm to locomote. Explain the action. _____

   _____

Figure 28.7 External anatomy of an earthworm.

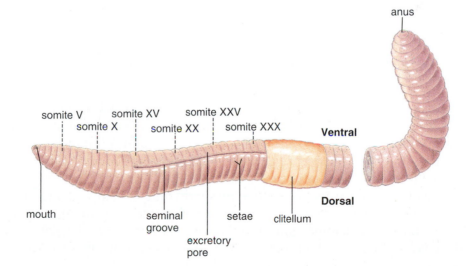

Internal Anatomy

1. Place a preserved earthworm on its ventral side in the dissecting pan. With a scalpel or razor blade, make a shallow incision slightly to the side of the blackish median dorsal blood vessel (Fig. 28.8*a*). Start your incision about 10 segments after the clitellum, and proceed anteriorly to the mouth. If you see black ooze, you have accidentally cut the intestine.
2. Identify the thin partitions, the **septa,** between segments.
3. Lay out the body wall, and pin every 10th segment to the wax in your pan. Add water to prevent drying out. While alive, the earthworm body wall is always moist and this facilitates gas exchange across the body wall. Notice the earthworm has no respiratory organ.
4. An earthworm feeds on **detritus** (organic matter) in the soil. Identify the digestive tract, which begins at the mouth and extends through body segments 1, 2, and 3. It opens into the swollen, muscular, thick-walled **pharynx,** which extends from segment 3 to segment 6. The **esophagus** extends from the pharynx through segment 14 to the **crop.** Next is the **gizzard,** which lies in segments 17 through 28.

   The intestine extends from the gizzard to the anus. Does the digestive system show specialization of parts?_____ Explain. _____
   _____
   _____

5. Identify the earthworm's circulatory system. The blood is always contained within vessels and never runs free. The **dorsal blood vessel** is readily seen along the dorsal side of the digestive tract. A series of "hearts" encircles the esophagus between segments 7 and 11, connecting the dorsal blood vessel with the ventral blood vessel. Does the earthworm have an open or closed circulatory system? _____ Explain. _____
   _____

6. Locate the earthworm's nervous system. The two-lobed brain is located on the dorsal surface of the pharynx in segment 3. Two nerves, one on each side of the pharynx, connect the brain to a ganglion that lies below the pharynx in segment 4. The **ventral nerve cord** then extends along the floor of the body cavity to the last segment.
7. Find the earthworm's excretory system, which consists of a pair of minute, coiled, white tubules, the **nephridia,** located in every segment except the first three and the last. Each nephridium opens to the outside by means of an excretory pore. Does the excretory system show that the earthworm is segmented? _____ Explain. _____
   _____
   _____

8. Identify the earthworm's reproductive system, including **seminal vesicles,** light-colored bodies in segments 9 through 12, which house maturing sperm that have been formed in two pairs of testes within them; **sperm ducts** that pass to openings in segment 15; and **seminal receptacles** (four small, white, special bodies that lie in segments 9 and 10), which store sperm received from another worm. **Ovaries** are located in segment 13 but are too small to be seen.
9. During mating, earthworms are arranged so that the sperm duct openings of one worm are just about, but not quite, opposite the seminal receptacle openings of the other worm. After being released, the sperm pass down a pair of seminal grooves on the ventral surface (see Fig. 28.7) and then cross over at the level of the seminal receptacles of the opposite worm. Once the worms separate, eggs and sperm are released into a cocoon secreted by the clitellum. Is the earthworm hermaphroditic? _____ Explain. _____
   _____

10. Does the earthworm have a respiratory system? _____ How does it exchange gases? _____

_____

11. Why would you expect an earthworm to lack an exoskeleton? (*Hint:* Review question 10.) _____

_____

Figure 28.8   Internal anatomy of an earthworm, dorsal view.
a. Drawing shows internal organs and a cross section.
b. Dissected specimen.

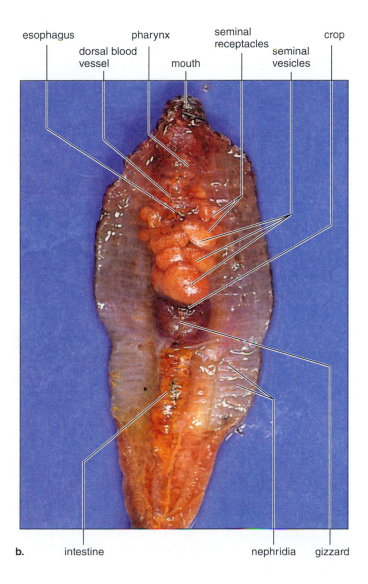

mouth

brain

pharynx

heart
esophagus

seminal
receptacle

seminal
vesicle

sperm
duct

crop

gizzard

nephridium

dorsal blood
vessel

intestine

septum

clitellum

typhlosole
lumen of
intestine
coelom

a.

Virtual Lab **Earthworm Dissection**
A virtual lab called Earthworm Dissection is available on the *Inquiry into Life* website **www.mhhe.com/maderInquiry14**. Follow the directions given to click and drag labels to external and internal illustrations of earthworm anatomy.

esophagus   pharynx   seminal   crop
                               receptacles
     dorsal blood   mouth   seminal
     vessel   vesicles

b.   intestine   nephridia   gizzard

1. Obtain a prepared slide of a cross section of an earthworm (Fig. 28.9). Examine the slide under the dissecting microscope and under the light microscope.
2. Identify the following structures.
   a. **Body wall:** A thick outer circle of tissue, consisting of the **cuticle** and the **epidermis.**
   b. **Coelom:** A relatively clear space with scattered fragments of tissue.
   c. **Intestine:** An inner circle with a suspended fold.
   d. **Typhlosole:** A fold that increases the intestine's surface area.
   e. **Ventral nerve cord:** A white, threadlike structure.
3. Does the typhlosole help in nutrient absorption? _____ Explain. _____

_____

Figure 28.9 Cross section of an earthworm.
Cross-section micrograph as it would appear under the microscope.

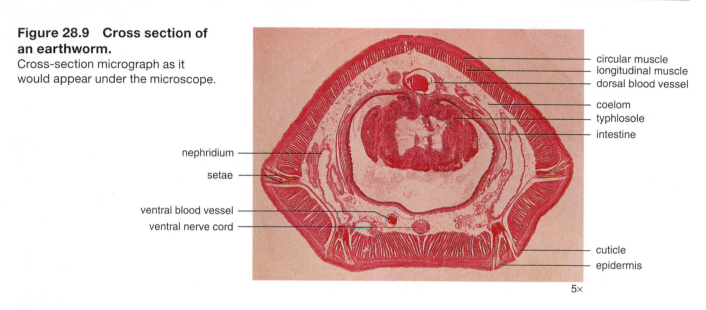

circular muscle
longitudinal muscle
dorsal blood vessel
coelom
typhlosole
intestine
nephridium
setae
ventral blood vessel
ventral nerve cord
cuticle
epidermis
5×

Conclusion: Comparison of Clam to Earthworm

- Complete Table 28.2 to compare the anatomy of a clam to that of an earthworm.

Table 28.2 Comparison of Clam to Earthworm

	Clam	Earthworm
Nervous system		
Digestion		
Skeleton		
Excretory organ		
Circulation		
Respiratory organ		
Locomotion		
Reproduction		

28.3 Arthropods

Arthropods (phylum Arthropoda) have paired, jointed appendages and a hard exoskeleton that contains chitin. The chitinous exoskeleton consists of hardened plates separated by thin, membranous areas that allow movement of the body segments and appendages.

Figure 28.10*a* features insects and relatives. **Insects** with three pairs of legs, with or without wings, and three distinct body regions comprise 95% of all arthropods. **Millipedes** have two pairs of legs per segment, while **centipedes** have one pair of legs per segment. Figure 28.10*b* features spiders and relatives. Spiders and scorpions have four pairs of legs, no antennae, and a cephalothorax (head and thorax are fused). The horseshoe crab is a living fossil. It has remained unchanged for thousands of years. The **crustaceans** (Fig. 28.10*c*), which include crabs, shrimp, and lobsters, have three to five pairs of legs, and two pairs of antennae. Barnacles are unusual, in that their legs are used to gather food.

For each animal in Figure 28.10, circle the obvious types of appendages.

Figure 28.10 Arthropod diversity.
a. Insects, millipedes, and centipedes are related. **b.** Spiders, scorpions, and horseshoe crabs are related. **c.** Crabs, shrimp, and barnacles, among others, are crustaceans.

a. Insects and relatives

Honeybee, *Apis mellifera*

Millipede, *Ophyiulus pilosus*

Centipede, *Scolopendra* sp.

b. Spider and relatives

Spider, *Argiope rafaria*

Scorpion, *Hadrurus hirsutus*

Horseshoe crab, *Limulus polyphemus*

c. Crustaceans

Crab, *Cancer productus*

Shrimp, *Stenopus* sp.

Barnacles, *Lepas anatifera*

Anatomy of a Crayfish

Crayfish also belong to the group of arthropods called crustaceans. Crayfish are adapted to an aquatic existence. They are known to be scavengers, but they also prey on other invertebrates. The mouth is surrounded by appendages modified for feeding, and there is a well-developed digestive tract. Dorsal, anterior, and posterior arteries carry **hemolymph** (blood plus lymph) to tissue spaces (hemocoel) and sinuses. In contrast to vertebrates, there is a ventral nerve cord.

Observation: Anatomy of a Crayfish

External Anatomy

1. In a preserved crayfish, identify the chitinous **exoskeleton.** With the help of Figure 28.11, identify the head, thorax, and abdomen. Together, the head and thorax are called the **cephalothorax;**

**Figure 28.11
Anatomy of a crayfish.**

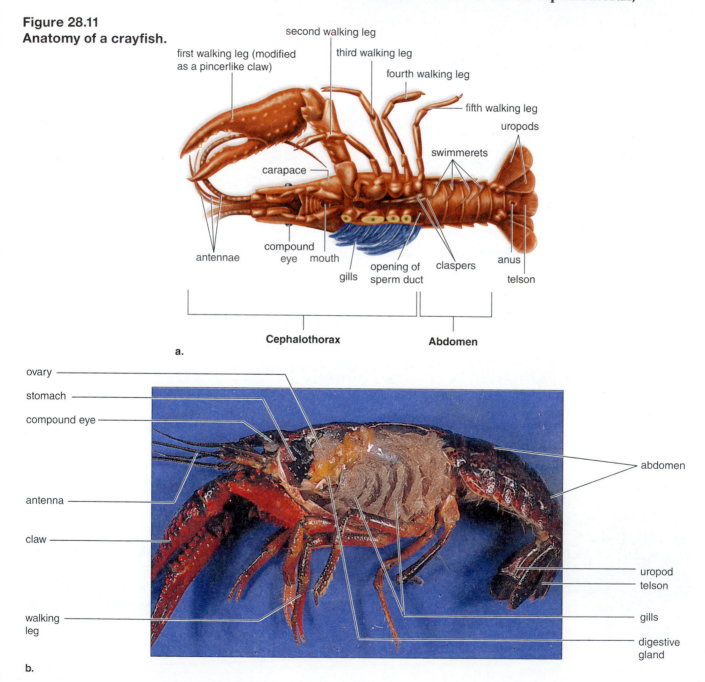

the cephalothorax is covered by the **carapace.** Has specialization of segments occurred? _____

Explain. _____

2. Find the **antennae,** which project from the head. At the base of each antenna, locate a small, raised nipple containing an opening for the **green glands,** the organs of excretion. Crayfish excrete a liquid nitrogenous waste.

3. Locate the **compound eyes,** composed of many individual units for sight. Do crayfish demonstrate

   cephalization? _____ Explain. _____

4. Identify the six pairs of appendages around the mouth for handling food.

5. Find the five pairs of walking legs attached to the cephalothorax. The most anterior pair is modified as pincerlike claws.

6. Locate the five pairs of **swimmerets** on the abdomen. In males, the anterior two pairs are stiffened and folded forward. They are claspers that aid in the transfer of sperm during mating.

7. In the female, identify the **seminal receptacles,** a swelling located between the bases of the third and fourth pairs of walking legs. Sperm from the male are deposited in the seminal receptacles. In the male, identify the opening of the sperm duct located at the base of the fifth walking leg. Find the last abdominal segment, which bears a pair of broad, fan-shaped **uropods** that, together with a terminal extension of the body, form a tail. Has specialization of appendages

   occurred? _____ Explain. _____

Internal Anatomy

1. Cut away the lateral surface of the carapace with scissors to expose the **gills** (Fig. 28.11*b*). Observe that the gills occur in distinct, longitudinal rows. How many rows of gills are there in

   your specimen? _____ The outer row of gills is attached to the base of certain appendages.

   Which ones? _____

2. Remove a gill with your scissors by cutting it free near its point of attachment, and place it in a watch glass filled with water. Observe the numerous gill filaments arranged along a central axis.

3. Carefully cut away the dorsal surface of the carapace with scissors and a scalpel. The epidermis that adheres to the exoskeleton secretes the exoskeleton. Remove any epidermis adhering to the internal organs.

4. Identify the diamond-shaped heart lying in the middorsal region. A crayfish has an open circulatory system. Carefully remove the heart.

5. Locate the **gonads** anterior to the heart in both the male and female. The gonads are tubular structures bilaterally arranged in front of the heart and continuing behind it as a single mass. In the male, the testes are highly coiled, white tubes.

6. Find the **mouth;** the short, tubular **esophagus;** and the two-part **stomach,** with the attached **digestive gland,** that precedes the intestine.

7. Identify the **green glands,** two excretory structures just anterior to the stomach, on the ventral segment wall.

8. Remove the thoracic contents previously identified.

9. Identify the **brain** in front of the esophagus. The brain is connected to the ventral nerve cord by a pair of nerves that pass around the esophagus.

10. Remove the animal's entire digestive tract, and float it in water. Observe the various parts, especially the connections of the digestive gland to the stomach.

11. Cut through the stomach, and notice in the anterior region of the stomach wall the heavy, toothlike projections, called the **gastric mill,** which grind up food. Do you see any grinding stones ingested by

    the crayfish? _____

    If possible, identify what your specimen had been eating. _____

Anatomy of a Grasshopper

The grasshopper is an **insect.** All insects have a head, a thorax, and an abdomen. Their appendages always include (1) three pairs of jointed legs and usually (2) two antennae as sensory organs. Grasshoppers are adapted to live on land. Wings and jumping legs are suitable for locomotion on land; **Malpighian tubules** save water by secreting a solid nitrogenous waste; the **tracheae** are tiny tubules that deliver air directly to the muscles; and the male has a penis with attached claspers to deliver sperm to the seminal receptacles of a female so they do not dry out.

Observation: Anatomy of a Grasshopper

External Anatomy

1. Obtain a preserved grasshopper *(Romalea),* and study its external anatomy with the help of Figure 28.12*a.* Identify the head, thorax, and abdomen.

2. Use a hand lens or dissecting microscope to examine the grasshopper's special sense organs of the **head.** Identify the **antennae** (a pair of long, jointed feelers), the **compound eyes,** and the three dotlike **simple eyes.** The labial palps, labeled in Figure 28.12*a,* have sense organs for tasting food.

3. Note the sturdy **mouthparts,** which are used for chewing plant material. A grasshopper's mouth parts are quite different from those of a piercing and sucking insect.

4. Locate the leathery **forewings** and the inner, membraneous **hindwings** attached to the **thorax.** Which pair of legs is used for jumping? _____ How many segments does each leg have? _____

5. Is locomotion in the grasshopper adapted to land? _____

    Explain. _____

    _____

6. In the **abdomen,** identify the **tympana** (sing., **tympanum**), one on each side of the first abdominal segment (Fig. 28.12*a*). The grasshopper detects sound vibrations with these membranes.

7. Locate the **spiracles,** along the sides of the abdominal segments. These openings allow air to enter the tracheae, which constitute the respiratory system.

8. Find the **ovipositors** (Figs. 28.12*a* and 28.13*a*), four curved and pointed processes projecting from the abdomen of the female. These are used to dig a hole in which eggs are laid. The male has a **penis** with **claspers** used during copulation (Fig. 28.13*b*).

Figure 28.12 Female grasshopper.
a. External anatomy. **b.** Internal anatomy.

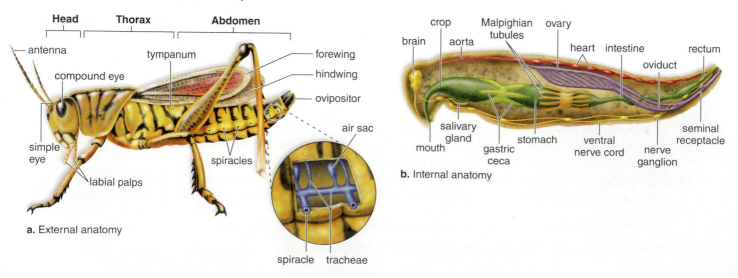

a. External anatomy

b. Internal anatomy

Figure 28.13 Grasshopper genitalia.
a. Females have an ovipositor, and **b.** males have claspers at the distal end of the penis.

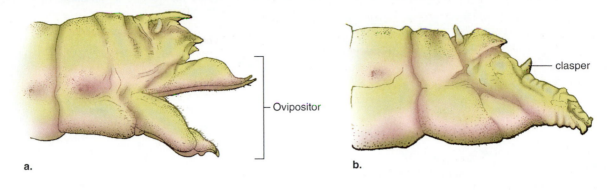

a.

clasper

Ovipositor

b.

Internal Anatomy

Observe a longitudinal section of a grasshopper if available on demonstration. Try to locate the structures shown in Figure 28.12*b*. What is the function of Malpighian tubules?_____

Conclusion: Comparison of Crayfish to Grasshopper

Compare the adaptations of a crayfish to those of a grasshopper by completing Table 28.3. Put a star beside each item that indicates an adaptation to life in the water (crayfish) and to life on land (grasshopper). Check with your instructor to see if you identified the maximum number of adaptations.

Table 28.3 Comparison of Crayfish to Grasshopper		
	Crayfish	Grasshopper
Locomotion		
Respiration		
Sense organs		
Nervous system		

External reproductive features
    Male
    Female		

Grasshopper Metamorphosis

Metamorphosis means a change, usually a drastic one, in form and shape. Grasshoppers undergo *incomplete metamorphosis,* a gradual change in form rather than a drastic change. The immature stages of the grasshopper are called **nymphs**, and they are recognizable as grasshoppers even though they differ somewhat in shape and form (Fig. 28.14*a*). Some insects undergo what is called *complete metamorphosis,* in which case they have three stages of development: **larvae**, **pupa**, and **adult** (Fig. 28.14*b*). Metamorphosis occurs during the pupa stage when the animal is enclosed within a hard covering. The animals best known for complete metamorphosis are the butterfly and the moth, whose larval stage is called a caterpillar and whose pupa stage is the cocoon; the adult is the butterfly or moth.

Figure 28.14 Metamorphosis.

a. During incomplete metamorphosis of a grasshopper, a series of **nymphs** leads to a full-grown grasshopper. **b.** During complete metamorphosis of a moth, a series of larvae lead to pupation. The **adult** hatches out of the pupa.

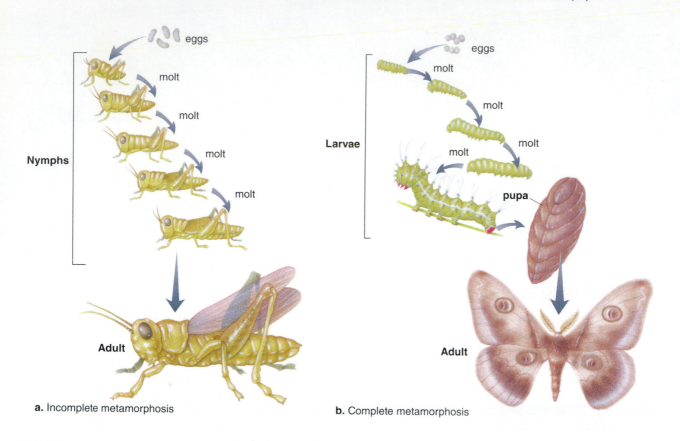

a. Incomplete metamorphosis

b. Complete metamorphosis

Observation: Grasshopper Metamorphosis

1. Use Figure 28.14 to add the grasshopper and the moth to Table 28.4.
2. Examine any specific life cycle displays or plastomounts that illustrate complete and incomplete metamorphosis and add these examples to Table 28.4.

Table 28.4 Insect Metamorphosis	
Specimen	**Complete or Incomplete Metamorphosis**

Conclusions: Insect Metamorphosis

- In insects with incomplete metamorphosis, do the nymphs or the adult have better developed wings? _____ What is the benefit of wings to an insect?_____
_____

- What stage is missing when an insect has incomplete metamorphosis? _____ What happens during this stage? _____
- What form, the larvae or the adult, disperses new individuals in flying insects that exhibit complete metamorphosis? _____
  How is this a benefit? _____
  _____
- In insects that undergo complete metamorphosis, the larvae and the adults utilize different food sources and habitats. Why might this be a benefit? _____
  _____

28.4 Echinoderms

The echinoderms (phylum Echinodermata) are the only invertebrate group that shares deuterostome development with the vertebrates (see Fig. 27.1). Unlike the vertebrates, all **echinoderms** are marine, and they dwell on the seabed, either attached to it, like sea lilies, or creeping slowly over it. The name *echinoderm* means "spiny-skinned," and most members of the group have defensive spines on the outside of their bodies. The spines arise from an **endoskeleton** composed of calcium carbonate plates. The endoskeleton supports the body wall and is covered by living tissue that may be soft (as in sea cucumbers) or hard (as in sea urchins).

Especially note that (1) adult echinoderms are radially symmetrical, with generally five points of symmetry arranged around the axis of the mouth but the larvae bilaterally symmetrical, and (2) the echinoderms' most unique feature is their **water vascular system.** In those echinoderms in which the arms make contact with the substratum, the **tube feet** associated with the water vascular system are used for locomotion. In other echinoderms, the tube feet are used for gas exchange and food gathering.

Echinoderms belong to one of five groups: sea lilies and feather stars; sea stars; brittle stars; sea urchins and sand dollars; and sea cucumbers (Fig. 28.15). *Where appropriate in Figure 28.15, write "ORS" for obvious radial symmetry or "RSNO" for radial symmetry not obvious on the lines provided.*

Figure 28.15 Echinoderm diversity.

a. Sea lily, *Comonthino* sp. _____   b. Sea star, *Pentaceraster cumingi* _____   c. Brittle stars, *Ophiopholis aculeata* _____

d. Sea urchin, *Stronglocentrotus pranciscanus* _____   e. Sand dollar, *Dendraster excentricus* _____   f. Sea cucumber, *Pseudocolochirus* sp. _____

Anatomy of a Sea Star

Sea stars (starfish) usually have five arms that radiate from a central disk. The mouth is normally oriented downward, and when sea stars feed on clams, they use the suction of their tube feet to force the shells open a crack. Then they evert the cardiac portion of the stomach, which releases digestive juices into the mantle cavity. Partially digested tissues are taken up by the pyloric portion of the stomach; digestion continues in this portion of the stomach and in the digestive glands found in the arms.

Observation: Anatomy of a Sea Star

External Anatomy

1. Place a preserved sea star in a dissecting pan so that the aboral side is uppermost.
2. With the help of Figure 28.16, identify the **central disk** and five arms. What type of symmetry does an adult sea star have? _____

Figure 28.16   Anatomy of a sea star.
a. Diagram and **(b)** image of dissected sea star. Both show the aboral side. **c.** Image of cut arm. **d.** Canals and tube feet of water vascular system. Seen from aboral side.

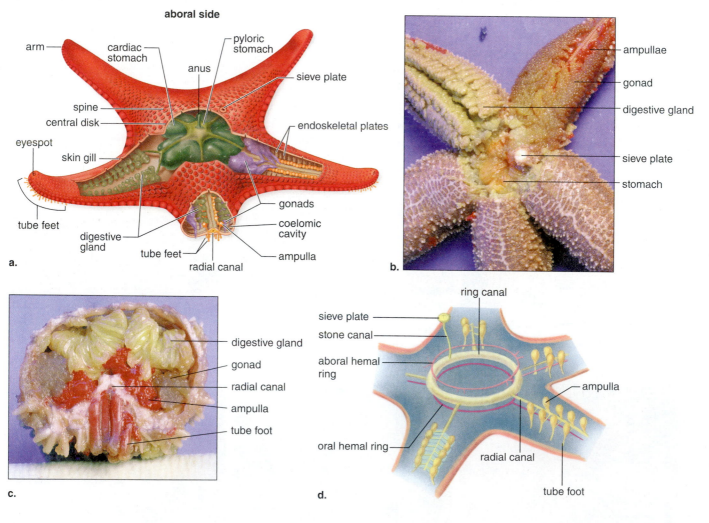

3. With the oral side uppermost, find the mouth, located in the center and protected by spines. Why is this side of the sea star called the oral side? _____

4. Locate the groove that runs along the middle of each arm and the **tube feet** (suctionlike disks) in rows on either side of the groove. Pluck away the tube feet from one area.

   How many rows of feet are there? _____ In the valley of each ambulacral groove, identify the **radial canal** that extends to the tip of each arm.

5. Turn the sea star over to its aboral side.

6. Locate the anal opening (Fig. 28.16). Why is this side of the sea star called the aboral side? _____

   _____

7. Lightly run your fingers over the spines extending from calcium carbonate plates that lie buried in the body wall beneath the surface. The plates form an endoskeleton of the animal.

8. Identify the **sieve plate (madreporite),** a brownish, circular spot between two arms where water enters the water vascular system.

Internal Anatomy

1. Place the sea star so that the aboral side is uppermost. Refer to Figure 28.16*a* and *b* as you dissect the sea star following these instructions:

2. Cut the tip of one of the arms and, with scissors, carefully cut through the body wall along each side of this arm.

3. Carefully lift up the upper body wall. Separate any internal organs that may be adhering so that all internal organs are left intact.

4. Cut off the body wall near the central disk, but leave the sieve plate (madreporite) in place.

5. Remove the body wall of the central disk, being careful not to injure the internal organs.

6. Identify the digestive system. The mouth leads into a short **esophagus,** which is connected to the saclike **cardiac stomach.** When a sea star eats, the cardiac stomach sticks out through the sea star's mouth and starts digesting the contents of a clam or oyster. Above the cardiac stomach is the **pyloric stomach,** which leads to a short intestine. Each arm contains one pair of **digestive glands.**

   To which stomach do the digestive glands attach? _____

7. Cut off a portion of an arm, and examine the cut edge (Fig. 28.16*c*). Identify the digestive glands, ambulacral groove, radial canal, ampullae, and tube feet.

8. Remove the digestive glands of one arm.

9. Identify the **gonads** extending into the arm. What is the function of gonads? _____

   _____

   It is not possible to distinguish male sea stars from females by this observation.

10. Remove both stomachs.

11. In the **water vascular system** (Fig. 28.16*d*), you have already located the sieve plate and tube feet. Now try to identify the following components:
    a. **Stone canal:** Takes water from the sieve plate to the ring canal.
    b. **Ring canal:** Surrounds the mouth and takes water to the radial canals.
    c. **Radial canals:** Send water into the ampulla. When the ampullae contract, water enters the tube feet. Each tube foot has an inner muscular sac called an ampulla. The ampulla contracts and forces water into the tube foot.

    What is the function of the water vascular system? _____

    _____

_____ 1. Jointed appendages and an exoskeleton are characteristic of what group of animals?

_____ 2. Crayfish belong to what group of arthropods?

_____ 3. A clam belongs to what group of molluscs?

_____ 4. Molluscs, annelids, and arthropods are all what type of animal?

_____ 5. A visceral mass, foot, and mantle are characteristic of what group of animals?

_____ 6. All the animals studied today have what type of coelom?

_____ 7. In a clam, which structure secretes the shell?

_____ 8. The clam is a filter feeder, but the squid is a(n) _____.

_____ 9. The annelids are the first of the animal phyla studied to have what general characteristic?

_____ 10. Which of the three types of annelids has suckers as an adaptation to its way of life?

_____ 11. What term indicates that earthworms have both male and female organs?

_____ 12. The arthropods are the first of the animal phyla to have what general characteristic?

_____ 13. What type of excretory organs are attached to the intestine of a grasshopper?

_____ 14. Contrast the respiratory organ of a crayfish with that of a grasshopper.

_____ 15. Identify the muscular organ used for locomotion in the clam.

_____ 16. An open circulatory system has what feature missing in a closed system?

_____ 17. Echinoderms have what type symmetry?

_____ 18. Sea stars locomote by means of what structures?

Thought Questions

19. Compare respiratory organs in the crayfish and the grasshopper. How are these suitable to the habitat of each?

20. For each of the following characteristics, name an animal with the characteristic, and state the characteristic's advantages:

   a. Closed circulatory system

   b. Jointed appendages

   c. Exoskeleton

   d. Segmentation

29

The Vertebrates

Introduction

Vertebrates are **chordates.** All chordates have (1) a dorsal tubular nerve cord; (2) a dorsal supporting rod, called a notochord, at some time in their life history; (3) a postanal tail (e.g., tailbone or coccyx); and, in chordates that breathe by means of gills, (4) pharyngeal pouches that become gill slits. In terrestrial chordates, these pouches are modified for other purposes.

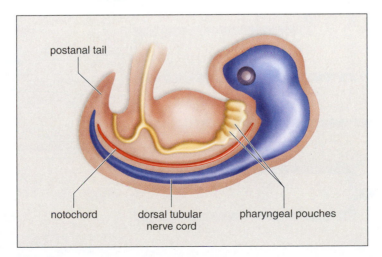

29.1 Evolution of Chordates

The evolutionary tree of the chordates (Fig. 29.1) shows that some chordates, notably the **tunicates** and **lancelets**, are not vertebrates. These chordates retain the notochord and are called the **invertebrate chordates**. Explain the term invertebrate chordates. _____

    The other animal groups in Figure 29.1 are **vertebrates** in which the notochord has been replaced by the vertebral column. Fishes include three groups: the **jawless fishes** were the first to evolve, followed by the **cartilaginous fishes** and then the **bony fishes**. The bony fishes include the **ray-finned** fishes (the largest group of vertebrates) and the **lobe-finned fishes**. The first lobe-finned fishes had a bony skeleton, fleshy appendages, and a lung. These lobe-finned fishes lived in shallow pools and gave rise to the amphibians. What three features called out in Figure 29.1 evolved among fishes? _____

_____

    The terrestrial vertebrates are all **tetrapods** because they have four limbs. The limbs of tetrapods are _____ appendages just like those of arthropods. Amphibians still return to the water to reproduce, but **reptiles** are fully adapted to life on land because among other features they produce an **amniotic egg**. The amniotic egg is so named because the embryo is surrounded by an amniotic membrane that encloses amniotic fluid. Therefore, amniotes develop in an aquatic environment of their own making.

Do all animals develop in a water environment? _____ Explain. _____

_____

In placental mammals, such as humans, the fertilized egg develops inside the female, where the unborn receives nutrients via the placenta. Reptiles (including birds) and mammals have many other adaptations that are suitable to living on land, as we will stress in later sections.

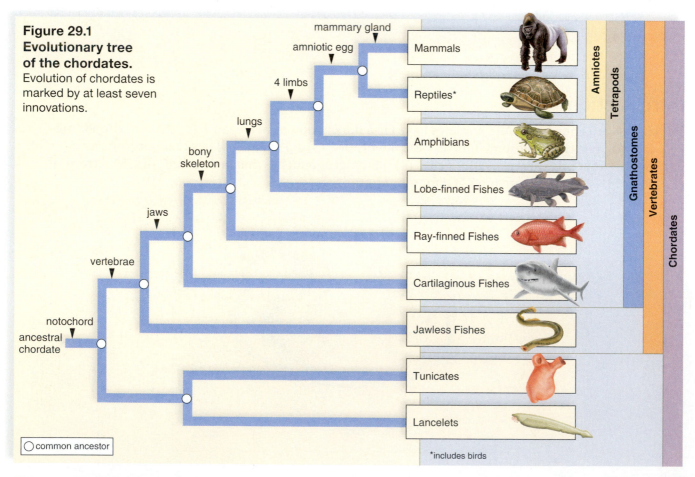

**Figure 29.1
Evolutionary tree
of the chordates.**
Evolution of chordates is
marked by at least seven
innovations.

*includes birds

29.2 Invertebrate Chordates

Among the chordates (phylum Chordata), two groups contain invertebrates, and the others contain the vertebrates.

Invertebrate Chordates

The two types of invertebrate chordates are urochordates and cephalochordates.

1. **Urochordates.** The tunicates, or sea squirts (Fig. 29.2), come in varying sizes and shapes, but all have incurrent and excurrent siphons. **Gill slits** are the only remaining chordate characteristic in adult tunicates. Examine any examples of tunicates on display.

2. **Cephalochordates.** Lancelets, also known as amphioxus *(Branchiostoma),* are small, fishlike animals that occur in shallow marine waters in most parts of the world. They spend most of their time buried in the sandy bottom, with only the anterior end projecting.

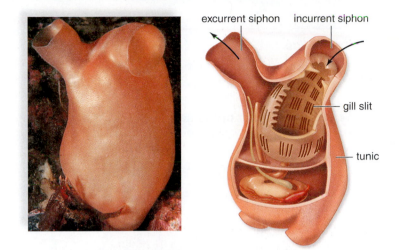

Figure 29.2 Urochordates.
The gill slits of a tunicate are the only chordate characteristic remaining in the adult.

Observation: Lancelet Anatomy

Preserved Specimen

1. Examine a preserved lancelet (Fig. 29.3).
2. Identify the **caudal fin** (enlarged tail) used in locomotion, the **dorsal fin,** and the short **ventral fin.**
3. Examine the lancelet's V-shaped muscles.

Figure 29.3 Anatomy of the lancelet, *Branchiostoma.*
Lancelets feed on microscopic particles filtered out of the constant stream of water that enters the mouth and exits through the gill slits into a protective atrium formed by body folds. The water exits at the atriopore.

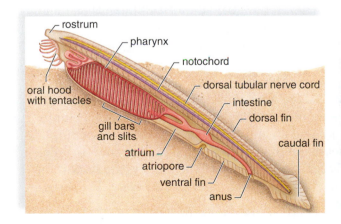

4. Find the tentacled **oral hood,** located anterior to the mouth and covering a vestibule. Water entering the mouth is channeled into the **pharynx,** where food particles are trapped before the water exits at the **atriopore.** Lancelets are filter feeders. Has cephalization occurred? _____

   Explain. _____

   _____

Prepared Slide

Examine a prepared cross section of a lancelet (Fig. 29.4), and note three chordate characteristics: notochord, dorsal tubular nerve cord, and gill slits.

Figure 29.4 Lancelet cross section.
Cross-section slide as it would appear under a microscope.

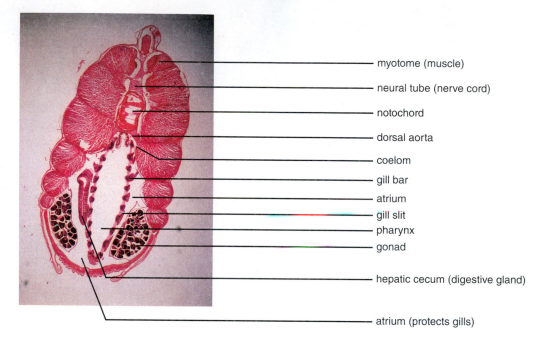

- myotome (muscle)
- neural tube (nerve cord)
- notochord
- dorsal aorta
- coelom
- gill bar
- atrium
- gill slit
- pharynx
- gonad
- hepatic cecum (digestive gland)
- atrium (protects gills)

29.3 Vertebrates

In vertebrates (subphylum Vertebrata) (Fig. 29.5), the embryonic notochord is replaced by a vertebral column composed of individual vertebrae that protect the nerve cord. The internal jointed skeleton consists not only of a vertebral column, but also of a skull that encloses and protects the well-developed brain. There is an extreme degree of cephalization with complex sense organs. The eyes develop as outgrowths of the brain. The ears are primarily equilibrium devices in aquatic vertebrates; and in land vertebrates, they function as sound wave receivers.

The vertebrates are extremely motile and have well-developed muscles and usually paired appendages. They have bilateral symmetry and are segmented, as witnessed by the vertebral column. There is a large body cavity, a complete gut with both a mouth and an anus (or instead, a cloacal opening), and the circulatory system consists of a well-developed heart and many blood vessels. They have an efficient means of extracting oxygen from water (gills) or air (lungs) as appropriate. The kidneys are important excretory and water-regulating organs that conserve or rid the body of water as necessary. The sexes are generally separate, and reproduction is usually sexual.

Figure 29.5 Vertebrate groups.

Cartilaginous fishes
Lack operculum and swim bladder; tail fin usually asymmetrical (sharks, skates, and rays)

Blue shark

Bony fishes
Operculum; swim bladder or lungs; tail fin usually symmetrical: lung-fishes, lobe-finned fishes, and ray-finned fishes (herring, salmon, sturgeon, eels, and sea horse)

Blueback butterflyfish

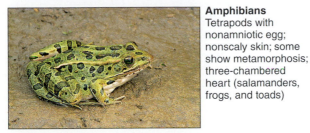

Amphibians
Tetrapods with nonamniotic egg; nonscaly skin; some show metamorphosis; three-chambered heart (salamanders, frogs, and toads)

Northern leopard frog

Reptiles
Tetrapods with amniotic egg; scaly skin (snakes, lizards, turtles, and tortoises)

Pearl River redbelly turtle

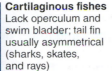

Birds
Now grouped with reptiles; tetrapods with feathers; bipedal with wings; double circulation (sparrows, penguins, and ostriches)

Scissor-tailed flycatcher

Mammals
Tetrapods with hair, mammary glands; double circulation; teeth differentiated: monotremes (spiny anteater and duckbill platypus), marsupials (opossum and kangaroo), and placental mammals (whales, rodents, dogs, cats, elephants, horses, bats, and humans)

Gray fox

Anatomy of the Frog

In this laboratory the anatomy of the frog will be considered typical of vertebrates in general. Frogs are amphibians, a group of animals in which metamorphosis occurs. Metamorphosis includes a change in structure, as when an aquatic tadpole becomes a frog with lungs and limbs (Fig. 29.6). Amphibians were the first vertebrates to be adapted to living on land; however, they typically return to the water to reproduce. Note every structure mentioned in the following Observations that represents an adaptation to a land environment.

Figure 29.6 External frog anatomy.

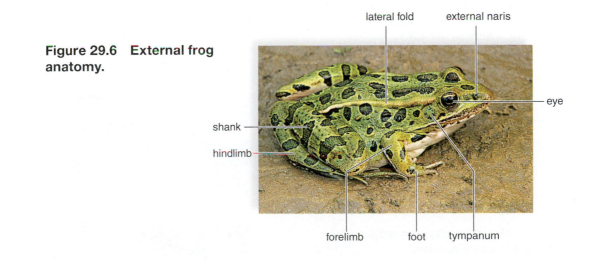

lateral fold   external naris   eye   shank   hindlimb   forelimb   foot   tympanum

Observation: External Anatomy of the Frog

1. Place a preserved frog *(Rana pipiens)* in a dissecting tray.
2. Identify the bulging eyes, which have a nonmovable upper and lower lid but can be covered by a **nictitating membrane** that serves to moisten the eye.
3. Locate the **tympanum** behind each eye (Fig. 29.6). What is the function of a tympanum? _____

   _____

4. Examine the external **nares** (sing., **naris,** or **nostril**). Insert a probe into an external naris, and observe that it protrudes from one of the paired small openings, the internal nares (Fig. 29.7), inside the mouth

   cavity. What is the function of the nares? _____
5. Identify the paired limbs. The bones of the fore- and hindlimbs are the same as in all tetrapods, in that the first bone articulates with a girdle and the limb ends in phalanges. The hind feet have five

   phalanges, and the forefeet have only four phalanges. Which pair of limbs is longest? _____

   How does a frog locomote on land? _____

   _____

   What is a frog's means of locomotion in the water? _____

   _____

Observation: Internal Anatomy of the Frog

Virtual Lab Frog Dissection   A virtual lab called Frog Dissection is available on the *Inquiry into Life* website **www.mhhe.com/mader inquiry14**. Allow ample time to complete this virtual lab with audio that allows a complete dissection of the frog systems. If highlighted buttons/arrows do not lead you to the next section, click on the menu provided.

Mouth

1. Open your frog's mouth very wide (Fig. 29.7), cutting the angles of the jaws if necessary.
2. Identify the tongue attached to the lower jaw's anterior end.
3. Find the **auditory (eustachian) tube** opening in the angle of the jaws. These tubes lead to the ears. Auditory tubes equalize air pressure in the ears.
4. Examine the **maxillary teeth** located along the rim of the upper jaw. Another set of teeth—**vomerine teeth**—is present just behind the midportion of the upper jaw.

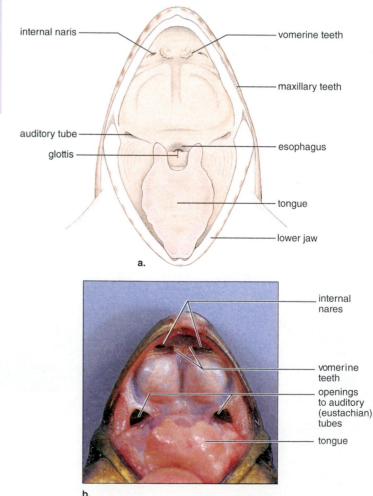

a.

internal naris — vomerine teeth — maxillary teeth — auditory tube — glottis — esophagus — tongue — lower jaw

internal nares — vomerine teeth — openings to auditory (eustachian) tubes — tongue

b.

Figure 29.7 Mouth cavity of a frog.
a. Drawing. **b.** Dissected specimen.

5. Locate the **glottis,** a slit through which air passes into and out of the **trachea,** the short tube from glottis to lungs. What is the function of a glottis? _____ _____

6. Identify the **esophagus,** which lies dorsal and posterior to the glottis and leads to the stomach.

Opening the Frog

1. Place the frog ventral side up in the dissecting pan. Lift the skin with forceps, and use scissors to make a large, circular cut to remove the skin from the abdominal region as close to the limbs as possible. Cut only skin, not muscle.
2. Now, remove the muscles by cutting through them in the same circular fashion. At the same time, cut through any bones you encounter. A vein, called the abdominal vein, will be slightly attached to the internal side of the muscles.
3. Identify the **coelom,** or body cavity. Recall from Laboratory 27 that vertebrates are deuterostomes in which the first embryonic opening becomes the anus and the second opening becomes the mouth.
4. If your frog is female, the abdominal cavity is likely to be filled by a pair of large, transparent **ovaries,** each containing hundreds of black and white eggs. Gently lift the left ovary with forceps, and find its place of attachment. Cut through the attachment, and remove the ovary in one piece.

Respiratory System and Liver

1. Insert a probe into the glottis, and observe its passage into the trachea. Enlarge the glottis by making short cuts above and below it. When the glottis is spread open, you will see a fold on either side; these are the vocal cords used in croaking.
2. Identify the **lungs,** two small sacs on either side of the midline and partially hidden under the liver (Fig. 29.8). Sequence the organs in the respiratory tract to trace the path of air from the external nares to the lungs. _____ _____

3. Locate the **liver,** the large, prominent, dark-brown organ in the midventral portion of the trunk (Fig. 29.8). Between the right half and left half of the liver, find the **gallbladder.**

Circulatory System

1. Lift the liver gently. Identify the **heart,** covered by a membranous covering (the **pericardium**). With forceps, lift the covering, and gently slit it open. The heart consists of a single, thick-walled **ventricle** and two (right and left) anterior, thin-walled **atria.**
2. Locate the three large veins that join together beneath the heart to form the **sinus venosus.** (To lift the heart, you may have to snip the slender strand of tissue that connects the atria to the pericardium.) Blood from the sinus venosus enters the right atrium. The left atrium receives blood from the lungs.
3. Find the **conus arteriosus,** a single, wide arterial vessel leaving the ventricle and passing ventrally over the right atrium. Follow the conus arteriosus forward to where it divides into three branches on each side. The middle artery on each side is the **systemic artery,** which fuses behind the heart to become the **dorsal aorta.** The dorsal aorta transports blood through the body cavity and gives off many branches. The **posterior vena cava** begins between the two kidneys and returns blood to the sinus venosus. Which vessel lies above (ventral to) the other? _____ _____

Digestive Tract

1. Identify the **esophagus,** a very short connection between the mouth and the stomach. Lift the left liver lobe, and identify the stomach, whitish and J-shaped. The **stomach** connects with the esophagus anteriorly and with the small intestine posteriorly.

2. Find the **small intestine** and the **large intestine,** which enters the **cloaca.** The cloaca lies beneath the pubic bone and is a general receptacle for the intestine, the reproductive system, and the urinary system. It opens to the outside by way of the anus. Sequence the organs in the digestive tract to trace the path of food from the mouth to the cloaca. _____

_____

Accessory Glands

1. You identified the liver and gallbladder previously. Now try to find the **pancreas,** a yellowish tissue near the stomach and intestine.
2. Lift the stomach to see the **spleen,** a small, pea-shaped body.

Figure 29.8 Internal organs of a female frog, ventral view.

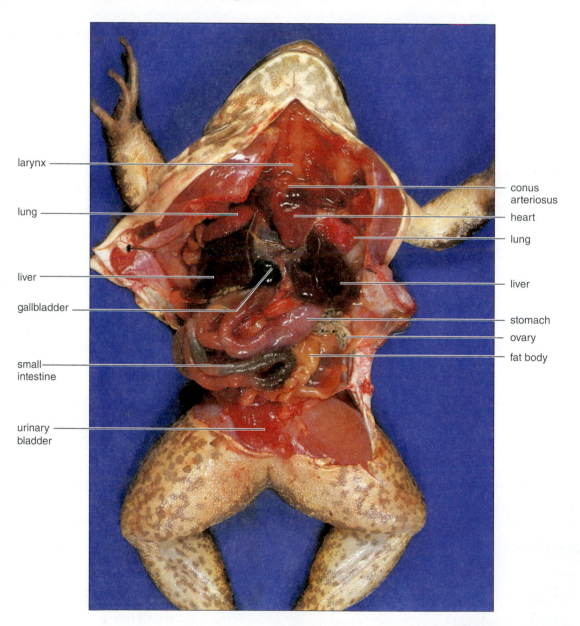

larynx

lung

liver

gallbladder

small intestine

urinary bladder

conus arteriosus

heart

lung

liver

stomach

ovary

fat body

Urogenital System

1. Identify the **kidneys,** long, narrow organs lying against the dorsal body wall (Fig. 29.9).
2. Locate the **testes** in a male frog (Fig. 29.9). Testes are yellow, oval organs attached to the anterior portions of the kidneys. Several small ducts, the **vasa efferentia,** carry sperm into kidney ducts that also carry urine from the kidneys. **Fat bodies,** which store fat, are attached to the testes.
3. Locate the ovaries in a female frog. The ovaries are attached to the dorsal body wall (Fig. 29.10). Fat bodies are also attached to the ovaries. Highly coiled **oviducts** lead to the cloaca. The ostium (opening) of the oviduct is dorsal to the liver.
4. Find the **mesonephric ducts**—thin, white tubes that carry urine from the kidney to the cloaca. In female frogs, you will have to remove the left ovary to see the mesonephric ducts.
5. Locate the **cloaca.** You will need to split through the bones of the pelvic girdle in the midventral line and carefully separate the bones and muscles to find the cloaca.
6. Identify the urinary bladder attached to the ventral wall of the cloaca. In frogs, urine backs up into the bladder from the cloaca.

7. Explain the term *urogenital system.* _____

   _____

8. The cloaca receives material from (1) _____,

   (2) _____, and (3) _____.

9. Compare the frog's urogenital system to the human urinary system, which in females has no connection to the genital system. *Beside each organ listed on the right, tell how the comparable frog organ differs from that of a human.*

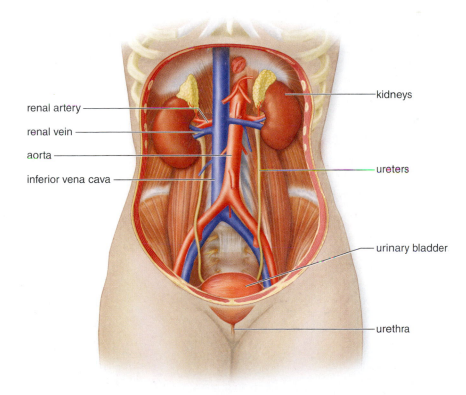

renal artery

renal vein

aorta

inferior vena cava

kidneys

ureters

urinary bladder

urethra

Figure 29.9   Urogenital system of a male frog.
a. Drawing. **b.** Dissected specimen.

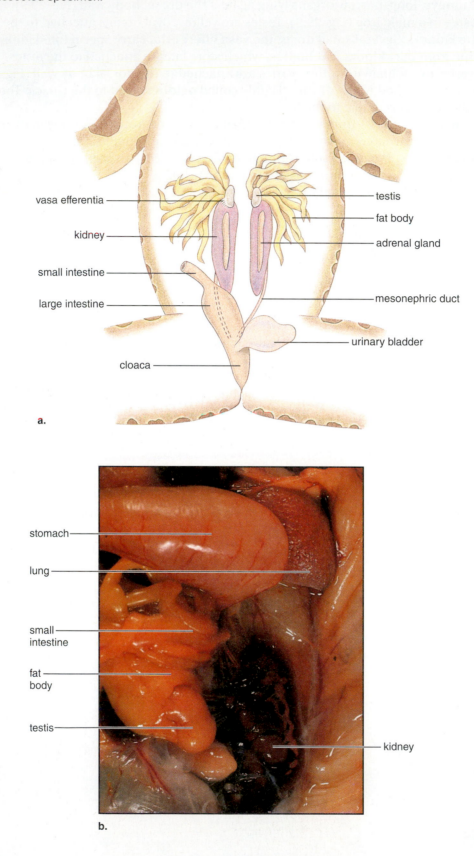

vasa efferentia

kidney

small intestine

large intestine

cloaca

testis

fat body

adrenal gland

mesonephric duct

urinary bladder

a.

stomach

lung

small
intestine

fat
body

testis

kidney

b.

Figure 29.10   Urogenital system of a female frog.
a. Drawing. b. Dissected specimen.

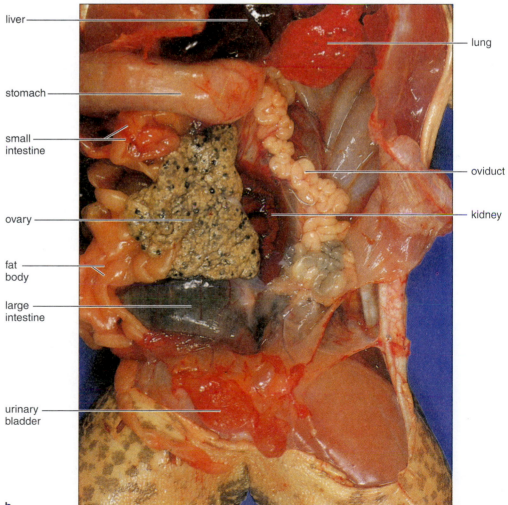

ostium

oviduct

kidney

ovary

fat body

adrenal gland

large intestine

mesonephric duct

uterus

urinary bladder

cloaca

a.

liver

lung

stomach

small intestine

oviduct

ovary

kidney

fat body

large intestine

urinary bladder

b.

In the frog demonstration dissection, identify the **brain,** lying exposed within the skull. With the help of Figure 29.11, find the major parts of the brain.

Figure 29.11   Frog brain, dorsal view.
a. Drawing. **b.** Dissected specimen.

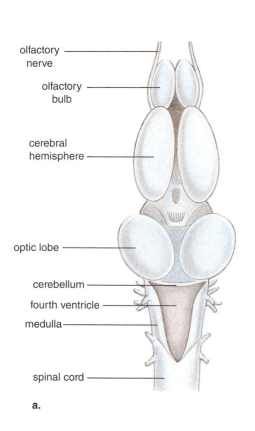

olfactory nerve
olfactory bulb
cerebral hemisphere
optic lobe
cerebellum
fourth ventricle
medulla
spinal cord

a.

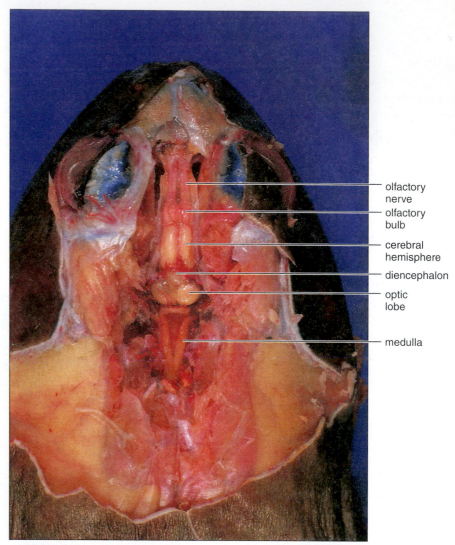

olfactory nerve
olfactory bulb
cerebral hemisphere
diencephalon
optic lobe
medulla

b.

29.4 Comparative Vertebrate Anatomy

In addition to the frog (amphibian), examine the perch (fish), pigeon (reptile), and rat (mammal) on display.

Observation: External Anatomy of Vertebrates

1. Compare the external features of the perch, frog, pigeon, and rat by examining specimens in the laboratory. Answer the following questions and record your observations in Table 29.1.
    a. Is the skin smooth, scaly, hairy, or feathery?
    b. Is there any external evidence of segmentation?
    c. Are all forms bilaterally symmetrical?
    d. Is the body differentiated into regions?
    e. Is there a well-defined neck?
    f. Is there a postanal tail?
    g. Are there nares (nostrils)?
    h. Is there a cloaca opening, or are there urogenital and anal openings?
    i. Are eyelids present? How many?
    j. How many appendages are there? (Fins are considered appendages.) (Fig. 29.12)

Table 29.1 Comparison of External Features

	Perch	Frog	Pigeon	Rat
a. Skin				
b. Segmentation				
c. Symmetry				
d. Regions				
e. Neck				
f. Postanal tail				
g. Nares				
h. Cloaca				
i. Eyelids				
j. Appendages				

Figure 29.12 Perch anatomy.

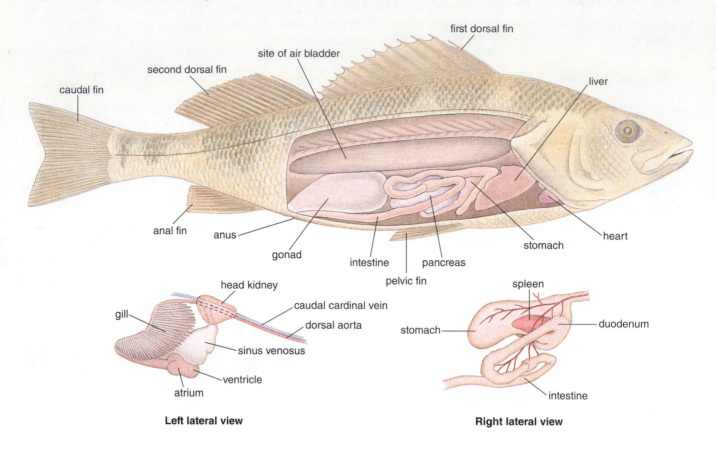

Left lateral view

Right lateral view

2. For evidence that birds are reptiles consider that birds
   a. have feathers, which are modified scales.
   b. have scales on their feet.
   c. and reptiles both lay eggs.
   d. and reptiles have similar internal organs.
   e. and reptiles also show some skeletal (skull) similarities.

   Which of these can you substantiate by external examination? _____

   _____

3. The perch, pigeon, and rat have a nearly impenetrable covering. Why is this an advantage in

   each case? _____

   _____

4. A frog uses its skin for breathing. Describe its skin in more detail. _____

   _____

Figure 29.13 Cardiovascular systems in vertebrates.

a. In a fish, the blood moves in a single loop. The heart has a single atrium and ventricle, which pumps the blood into the gill region, where gas exchange takes place. **b.** Amphibians have a double-loop system in which the heart pumps blood to both the gills and the body. **c.** In birds and mammals, the right side pumps blood to the lungs, and the left side pumps blood to the body.

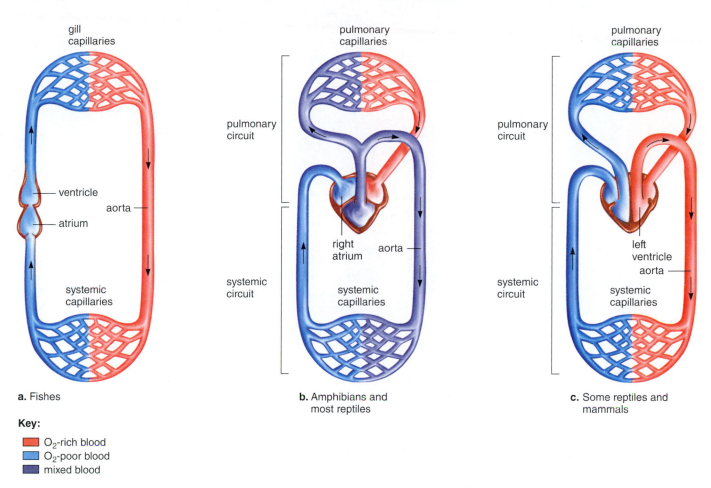

a. Fishes

b. Amphibians and most reptiles

c. Some reptiles and mammals

Key:
- O_2-rich blood
- O_2-poor blood
- mixed blood

Observation: Internal Anatomy of Vertebrates

1. Examine the internal organs of the perch, frog, pigeon (see Fig. 29.14), and rat (see Fig. 29.15).
2. If necessary, make a median longitudinal incision in the ventral body wall, from the jaws to the anus. The body cavity is called a coelom because it is completely lined by mesoderm.
3. Which of these animals has a **diaphragm** dividing the body cavity into **thorax** and **abdomen?**

_____

Circulatory Systems

Study heart models for a fish, amphibian, bird, and mammal. Trace the path of the vessel that leaves the ventricle(s), and determine whether the animals have a **pulmonary system** (Fig. 29.13). The word *pulmonary* comes from the Latin *pulmonarius,* meaning "lungs."

Figure 29.14   Pigeon anatomy.
Note that the lungs of the pigeon,
like other birds, have attached air
sacs which make the lungs of birds
more efficient.

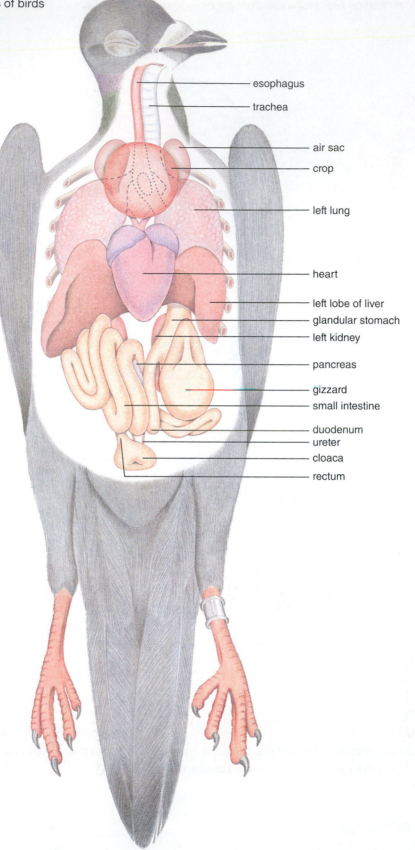

esophagus

trachea

air sac

crop

left lung

heart

left lobe of liver

glandular stomach

left kidney

pancreas

gizzard

small intestine

duodenum

ureter

cloaca

rectum

Complete Table 29.2.

Table 29.2 Comparative Circulatory Systems

Animal	Number of Heart Chambers	Pulmonary Circuit (Yes or No)
Perch		
Frog		
Pigeon		
Rat		

1. Do fish have a blood vessel that returns blood from the gills to the heart? _____ Would you expect blood pressure to be high or low after blood has moved through the gills? _____

2. What animals studied have pulmonary vessels that take blood from the heart to the respiratory organ and back to the heart? _____ What is the advantage of a pulmonary circuit? _____

3. Which of these animals has a four-chambered heart? _____ What is the advantage of having separate ventricles? _____
_____

4. The circulatory system distributes the heat of muscle contraction in birds and mammals. Is the anatomy of birds and mammals conducive to maintaining a warm internal temperature? _____ Explain. _____
_____

Respiratory Systems

Compare the respiratory systems of the perch, frog, pigeon, and rat, and complete Table 29.3 by checking the anatomical features that appear in each animal.

Table 29.3 Respiratory Systems

	Gills	Trachea	Lungs	Rib Cage*	Diaphragm	Air Sacs
Perch						
Frog						
Pigeon						
Rat						

*A rib cage consists of ribs plus a sternum. Some ribs are connected to the sternum, which lies at the midline in the anterior portion of the rib cage.

1. Among the animals studied, only a perch breathes by _____. Can the particular respiratory organ be related to the environment of the animals? _____ Explain. _____
_____

2. Knowing that gills are attached to the pharynx (throat in humans), explain why fish have no trachea. _____

_____

3. A rib cage is present in the rat and pigeon but missing in the frog. Can this difference be related to the fact that frogs breathe by positive pressure, while birds and mammals breathe by negative pressure? _____ (A frog swallows air and then pushes the air into its lungs; in birds and mammals, the thorax expands first, and then the air is drawn in.) Explain. _____

_____

4. A diaphragm is present only in mammals (e.g., rat). Of what benefit is this feature to the expansion of lungs in mammals? _____

_____

_____

5. Air sacs (not shown in Fig. 29.14) are present only in birds. This feature allows air to pass one way through the lungs of a bird and greatly increases the bird's ability to extract oxygen from the air.

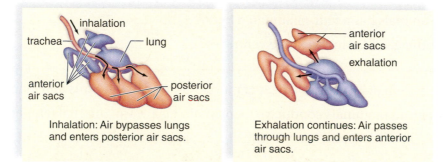

Inhalation: Air bypasses lungs and enters posterior air sacs.

Exhalation continues: Air passes through lungs and enters anterior air sacs.

Digestive Systems and Urogenital Systems

1. All the vertebrates have a stomach, small intestine, and large intestine where food is processed. They also have a liver and pancreas. Which is the larger, more prominent organ (liver or pancreas) in the pigeon and rat? _____

2. All vertebrates have an anus. As noted in Table 29.1, which vertebrates studied have a cloaca (receptacle for the urogenital and digestive systems)? _____

In the other vertebrates studied, the urogenital and digestive systems are separate.

3. Urogenital systems. All the animals have gonads and kidneys. Which of these organs is involved in urine production? _____

As we shall learn later, the kidneys help maintain the proper balance of fluid and salts in the blood.

Which of these organs is involved in reproduction? _____

The sexes are separate in vertebrates: females have ovaries and males have testes. In reptiles and mammals, males usually have a penis to pass sperm to the female. What is the chief biological benefit of the penis in terrestrial animals? _____

Figure 29.15 Rat anatomy.

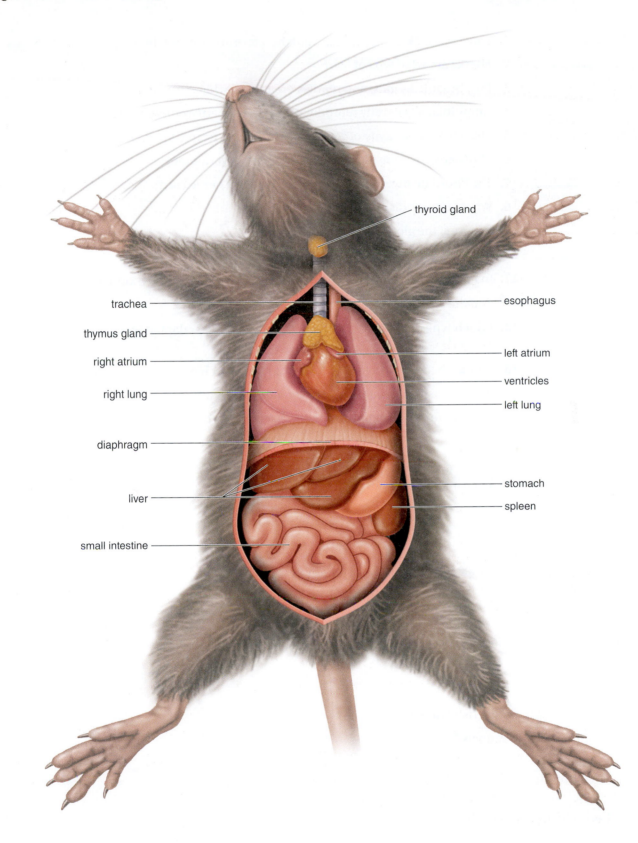

thyroid gland

trachea

thymus gland

right atrium

right lung

diaphragm

liver

small intestine

esophagus

left atrium

ventricles

left lung

stomach

spleen

_____ 1. Lancelets lack _____ and therefore are not vertebrates.

_____ 2. Sharks are what type of fish?

_____ 3. Fish breathe by gills, and reptiles breathe by what structures?

_____ 4. Amphibians generally reproduce in what type environment?

_____ 5. The abdominal cavity of vertebrates is derived from what embryonic structure?

_____ 6. Birds are now considered to be what type of vertebrate?

_____ 7. The chordate invertebrates have what type of nerve cord?

_____ 8. What features distinguish a mammal?

_____ 9. In a frog, the glottis allows air to enter the _____.

_____ 10. In a frog, the esophagus allows food to enter the _____.

_____ 11. In a frog, the cloaca receives material from the intestine, the kidneys,

and the _____.

_____ 12. Which type of vertebrate is the first type to be fully adapted for reproduction on land?

_____ 13. A frog's heart has one ventricle, but a rat's heart has how many ventricles?

_____ 14. Birds have what anatomical feature not seen in any other type of vertebrate?

_____ 15. Vertebrates are (deuterostomes or protostomes) _____.

Thought Questions

16. What is the major difference in the respiratory system of a perch compared with that of a frog, a pigeon, and a rat?

17. Associate three of the chordate characteristics with a system of the body.

18. How does the major difference between the heart of a frog and that of a rat affect the blood?

19. What anatomical characteristics are associated with

   a. Respiration in mammals but not birds?

   b. Locomotion in fish but not amphibians?

   c. Reproduction in birds but not fish?

L A B O R A T O R Y

30

Sampling Ecosystems

Learning Outcomes

Introduction
- Define ecology and an ecosystem. Identify the abiotic and biotic components of an ecosystem.   443–44

30.1 Terrestrial Ecosystems
- Define and give examples of producers in terrestrial ecosystems.   444–47
- Define and give examples of consumers in terrestrial ecosystems.   444–47
- Define and give examples of decomposers in terrestrial ecosystems.   444–47

30.2 Aquatic Ecosystems
- Define aquatic ecosystem.   448
- Give examples of producers, consumers, and decomposers in aquatic ecosystems.   448–51

Introduction

Ecology is the study of interactions between organisms and their physical environment within an **ecosystem.** The **abiotic** (nonliving) components of an ecosystem include soil, water, light, inorganic nutrients, and weather variables. The **biotic** (living) components can be organized according to the **trophic** (feeding) level in which each organism belongs. This includes producers, consumers, and decomposers.

   Producers are autotrophic organisms with the ability to carry on photosynthesis and to make food for themselves (and indirectly for the other populations as well). In terrestrial ecosystems, the predominant producers are green plants, while in freshwater and saltwater ecosystems, the dominant producers are various species of algae.

   Consumers are heterotrophic organisms that eat available food. Three types of consumers can be identified, according to their food source:

1. **Herbivores** feed directly on green plants and are termed primary consumers. A caterpillar feeding on a leaf is a herbivore.

2. **Carnivores** feed on other animals and are therefore secondary or tertiary consumers. A blue heron feeding on a fish is a carnivore.

3. **Omnivores** feed on both plants and animals. A human who eats both leafy green vegetables and beef is an omnivore.

   Decomposers and **detritivores** are organisms of decomposition, such as bacteria, fungi, and millipedes, that break down **detritus** (nonliving organic matter) to inorganic matter, which can be used again by producers. In this way, the same chemical elements are constantly recycled in an ecosystem.

   The trophic structure (feeding relationships) of an ecosystem is represented in the form of a pyramid, such as the one shown in Figure 30.1. **Biomass** is the weight of all the organisms at each trophic level.

Figure 30.1   Ecological pyramid.
Organisms at lower trophic levels are higher in number and have greater biomass than organisms at higher trophic levels. The examples of organisms on the left form an aquatic food chain in which herons eat fish, which eat zooplankton, which eat phytoplankton. The examples of organisms on the right form a terrestrial food chain in which owls eat shrews, which eat beetles, which eat plants.

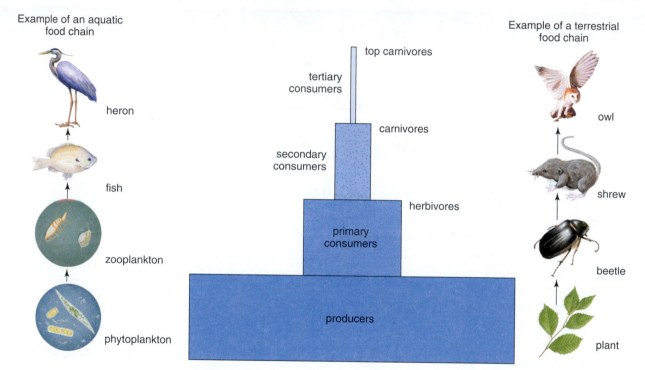

30.1   Terrestrial Ecosystems

Examining an ecosystem in a scientific manner requires concentrating on a representative portion of that ecosystem and recording as much information about it as possible. Representative areas or plots should be selected randomly. For example, random sampling of a terrestrial ecosystem often involves tossing a meterstick gently into the air in the general area to be sampled and then sampling the square meter where the stick lands. A **terrestrial ecosystem** sampling should include samples from the air above and from the various levels of plant materials growing on and in the ground beneath the selected plot.

Study of Terrestrial Sampling Site

The objective of the next Experimental Procedure is to characterize the abiotic and biotic components of a terrestrial ecosystem and to determine how those factors affect the trophic structure of an ecosystem. Although terrestrial ecosystems include deciduous forest, prairie, scrubland, and desert, a weedy field, if dominated by annual and perennial herbaceous plants up to a meter or so in height, is also a good site choice.

1. Gather all necessary equipment, such as metersticks, jars and bags (brown paper and plastic) with labels for collecting specimens, nets, pH paper, thermometers, and other testing equipment, to take with you to the site. Do not forget data-recording materials. Number your collection jars and bags so that you can use these numbers when recording data later (see Table 30.2).

2. When you arrive at the site, take several minutes to observe the general area. Describe what you observe, including weather conditions.

_____

_____

3. Choose two sampling locations at the site. The two locations should differ in significant features (e.g., northern or southern exposure, high- or low-slope position, time since last disturbance, native versus exotic species composition). Formulate hypotheses about differences between the two sampling locations in regard to the following variables:

   a. Air temperature, humidity, and light intensity _____

   _____

   b. Soil temperature, moisture, and pH _____

   _____

   c. Producer biomass and diversity (variety of producers and how many of each type) _____

   _____

   d. Consumer biomass and diversity _____

   _____

Experimental Procedure: Terrestrial Ecosystems

Your instructor will organize class members into teams, and each team will be assigned specific tasks at each of the two chosen sampling locations. At each sampling location (e.g., shaded versus not shaded), you will randomly choose three 1 m square plots. Within each of these plots, randomly choose one 0.1 m square area (subplot) for all of the following abiotic variables and record your measurements in Table 30.1. For each of the biotic variables, randomly choose separate 0.1 m square subplots to sample and complete Table 30.2. Keep outside the 1 m square plots when sampling to minimize disturbance.

Abiotic Components

1. Measure air temperature, relative humidity, and light intensity at 0.5 m above the surface of the ground. Calculate the average for all three subplots (replicates).
2. Measure soil temperature, soil moisture, and soil pH (and any standing water) at 0.2 m below the surface using a soil corer. Calculate the average for all three replicates.

Biotic Components—Plants

In each subplot:

1. Count the total number of live plants.
2. Harvest each entire living plant and wash any soil from the roots. Place plant material in labeled brown paper bags.

Biotic Components—Animals

In each subplot:

1. Sweep with a net as thoroughly as possible about three to five times to capture different organisms on the vegetation in each plot. Empty the contents of the net into a labeled jar of alcohol for later sorting and identification.
2. Collect leaf litter samples in labeled plastic bags for Berlese (or Tullgren) funnel analysis.

Table 30.1 Abiotic Components of a Terrestrial Ecosystem

	Abiotic Factor	Location 1				Location 2			
		a	b	c	avg	a	b	c	avg
Air	Temperature								
	Humidity								
	Light intensity								
Soil	Temperature								
	Moisture								
	pH								

Experimental Procedure: Terrestrial Ecosystem cont'd

Laboratory Work

1. Examine collected plants and animals using a stereomicroscope, or a compound microscope, if appropriate. Group organisms into different types based on morphological features (morphotypes) and classify them as producers, consumers, or decomposers. Further classify invertebrates into herbivores, detritivores, and carnivores. Complete Table 30.2.
2. Determine the dry biomass (weight) of the plant material, or wet biomass if a drying oven is not available. Dry biomass, although more time consuming to measure, is preferable when comparing biomasses among sites and between trophic levels.
3. Determine the dry biomass of the animal material (or wet weight if plant wet weight was used).
4. Construct graphs comparing the abiotic conditions of each terrestrial sampling location.
5. Construct a pyramid for biomass, morphotype, and the total number of producers, herbivores (including detritivores), and carnivores.
6. Select any three producers and any three consumers from the organisms collected, and explain how each has adapted to its terrestrial environment.

   Producer 1 _____

   Producer 2 _____

   Producer 3 _____

   Consumer 1 _____

   Consumer 2 _____

   Consumer 3 _____

7. Return all living creatures and litter samples to their respective collection sites, as explained by your instructor. If any organisms were preserved, ask your instructor what to do with them.

Table 30.2 Biotic Components of a Terrestrial Ecosystem

	Biotic Factor		Location 1				Location 2			
			a	b	c	avg	a	b	c	avg
Plants	Producers	Total number								
		Number of morphotypes								
		Biomass								
Animals on vegetation	Herbivores	Total number								
		Number of morphotypes								
		Biomass								
	Carnivores	Total number								
		Number of morphotypes								
		Biomass								
Animals in litter	Herbivores	Total number								
		Number of morphotypes								
		Biomass								
	Carnivores	Total number								
		Number of morphotypes								
		Biomass								
	Detritivores	Total number								
		Number of morphotypes								
		Biomass								

Conclusions: Terrestrial Ecosystems

Compare the results of this terrestrial ecosystem analysis with the hypotheses you formulated (see page 445).

- Air temperature, humidity, and light intensity _____

  _____

- Soil temperature, moisture, and pH _____

  _____

- Producer biomass and diversity _____

  _____

- Consumer biomass and diversity _____

  _____

30.2 Aquatic Ecosystems

An **aquatic ecosystem** sampling should include samples from the air above the water column, the column of water itself, and the soil beneath the water column.

Study of Aquatic Sampling Site

The objective of the next Experimental Procedure is to characterize the abiotic and biotic components of an aquatic ecosystem and to determine how those factors affect the trophic structure of an ecosystem. Aquatic ecosystems consist of freshwater ecosystems (e.g., lakes, ponds, rivers, and streams) and marine ecosystems (e.g., oceans). A good site for this study is a large pond, small lake, or reservoir (with a shallow margin having rooted aquatic plants and a deeper zone with water 1–2 m deep).

1. Gather all necessary equipment, such as metersticks, collection jars and bags with labels, nets, pH paper, thermometers, and other testing equipment, to take with you to the site. Do not forget data-recording materials. Number your collection jars and bags so that you can use these numbers when recording data in Table 30.3.

2. When you arrive at the site, take several minutes to observe the general area. Describe what you observe, including weather conditions.

   _____

   _____

3. Choose two sampling locations that differ in significant features (e.g., sheltered by trees versus unsheltered, near stream inflow versus far from stream inflow). Plan to sample conditions near shore (shallow-water zone) and in deeper water away from shore (or only one or the other, if logistics are limiting).

   Formulate hypotheses about differences between the two sampling locations in

   a. Temperature, humidity, and light intensity above the surface _____

   _____

   b. Temperature, dissolved oxygen, pH, and visibility below the surface _____

   _____

   c. Producer biomass and diversity (variety of producers and how many of each type) _____

   _____

   d. Consumer biomass and diversity _____

   _____

Table 30.3 Abiotic Components of an Aquatic Ecosystem

		Location 1				Location 2			
	Abiotic Factor								
		a	b	c	avg	a	b	c	avg
Air	Temperature								
	Humidity								
	Light intensity								
Water	Temperature								
	Dissolved oxygen								
	pH								
	Visibility								

Experimental Procedure: Aquatic Ecosystem

Your instructor will organize class members into teams, and each team will be assigned specific tasks at each of the two sampling locations. At each sampling location (e.g., shaded versus not shaded), you will randomly choose three 1 m square plots. Within each of these plots, randomly choose one 0.1 m square area (subplot) for all of the following abiotic variables and record your measurements in Table 30.3. For each of the biotic variables, randomly choose separate 0.1 m square subplots to sample and complete Table 30.4. Keep outside the 1 m square plots when sampling to minimize disturbance.

Abiotic Components

1. Measure air temperature, relative humidity, and light intensity at 0.5 m above the surface of the water.
2. Measure water temperature, dissolved oxygen, pH, and visibility at 0.5 m below the surface.

Biotic Components: Plants

In each subplot:

1. Count the total number of live plants in each subplot.
2. Harvest each entire plant and wash any sediment from the roots. Place plant material in labeled brown paper bags.

Biotic Component: Plankton

Lower a plankton net with attached collecting bottle into the water to a depth of 0.5 m. Slowly raise the net vertically two to three times to collect plankton in the water column. Pour the sample into a labeled collecting jar and preserve with Lugol's solution for later sorting and identification.

Biotic Component: Animals

Choose an area of the plot not disturbed by other sampling but near emergent vegetation. Sweep through the water inside the plot three to five times. Be sure to sample around the vegetation and bump the substrate several times to dislodge benthos from the sediment. Empty the contents of the net into a sieve placed over a bucket of water. Examine, wash, and collect any macroinvertebrates and place them in the bucket. Transfer collected organisms into a labeled jar of alcohol for later sorting and identification.

Laboratory Work

1. Examine collected plants, plankton, and animals using a stereomicroscope, or a compound microscope, if appropriate. Group organisms into different types based on morphological features (morphotypes) and classify them as producers, consumers, or decomposers. Further classify plankton and invertebrates into herbivores, detritivores, and carnivores. Complete Table 30.4.
2. Determine the dry biomass (weight) of the plant material, or wet biomass if a drying oven is not available.

Table 30.4   Biotic Components of an Aquatic Ecosystem

	Biotic Factor		Location 1				Location 2			
			a	b	c	avg	a	b	c	avg
Plants	Producers	Total number								
		Number of morphotypes								
		Biomass								
Plankton	Producers	Total number								
		Number of morphotypes								
		Biomass								
	Herbivores	Total number								
		Number of morphotypes								
		Biomass								
	Carnivores	Total number								
		Number of morphotypes								
		Biomass								
Animals	Herbivores	Total number								
		Number of morphotypes								
		Biomass								
	Carnivores	Total number								
		Number of morphotypes								
		Biomass								
	Detritivores	Total number								
		Number of morphotypes								
		Biomass								

3. Determine the dry biomass of the plankton and animal material (or wet weights if plant wet weight was used).
4. Construct graphs comparing the abiotic conditions of each aquatic sampling location.
5. Construct a pyramid for biomass, morphotype, and the total number of individuals of producers, herbivores (including detritivores), and carnivores.
6. Select any three producers and any three consumers from the organisms collected, and explain how each has adapted to its aquatic environment.

Producer 1 _____

Producer 2 _____

Producer 3 _____

Consumer 1 _____

Consumer 2 _____

Consumer 3 _____

7. Return all living organisms and water samples to their respective collection sites, as explained by your instructor. If any organisms were preserved, ask your instructor what to do with them. Place any exposed petri dishes in a designated area for incubation until the next laboratory.

Conclusions: Aquatic Ecosystems

Compare the results of this aquatic ecosystem analysis with the hypotheses you formulated (see page 448).

- Temperature, humidity, and light intensity above the surface _____
  _____

- Temperature, dissolved oxygen, pH, and visibility below the surface _____
  _____

- Producer biomass and diversity _____
  _____

- Consumer biomass and diversity _____
  _____

Laboratory Review 30

_____ 1. This laboratory studies organisms at what level of organization?

_____ 2. Give an example of an abiotic component of an ecosystem.

_____ 3. Give an example of a biotic component of an ecosystem.

_____ 4. Which type of consumer feeds directly on green plants?

_____ 5. Which type of consumer feeds on both plants and animals?

_____ 6. If your assigned ecosystem contains plankton, it is most likely which type of ecosystem?

_____ 7. Which type of ecosystem contains decomposers?

_____ 8. In a forest ecosystem, what type of organisms are the predominant producers?

_____ 9. What is the role of a producer?

_____ 10. Give an example of a consumer in a terrestrial ecosystem.

_____ 11. In an aquatic ecosystem, what type of organisms are the predominant producers?

_____ 12. Give an example of a consumer in an aquatic ecosystem.

_____ 13. What type of organisms are the predominant decomposers in an ecosystem?

_____ 14. What does a decomposer take in for nutrients?

_____ 15. How do phytoplankton differ from zooplankton?

Thought Questions

16. What is an ecosystem?

17. What is the role of decomposers in an ecosystem?

18. Explain how a significant change in temperature or pH might affect the biotic components of an ecosystem.

31

Effects of Pollution on Ecosystems

Introduction

This laboratory will consider three causes of aquatic pollution: thermal pollution, acid pollution, and cultural eutrophication. **Thermal pollution** occurs when water temperature rises above normal. As water temperature rises, the amount of oxygen dissolved in water decreases, possibly depriving organisms and their cells of an adequate supply of oxygen. Deforestation, soil erosion, and the burning of fossil fuels contribute to thermal pollution, but the chief cause is use of water from a lake or the ocean as a coolant for the waste heat of a power plant.

When sulfur dioxide and nitrogen oxides enter the atmosphere, usually from the burning of fossil fuels, they are converted to acids, which return to Earth as **acid deposition** (acid rain or snow). Acid deposition kills plants, aquatic invertebrates, and also decomposers, threatening the entire ecosystem.

Figure 31.1 Cultural eutrophication.
Eutrophic lakes tend to have large populations of algae and rooted plants.

Cultural eutrophication, or overenrichment, is due to runoff from agricultural fields, wastewater from sewage treatment plants, and even excess detergents. These sources of excess nutrients cause an algal bloom seen as a green scum on a lake (Fig. 31.1). When algae overgrow and die, decomposition robs the lake of oxygen, causing a fish die-off.

31.1 Studying the Effects of Pollutants

We are going to study the effects of pollution by observing its effects on hay infusion organisms, on seed germination, and on an animal called *Gammarus*.

Study of Hay Infusion Culture

A hay infusion culture (hay soaked in water) contains various microscopic organisms, but we will be concentrating on how the pollutants in our study affect the protozoan populations in the culture. We will consider both of these aspects:

species composition: number of different types of protozoans
species diversity: diversity increases as the relative abundance of each type protozoan.

Experimental Procedure: Effect of Pollutants on a Hay Infusion Culture

During this Experimental Procedure you will examine, by preparing a wet mount, hay infusion cultures that have been treated in the following manner.

1. Control culture. This culture simulates the species composition and diversity of an untreated culture. Prepare a wet mount and answer the following questions:
   With the assistance of Figure 31.2 and any guides available in the laboratory, identify as many different types of protozoans as possible in the hay infusion culture. State whether species composition is high, medium, or low. Record your estimation in the second column of Table 31.1. Do you judge species diversity to be high, medium, or low? Record your estimation in the third column of Table 31.1.
2. Oxygen-deprived culture. Thermal pollution causes water to be oxygen deprived; therefore, when we study the effects of low oxygen on a hay infusion culture, we are studying an effect of thermal pollution. Prepare a wet mount of this culture and determine if there is a change in species composition and diversity. Again record the species composition and species diversity as high, medium, or low in Table 31.1.

Figure 31.2 Microorganisms in hay infusion cultures.
Organisms are not to size.

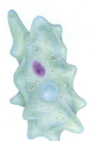

a. Amoeboid (*Amoeba*) **b.** Flagellate (*Euglena*) **c.** Ciliate (*Paramecium*) **d.** Ciliate (*Vorticella*) **e.** Ciliate (*Euplotes*)

f. Rotifer (*Philodina*) **g.** Ciliate (*Tetrahymena*) **h.** Ciliate (*Stentor*) **i.** Ciliate (*Colpoda*) **j.** Ciliate (*Didinium*)

Table 31.1 Effect of Pollution on a Hay Infusion Culture

Type of Culture	Species Composition (High, Medium, or Low)	Species Diversity (High, Medium, or Low)	Explanation
Control			
Oxygen-deprived			
Acidic			
Enriched			

3. **Acidic culture.** In this culture, the pH has been adjusted to 4 with sulfuric acid (H_2SO_4). This simulates the effect of acid rain on a hay infusion culture. Prepare a wet mount of this culture and determine if there is a change in species composition and diversity. Again record the species composition and species diversity as high, medium, or low in Table 31.1.

4. **Enriched culture.** More organic nutrients have been added to this culture. These nutrients will cause the algae population, which is food for most protozoans, to increase. In the short term, their species composition should increase. Eventually, as the algae die off decomposition will rob the water of oxygen and the protozoans may start to die off. Prepare a wet mount of this culture and determine if there is a change in species composition and diversity. Again record the species composition and species diversity as high, medium, or low in Table 31.1.

Conclusions

- What could be a physiological reason for the adverse effects of oxygen deprivation on a hay infusion culture? If consistent with your results, enter this explanation in the last column of Table 31.1.
- What could be a physiological reason for the adverse effects of a low pH on a hay infusion culture? If consistent with your results, enter this explanation in the last column of Table 31.1.
- What could be an environmental reason for the adverse affects of an enriched culture? If consistent with your results, enter this explanation in the last column of Table 31.1.

Effect of Acid Rain on Seed Germination

Seeds depend on favorable environmental conditions of temperature, light, and moisture to germinate, grow, and reproduce. Like any other biological process, germination requires enzymatic reactions that can be adversely affected by an unfavorable pH.

Experimental Procedure: Effect of Acid Rain on Seed Germination

In this Experimental Procedure we will test whether there is a negative correlation between acid concentration and germination. In other words, it is hypothesized that as acidity increases, the more likely seeds will _____.

Your instructor has placed 20 sunflower seeds in each of five containers with water of increasing acidity: 0% vinegar (tap water), 1% vinegar, 5% vinegar, 20% vinegar, and 100% vinegar.

1. Test and record the pH of solutions having the vinegar concentrations noted above. Record the pH of each solution in Table 31.2.
2. Count the number of germinated sunflower seeds in each container, and complete Table 31.2.

Table 31.2 Effect of Increasing Acidity on Germination of Sunflower Seeds

Concentration of Vinegar	pH	Number of Seeds that Germinated	Percent Germination
0%			
1%			
5%			
20%			
100%			

Conclusions: Effect of Increasing Acidity on Germination of Sunflower Seeds

- As you know, each enzyme has an optimum pH. Explain why acid rain is expected to inhibit metabolism, and therefore, seedling development. _____
- Do the data support or falsify your hypothesis? _____

Study of *Gammarus*

A small crustacean called *Gammarus* lives in ponds and streams (Fig. 31.3) where it feeds on debris, algae, or anything smaller than itself, such as some of the protozoans in Figure 31.1. In turn, fish like to feed on *Gammarus*.

Experimental Procedure: Gammarus

- Add 25 ml of spring water to a beaker and record the pH of the water. _____ pH
- Add four *Gammarus* to the container. Do they all use their legs in swimming? _____
- Which legs are used in jumping and climbing? _____
- What do *Gammarus* do when they "bump" into each other? _____

Control Sample

After observing *Gammarus,* decide what behaviors are most often observed. During a 5-minute time span, total the amount of time spent doing each of these behaviors.

Behaviors	Amount of Time	Total Time
1. _____	_____	_____
2. _____	_____	_____
3. _____	_____	_____

Test Sample

If so directed by your instructor, put a *Gammarus* in a beaker of spring water adjusted to pH 4 by adding vinegar. During a 5-minute time span, total the amount of time spent doing each of these behaviors.

Behaviors	Amount of Time	Total Time
1. _____	_____	_____
2. _____	_____	_____
3. _____	_____	_____

Figure 31.3 *Gammarus.*
Gammarus is a type of crustacean classified in a subphylum that also includes shrimp.

Conclusion

- Draw a conclusion from this study: _____

  _____

- Create a food chain that shows who eats whom when the food chain includes algae, protozoans,

  Gammarus, fish, and humans. _____

  a. Predict what would happen to this food chain if the water was

  oxygen deprived _____

  acidic _____

  enriched with inorganic nutrients (short term and long term) _____

  _____

Conclusions: Studying the Effects of Pollutants

- Give an example to show that the hay infusion study pertains to real ecosystems. _____

  _____

- What are the potential consequences of acid rain on crops that reproduce by seeds? _____

  _____

  On the food chains of the ocean? _____

- How does the addition of nutrients affect species composition and species diversity of an ecosystem

  over time? _____

31.2 Studying the Effects of Cultural Eutrophication

Chlorella, the green alga used in this study, is considered to be representative of algae in bodies of fresh water. The protozoan *Daphnia* feeds on green algae such as *Chlorella* (Fig. 31.4). First, you will observe how *Daphnia* feeds, and then you will determine the extent to which *Daphnia* could keep the effects of cultural eutrophication from occurring in a hypothetical example. Keep in mind that this case study is an oversimplification of a generally complex problem.

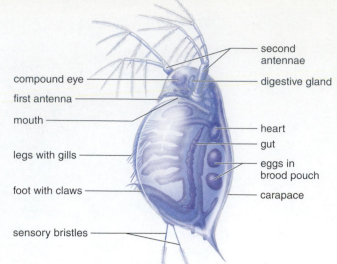

Figure 31.4   Anatomy of *Daphnia.*

Observation: Daphnia Feeding

1. Place a small pool of petroleum jelly in the center of a small petri dish.
2. Use a dropper to take a *Daphnia* from the stock culture, place it on its back (covered by water) in the petroleum jelly, and observe it under the stereomicroscope (see Fig. 31.4).
3. Note the clamlike carapace and the legs waving rapidly as the *Daphnia* filters the water.
4. Add a drop of *carmine solution,* and observe how the *Daphnia* filters the "food" from the water and passes it through the gut. The gut is more visible if you push the animal onto its side. In this position, you may also observe the heart beating in the region above the gut and just behind the head.
5. Allow the *Daphnia* to filter-feed for up to 30 minutes, and observe the progress of the carmine particles through the gut. Does the carmine travel completely through the gut in 30 minutes?

Experimental Procedure: Daphnia *Feeding on* Chlorella

This exercise requires the use of a spectrophotometer. Absorbance will be a measure of the algal population level; the greater the number of algal cells, the greater the absorbance. The higher the absorbance, the greater the amount of light absorbed and *not* passed through the solution.

1. Obtain two spectrophotometer tubes (cuvettes) and a Pasteur pipette.
2. Fill one of the cuvettes with distilled water, and use it to zero the spectrophotometer. Save this tube for step 6.
3. Use the Pasteur pipette to fill the second cuvette with *Chlorella.* Gently aspirate and expel the sample several times (without creating bubbles) to give a uniform dispersion of the algae.
4. Add ten hungry *Daphnia,* and following your instructor's directions, immediately measure the absorbance with the spectrophotometer. If a *Daphnia* swims through the beam of light, a strong deflection should occur; do not use any such higher readings—instead, use the lower figure for the absorbance. Record your reading in the first column of Table 31.3.
5. Remove the cuvette with the *Daphnia* to a safe place in a test tube rack. Allow the *Daphnia* to feed for 30 minutes.
6. Rezero the spectrophotometer with the distilled water cuvette.
7. Measure the absorbance of the experimental cuvette again. Record your data in the second column of Table 31.3, and explain your results in the third column.

Table 31.3 Spectrophotometer Data/*Daphnia* Feeding on *Chlorella*		
Absorbance Before Feeding	Absorbance After Feeding	Explanation

The following problem will test your understanding of the ecological value of a single species—in this case, *Daphnia.* Please realize that this is an oversimplification of a generally complex problem.

1. Assume that developers want to build condominium units on the shores of Silver Lake. Homeowners in the area have asked the regional council to determine how many units can be built without altering the nature of the lake. As a member of the council, you have been given the following information:

   The current population of *Daphnia,* 10 animals/liter, presently filters 24% of the lake per day, meaning that it removes this percentage of the algal population per day. This is sufficient to keep the lake essentially clear. Predation—the eating of the algae—will allow the *Daphnia* population to increase to no more than 50 animals/liter. Therefore, 50 *Daphnia*/liter will be available for feeding on the increased number of algae that would result from building the condominiums.

   Using this information, complete Table 31.4.

Table 31.4 *Daphnia* Filtering

Number of *Daphnia*/liter	Percent of Lake Filtered
10	24%
50	

2. The sewage system of the condominiums will add nutrients to the lake. Phosphorus output will be 1 kg per day for every 10 condominiums. This will cause a 30% increase in the algal population. Using this information, complete Table 31.5.

Table 31.5 Cultural Eutrophication

Number of Condominiums	Phosphorus Added	Increase in Algal Population
10	1kg	30%
20		
30		
40		
50		

Conclusion: Cultural Eutrophication

- Assume that phosphorus is the only nutrient that will cause an increase in the algal population and that *Daphnia* is the only type of zooplankton available to feed on the algae. How many condominiums would you allow the developer to build? _____

  _____

- What other possible impacts could condominium construction have on the condition of the lake?

  _____

  _____

_____ 1. Any effect on seeds would typify an effect on what trophic level in an ecosystem?

_____ 2. Any effect on *Gammarus* would typify an effect on what type of population in an ecosystem?

_____ 3. As water temperatures rise, water contains less _____.

_____ 4. What type of pollution results when water from rivers and ponds is used to cool industrial processes?

_____ 5. In your experiment, did you add acid or base to adjust the hay infusion culture to pH 4?

_____ 6. What condition does acid deposition cause that can be harmful to organisms?

_____ 7. What other portions of a food chain are threatened when plants die due to acid rain?

_____ 8. Cultural eutrophication begins with an excess of what type of substances?

_____ 9. Overenrichment causes which types of populations to increase in size beyond the ordinary?

_____ 10. In the case study, what two factors caused the producer population *(Chlorella)* to increase?

_____ 11. Does the absorbance increase or decrease after *Daphnia* feeds?

_____ 12. Why do excess nutrients alter some aquatic ecosystems?

Thought Questions

13. Does pollution affect all living things, including humans? How?

14. When pollutants enter the environment, they have far-ranging effects. Give an example from this laboratory.

15. Why might it be important to monitor predator/prey interactions when studying ecosystems?

A

Preparing a Laboratory Report/ Laboratory Report Form

A laboratory report has the sections noted in the outline that follows. Use this outline and a copy of the Laboratory Report Form on page A–3 to help you write a report assigned by your instructor. In general, do not use the words *we, my, our, your, us,* or *I* in the report. Use scientific measurements and their proper abbreviations. (For example, cm is the proper notation for centimeter and sec is correct for seconds.)

1. **Introduction:** Tell the reader what the experiment was about.
   a. **Background information:** Begin by giving an overview of the topic. Look at the Introduction to the Laboratory (and/or at the introduction to the section for which you are writing the report). Do not copy the information, but use it to get an idea about what background information to include.

   For example, suppose you are doing a laboratory report on "Solar Energy" in Laboratory 8 (Photosynthesis). You might give a definition of photosynthesis and explain the composition of white light.

   b. **Purpose:** Think about the steps of the experiment and state what the experiment was about. Tell the independent and dependent variable.

   For example, you might state that the purpose of the photosynthesis experiment was to determine the effect of white light versus green light on the photosynthetic rate. The independent variable was the color of light and the dependent variable was the rate of photosynthesis.

   c. **Hypothesis:** Consider the expected results of the experiment in order to state the hypothesis. It's possible that the Introduction to the Laboratory might hint at the expected results. State this in the form of a hypothesis.

   For example, you might state: It was hypothesized that white light would be more effective than green light for photosynthesis.

2. **Method:** Tell the reader how you did the experiment.
   a. **Equipment and sample used:** Use any illustrations in the laboratory manual that show the experimental setup to describe the equipment and the sample (subject) used.

   For example, for the photosynthesis experiment look at Figure 8.4 and describe what you see. You might state that a 150-watt lamp was the source of white light directed at *Elodea*, an aquatic plant, placed in a test tube filled with a solution of sodium bicarbonate ($NaHCO_3$). A beaker of water placed between the lamp and the test tube was a heat absorber.

   b. **Collection of data:** Think about what you did during the experiment such as what you observed or what you measured. Look at any tables you filled out in order to recall how the data were collected and what control(s) were.

   For example, for the photosynthesis experiment, you might state that the rate of photosynthesis was determined by the amount of oxygen released and was measured by how far water moved in a side arm placed in a stopper of the test that held *Elodea*. A control was the same experimental setup, except the test tube lacked *Elodea*.

3. **Results:**   Present the data in a clear manner.
   a. **Graph or table:**   If at all possible, show your data in table or graph form. You could reproduce a table you filled in or a graph you drew to show the results of the experiment. Be sure to include the title of the table; do not include any interpretation of the data column in the table.

For example, for the photosynthesis experiment you might reproduce Table 8.3.

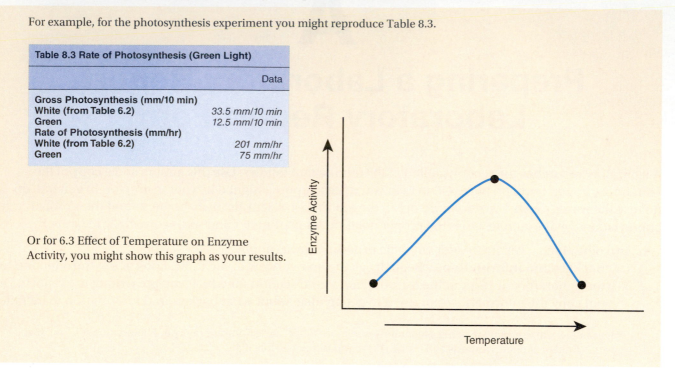

Table 8.3 Rate of Photosynthesis (Green Light)	
	Data
Gross Photosynthesis (mm/10 min)	
White (from Table 6.2)	*33.5 mm/10 min*
Green	*12.5 mm/10 min*
Rate of Photosynthesis (mm/hr)	
White (from Table 6.2)	*201 mm/hr*
Green	*75 mm/hr*

Or for 6.3 Effect of Temperature on Enzyme Activity, you might show this graph as your results.

   b. **Description of data:**   Examine your data, and decide what they tell you. Then, below any table or graph, add a description to help the reader understand what the table or graph is showing. Define any terms in the table that are not readily understandable.

For example, below Table 8.3 you might state that these data indicate that the rate of photosynthesis with white light is faster than with green light. Also, you should define gross photosynthesis. Or below the graph that shows the effect of temperature on enzyme activity, you might state that these data show that the rate of enzymatic activity speeds up until boiling occurs and then it drops off.

4. **Conclusion:**   Tell if the data support or do not support the hypothesis.
   a. **Compare the hypothesis with the data:**   Do your data agree or disagree with the hypothesis?

For example, for the photosynthesis experiment you might state: These results support the hypothesis that white light is more effective for photosynthesis than green light.

   b. **Explanation:**   Explain why you think you obtained these results. Look at any questions you answered while in the laboratory, and use them to help you decide on an appropriate explanation.

For example, the answers to the questions in 8.2 Solar Energy might help you state that white light gives a higher rate of photosynthesis because it contains all the visible light rays. Green light gives a lower rate because green plants such as *Elodea* do not absorb green light.

If your results do not support the hypothesis, explain why you think this occurred.

In this instance you might state that while white light contains all visible light rays and green light is not absorbed by a green plant, the experiment did not support the hypothesis because of failure to use a heat absorber when doing the green light experiment.

Laboratory Report for _____

1. **Introduction**
   a. Background information

   b. Purpose

   c. Hypothesis

2. **Method**
   a. Equipment used

   b. Collection of data

3. **Results**
   a. Graph or table
   (Place these on attached sheets.)

   b. Description of data

4. **Conclusion**
   a. Compare the hypothesis with the data

   b. Explanation

   c. Conclusion

Metric System

Unit and Abbreviation	Metric Equivalent	Approximate English-to-Metric Equivalents	Units of Temperature

Length

nanometer (nm) $= 10^{-9}$ m $(10^{-3}$ μm)
micrometer (μm) $= 10^{-6}$ m $(10^{-3}$ mm)
millimeter (mm) $= 0.001\ (10^{-3})$ m
centimeter (cm) $= 0.01\ (10^{-2})$ m

1 inch $= 2.54$ cm
1 foot $= 30.5$ cm
1 foot $= 0.30$ m

meter (m) $= 100\ (10^{2})$ cm
$= 1{,}000$ mm

1 yard $= 0.91$ m

kilometer (km) $= 1{,}000\ (10^{3})$ m

1 mi $= 1.6$ km

Weight (mass)

nanogram (ng) $= 10^{-9}$ g
microgram (μg) $= 10^{-6}$ g
milligram (mg) $= 10^{-3}$ g
gram (g) $= 1{,}000$ mg

1 ounce $= 28.3$ g
1 pound $= 454$ g
$= 0.45$ kg

kilogram (kg) $= 1{,}000\ (10^{3})$ g
metric ton (t) $= 1{,}000$ kg

1 ton $= 0.91$ t

Volume

microliter (μl) $= 10^{-6}$ l $(10^{-3}$ ml)
milliliter (ml) $= 10^{-3}$ l
$= 1$ cm^3 (cc)
$= 1{,}000$ mm^3

1 tsp $= 5$ ml
1 fl oz $= 30$ ml

liter (l) $= 1{,}000$ ml

1 pint $= 0.47$ l
1 quart $= 0.95$ l
1 gallon $= 3.79$ l

kiloliter (kl) $= 1{,}000$ l

Units of Temperature

°F °C

230 — 110
220
212° — 210 — 100 — 100°
200
190 — 90
180 — 80
170
160° — 160 — 70 — 71°
150
140 — 60
134° — 130 — 57°
120 — 50
110
105.8° — 110 — 40 — 41°
98.6° — 100 — 37°
90
80 — 30
70
60 — 20
56.66° — 60 — 13.7°
50 — 10
40
32° — 30 — 0 — 0°
20
10 — −10
0
−10 — −20
−20
−30 — −30
−40 — −40

Common Temperatures*

°C	°F	
100	212	Water boils at standard temperature and pressure.
71	160	Flash pasteurization of milk
57	134	Highest recorded temperature in the United States, Death Valley, July 10, 1913
41	105.8	Average body temperature of a marathon runner in hot weather
37	98.6	Human body temperature
13.7	56.66	Human survival is still possible at this temperature.
0	32.0	Water freezes at standard temperature and pressure.

* See temperature scale

To convert temperature scales:

$$°C = \frac{5}{9}\,(°F - 32)$$

$$°F = \frac{9}{5}\,(°C + 32)$$

Tree of Life

The tree of life in Figure C-1 depicted in this appendix shows how the three domains of life—Bacteria, Archaea, and Eukarya—are related, and indeed, how all organisms may be related to one another through the evolutionary process. The tree of life contains the following groups of organisms.

PROKARYOTES

Laboratories 4 and 24 review the prokaryotes in domains Bacteria and Archaea. Prokaryotic organisms are characterized by their simple structure but a complex metabolism. The chromosome of a prokaryote is not bounded by a nuclear envelope, and therefore, these organisms do not have a nucleus. Prokaryotes carry out all the metabolic processes performed by eukaryotes and many others besides. However, they do not have organelles, except for plentiful ribosomes.

DOMAIN BACTERIA

Laboratory 24 covers the diversity of bacteria, which are the most plentiful of all organisms. Bacteria are capable of living in most habitats, and carry out many different metabolic processes. Most bacteria are aerobic heterotrophs, but some are photosynthetic, and some are chemosynthetic. Motile forms move by flagella consisting of a single filament. Their cell wall contains peptidoglycan, and they have distinctive RNA sequences.

DOMAIN ARCHAEA

Domain Archaea is briefly mentioned in Laboratory 24. Archaean cell walls lack peptidoglycan, their lipids have a unique branched structure, and their ribosomal RNA sequences are distinctive. Examples are methanogens and the extremophiles.

DOMAIN EUKARYA

Laboratories 25–29 explore the eukaryotes. Eukaryotes have a complex cell structure with a nucleus and several types of organelles that compartmentalize the cell. Mitochondria that produce ATP and chloroplasts that produce carbohydrate are derived from prokaryotes that took up residence in a larger nucleated cell. They may be unicellular (the majority of the Protists), or multicellular (the plants, fungi, animals). Each multicellular group is characterized by a particular mode of nutrition. Flagella, if present, have a 9 + 2 organization.

PROTISTS

The protists (Laboratory 24) are a catchall group for any eukaryote that is not a plant, fungus, or animal. The protists are currently divided among six supergroups whose evolutionary relationships are being actively investigated. A supergroup is a major eukaryotic group, and six supergroups encompass all members of the domain Eukarya, including all protists, plants, fungi, and animals. Examples of protists include amoebas, the green algae, and the paramecium.

PLANTS

Plants (Laboratories 25 and 26) are photosynthetic eukaryotes that became adapted to living on land. This group includes aquatic green algae called charophytes, which have a haploid life cycle and share certain traits with the land plants.

The land plants exhibit an alternation-of-generations life cycle; protect a multicellular sporophyte embryo; produce gametes in gametangia; possess apical tissue that produces complex tissues; and possess a waxy cuticle that prevents water loss. Examples include mosses, ferns, the conifers, and the flowering plants.

FUNGI

Fungi (Laboratory 24) have multicellular bodies composed of hyphae; usually absorb food and lack flagella; and produce nonmotile spores during both asexual and sexual reproduction. Examples include the mushrooms, cup fungi, and molds.

ANIMALS

Animals (Laboratories 27–29) are multicellular, usually with specialized tissues and a digestive cavity; they ingest or absorb food; and they have a diploid life cycle. This diverse group includes invertebrate organisms such as sponges, worms, mollusks, and insects, and vertebrates such as fishes, reptiles, and humans.

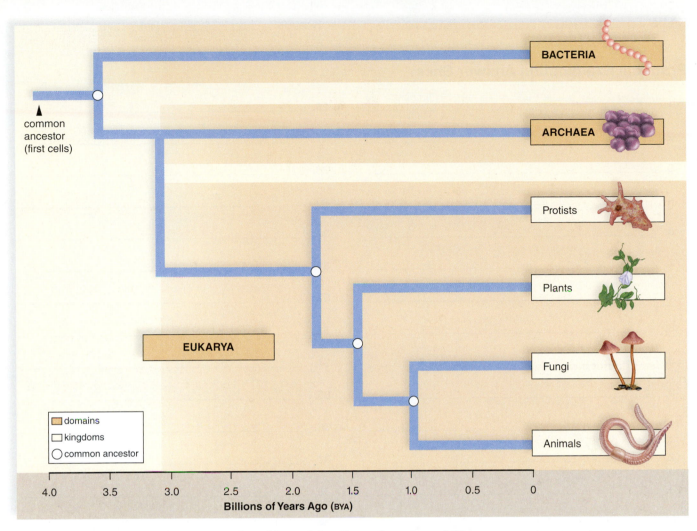

Figure C-1 The evolutionary relationships of the three domains of life.
Living organisms are classified into three domans: Bacteria, Archaea, and Eukarya. A time scale is provided on the bottom for reference.

Name _____ Section _____ Date _____

Practical Examination Answer Sheet

1. _____
2. _____
3. _____
4. _____
5. _____
6. _____
7. _____
8. _____
9. _____
10. _____
11. _____
12. _____
13. _____
14. _____
15. _____
16. _____
17. _____
18. _____
19. _____
20. _____
21. _____
22. _____
23. _____
24. _____
25. _____

26. _____
27. _____
28. _____
29. _____
30. _____
31. _____
32. _____
33. _____
34. _____
35. _____
36. _____
37. _____
38. _____
39. _____
40. _____
41. _____
42. _____
43. _____
44. _____
45. _____
46. _____
47. _____
48. _____
49. _____
50. _____

Photo Credits

Phototake; 19.10: © Martin Rotker/Phototake; p. 272(both): © Petit format/Photo Researchers; 19.12(both): © Ralph Hutchings/Visuals Unlimited.

Laboratory 20
Figure 20.4: © Evelyn Jo Johnson; 20.5: © Carolina Biological Supply/Phototake.

Laboratory 21
Figure 21.2a: © Superstock; 21.2b: © HFPA, 63rd Golden Globe Awards; 21.2c–f, h: © The McGraw-Hill Companies, Inc. Bob Coyle, photographer; 21.2g: © Corbis RF; p. 300(Turner): Courtesy UNC Medical Illustration and Photography; p. 300(Poly X): Courtesy The McElligott Family; p. 300(Klinefelter): Courtesy Stefan D. Schwarz, http://klinefeltersyndrome.org; p. 300(Jacob): Courtesy The Giles Family; 21.6(all): © CNRI/SPL/Photo Researchers.

Laboratory 22
Figures 22.9(both): © Bill Longcore/Photo Researchers.

Laboratory 23
Page 319(fossil dig): © Annie Griffiths Belt/Corbis; 23.2(trilobites): © Danita Delimont/Getty Images; 23.2(pseudoscorpion, mantis): © John Cancalosi/Getty Images; 23.2(snails): © Ed Reschke/Getty Images; 23.2(echinoderms): © DEA/G. Nimatallah/Getty Images; 23.2(ammonite): © Carl Pendle/Getty Images; 23.3(frog): © DEA/G. Cigolini/Getty Images; 23.3(snake, fish): © John Cancalosi/Getty Images; 23.3duckbill dinosaur): © Kevin Schafer/Getty Images; 23.3(bird): © WaterFrame/Alamy; 23.3(deerlike mammal): © Gary Ombler/Getty Images; 23.4(sassafras leaf): © Jonathan Blair/Getty Images; 23.4(ferns): © John Cancalosi/Getty Images; 23.4(seed plant leaves): © Sinclair Stammers/Photo Researchers; 23.4(maple leaf): © Biophoto Associates/Photo Researchers; 23.4poplar leaf): © James L. Amos/Photo Researchers; 23.4(flower): © Barbara Strnadova/Photo Researchers; 23.6(bird): © The McGraw-Hill Companies, Inc.; 23.6(man): © The McGraw-Hill Companies, Inc. Eric Wise, photographer; 23.5(bat): © Jack Milchanowski/Getty Images; 23.5(cat): © Marc Henrie/Getty Images; 23.5(lizard): © Mauricio Handler/Getty Images; 23.9(S. tchadensis): © HO/AFP/Newscom; 23.9(A. afarensis): © Friedrich Saurer/Alamy; 23.9(A. boisei): © Science VU/NMK/Visuals Unlimited; 23.9(H. habilis): © Herve Conge/ISM/Phototake; 23.9(H. sapiens): © 2007 Educational Images Ltd./Custom Medical Stock Photo.

Laboratory 24
Figure 24.1(bacteria): © Dr. Dennis Kunkel/Phototake; 24.1(paramecium): © Michael Abbey/Visuals Unlimited; 24.1(morel): © Corbis RF; 24.1(sunflower): © Photodisc Green/Getty RF; 24.1(snowgoose): © Charles Bush Photography; 24.2a: © Henry Aldrich/Visuals Unlimited; 24.3(both): © Kathy Park Talaro; 24.4a: © Dr. Richard Kessel & Dr. Gene Shih/Visuals Unlimited; 24.4b: © Gary Gaugler/Visuals Unlimited; 24.4c: © SciMAT/Photo Researchers; 24.5a: Courtesy Steven R. Spilatro, Marietta College, Marietta, OH; 24.5b: © Sherman Thomas/Visuals Unlimited; 24.6: © M.I. Walker/Photo Researchers; 24.7: © R. Knauft/Photo Researchers; 24.9b: © M.I. Walker/Photo Researchers; 24.10(left): © John D. Cunningham/Visuals Unlimited; 24.10(right): © Carolina Biological Supply/Visuals Unlimited; 24.11(rockweed): © D.P Wilson/Eric & David Hosking/Photo Researchers; 24.12a: © Steven P. Lynch; 24.12b: © Gary R. Robinson/Visuals Unlimited; 24.13: © Dennis Kunkel Microscopy, Inc./Phototake; 24.14: © Biophoto Assoc./Photo Researchers; 24.15c: © Eye of Science/Photo Researchers; 24.15d(left): © London School of Hygiene and Tropical Medicine/Photo Researchers; 24.15d(right): Courtesy CDC/Dr. Mae Melvin; 24.16b: © Tom Adams/Visuals Unlimited; 24.18a: © Biophoto Assoc./Photo Researchers; 24.18b: © Bill Keogh/Visuals Unlimited; 24.18c: © Gary R. Robinson/Visuals Unlimited; 24.18d: © Carol Wolfe/photographer; 24.19a: © Gary T. Cole/Biological Photo Service; 24.20: © Garry DeLong/Getty Images; 24.21a: © Carolina Biological Supply Co./Visuals Unlimited; 24.21b: © Carolina Biological Supply Co./Visuals Unlimited; 24.23a: © Everett S. Beneke/Visuals Unlimited; 24.23b: © John Hadfield/SPL/Photo Researchers; 24.23c: © P. Marazzi/SPL/Photo Researchers.

Laboratory 25
Figure 25.2(left): © Dr. John D. Cunningham/Visuals Unlimited; 25.2(right): © Kingsley Stern; 25.4(sporophyte): © Heather Angel/Natural Visions; 25.4(gametophyte): © Steven P. Lynch; 25.5–25.6: © 2007 J.R. Waaland/Biological Photo Service; 25.7b–25.8a(thallus): © Ed Reschke; 25.8b: © J.M. Conrarder/Nat'l Audubon Society/Photo Researchers; 25.8c: © R. Calentine/Visuals Unlimited; 25.9: © Steve Solum/Photoshot; 25.10a: © Carolina Biological Supply/Phototake; 25.11: © Robert P. Carr/Photoshot; 25.12: © Steven P. Lynch; 25.13: © The McGraw-Hill Companies, Inc. Carlyn Iverson, photographer; 25.14: © Ken Wagner/Visuals unlimited; 25.15(top): © Robert Knauft/Biology Pics/Photo Researchers; 25.15(bottom): © Carolina Biological Supply/Phototake.

Laboratory 26
Page 371(ovule): © Ed Reschke; 26.3a: © D. Cavagnaro/Visuals Unlimited; 26.3b: © Kingsley Stern; 26.3c: © D. Giannechini/Photo Researchers; 26.4(pollen grains): © Carolina Biological Supply/Phototake; 26.7a: © 2007 J.R. Waaland/Biological Photo Service; 26.7b: © Ed Reschke; 26.8: © Carolina Biological Supply/Phototake; 26.11: © W. P. Armstrong 2004.

Laboratory 27
Figure 27.3a: © Amar and Isabelle Guillen, Guillen Photography/Alamy; 27.3b: © Andrew J. Martinez/Photo Researchers; 27.3c: © Kenneth M. Highfill/Photo Researchers; 27.5: © Carolina Biological Supply/Visuals Unlimited; 27.6(both): © Kim Taylor/npl/Minden Pictures; 27.7a: © Azure Computer & Photo Services/Animals Animals; 27.7b: © Ron & Valerie Taylor/Photoshot; 27.7c: © NHPA/Charles Hood/Photoshot RF; 27.7d: © Amos Nachoum/Corbis; 27.9: © Carolina Biological Supply/Phototake; 27.10a: Photography by Marc C. Perkins, Orange Coast College, Costa Mesa, CA; image blending by Heather Bartell, Huntington Beach, CA; 27.10b: © Tom E. Adams/Visuals Unlimited; 27.12a: © R. Calentine/Visuals Unlimited; 27.13(both): © Larry Jenson/Visuals Unlimited; 27.14: © Carolina Biological Supply/Phototake; p. 398(elephantiasis): © Vanessa Vick/The New York Times/Redux; 27.16a: © Wim van Egmond/Visuals Unlimited.

Laboratory 28
Figure 28.1a: © Fred Bavendam/Minden; 28.1b: © Andrew J. Martinez/Photo Researchers; 28.1c: © IT Stock Free/Alamy RF; 28.1d: © Douglas Faulkner/Photo Researchers; 28.3b–28.4b: © Ken Taylor/Wildlife Images; 28.6a: © 2007 Roger K. Burnard/Biological Photo Service; 28.6b: © R. DeGoursey/Visuals Unlimited; 28.6c: © Diane R. Nelson; 28.6d: © C.P. Hickman/Visuals Unlimited; 28.8b: © Ken Taylor/Wildlife Images; 28.9: © John Cunningham/Visuals Unlimited; 28.10a(honeybee): © R. Williamson/Visuals Unlimited; 28.10a(millipede): © Bill Beatty/Visuals Unlimited; 28.10a(centipede): © Adrian Wenner/Visuals Unlimited; 28.10b(spider): © W.J. Weber/Visuals Unlimited; 28.10b(scorpion): © David M. Dennis; 28.10b(horseshoe crab): © E.R. Degginger/Photo Researchers; 28.10c(crab): © Tom McHugh/Photo Researchers; 28.10c(shrimp): © Alex Kerstitch/Visuals Unlimited; 28.10c(barnacles): © Kjell Sandved/Visuals Unlimited; 28.11b: © Ken Taylor/Wildlife Images; 28.15a: © Daniel Gotshall/Visuals Unlimited; 28.15b: © Hal Beral/Visuals Unlimited; 28.15c: © Robert Dunne/Photo Researchers; 28.15d: © Neil McDan/Photo Researchers; 28.15e: © Robert Clay/Visuals Unlimited; 28.15f: © Alex Kerstitch/Visuals Unlimited; 28.16(both): © BiologyImaging.com.

Laboratory 29
Figure 29.2: © Rick Harbo; 29.3: © Heather Angel/Natural Visions; 29.4: © Stan Sims/Visuals Unlimited; 29.5(shark): © Hal Beral/Visuals Unlimited; 29.5(fish): © Patrice/Visuals Unlimited; 29.5(frog): © Rod Planck/Photo Researchers; 29.5(turtle): © Suzanne and Joseph Collins/Photo Researchers; 29.5(bird): © Robert and Linda Mitchell Photography; 29.5(fox): © Craig Lorenz/Photo Researchers; 29.6: © Rod Planck/Photo Researchers; 29.7b: © Carolina Biological Supply/Phototake; 29.8–29.11b: © Ken Taylor/Wildlife Images.

Laboratory 31
Figure 31.1: © Michael Gadomski/Animals Animals; 31.3: © NOAA/Visuals Unlimited.

Index

Note: Page references followed by *f* and *t* refer to figures and tables, respectively.

internal anatomy of, 428–434
Fronds, of ferns, 365, 365*f*
Frontal bone, 248*f*, 249, 249*f*
    chimpanzee and human, comparison of, 328, 328*f*
Frontalis muscle, 251, 252*f*, 253*t*
Frontal lobe, 230, 231*f*, 233*f*
Fructose
    structure of, 91
    and yeast fermentation, 91–92
Fruit(s), 126, 131, 136–138, 136*f*, 377, 378*f*, 381
    aggregate, 137
    dry, 137
    fleshy, 137
        complex, 137
        simple, 137
    types of, 137
Fruit flies. *See Drosophila*
Fucus, 343–344, 344*f*
Fungi (sing., fungus), 335, 335*f*, 350–353
    diversity of, 350, 350*f*
    and human disease, 353, 353*f*
Funnel, of squid, 405, 406*f*

Gallbladder
    of fetal pig, 167*f*, 168, 169*f*, 170
    of frog, 429, 430*f*
    human, 171*f*, 174*f*
Gamete(s). *See also* Egg(s); Sperm
    chromosome number in, 65
    definition of, 57
    formation of, 65, 73. *See also* Oogenesis; Spermatogenesis
    of plants, 128, 128*f*, 357, 357*f*
Gametogenesis, 57
    in animals, 73–75, 73*f*
    definition of, 73
    segregation of alleles in, 276
Gametophyte(s), 128, 128*f*, 357, 357*f*, 358, 369
    female, 129, 129*f*–130*f*
    of ferns, 364, 364*f*, 366–367
    of flowering plants, 378*f*, 379–380, 380*f*
    male, 129, 129*f*–130*f*
    of moss, 358*f*, 359
    of pines, 372*f*, 373, 375
    of seedless vascular plants, 361
Gammarus, 456, 457*f*
    effects of pollutants on, 456–457
Ganglia (sing., ganglion), of clam, 405
Gas exchange, pulmonary, 188, 214, 214*f*
Gastric gland(s), 175*f*
Gastric mill, of crayfish, 415
Gastric pit(s), 175*f*
Gastrocnemius muscle, 252*f*, 253*t*
Gastrodermis, of *Hydra*, 389, 389*f*
Gastropods, 402, 402*f*
Gastrovascular cavity
    of Cnidarians, 388*f*
    of *Hydra*, 389, 389*f*
    of planarians, 392, 392*f*, 393, 393*f*–394*f*
Gastrula, 259
    early, 261
    late, 261–262
Gastrulation
    in frog, 261–262, 261*f*–262*f*
    in humans, 262
Gel electrophoresis
    in detection of sickle cell disease, 315, 315*f*
    equipment for, 312, 313*f*
    procedure for, 312, 313*f*, 315

Gemma (pl., gemmae), of liverworts, 361, 361*f*
Gemma cups, of liverworts, 361, 361*f*
Gene(s), 58*t*, 275. *See also* Allele(s)
Generative cell, of pollen grain, 378*f*, 379
Genetic counseling, 297–301
Genetic disorders
    detection of, 313–315
    inheritance of, 292–295
Genetics, 275
    human, 289–302
Genital pore, of planarians, 392, 392*f*
Genome
    definition of, 313
    human, 313
Genomic sequencing, for detection of sickle cell disease, 315
Genotype
    definition of, 275, 289
    heterozygous, 275, 275*f*, 289*f*
    homozygous dominant, 275, 275*f*, 289*f*
    homozygous recessive, 275, 275*f*, 289*f*
    and phenotype, 275*f*
Geologic timescale, 318–320
    dating within, 318
    divisions of, 318, 319*f*
    limitations of, 320
Germination, 141–142
    acid rain and, 455–456
    eudicot, 139*f*
    monocot, 140*f*
Germ layers, 259
    derivatives of, in humans, 262, 263*f*
    of planarians, 392
    of roundworms, 396
Gill(s), 325, 423
    of clam, 404*f*, 405
    of crayfish, 414*f*, 415
    of mushrooms, 352, 352*f*, 353
    of squid, 405, 406*f*
    in vertebrates, 426
Gill capillary(ies), in vertebrates, 437*f*
Gill mushroom, 352
Gill slits, 423
    dorsal tubular, of lancelet, 425*f*, 426, 426*f*
    of tunicates, 425, 425*f*
Gingko(es), 371, 371*f*
Gizzard, of earthworm, 410, 411*f*
Gladiolus, corm of, 116, 116*f*
Glans penis, 204*f*
Gloeocapsa, 340, 340*f*
Glomerular capsule (Bowman's capsule), 223, 224*f*
Glomerular filtrate, 225
Glomerular filtration, 223–224, 224*f*, 225
Glomerulus (pl., glomeruli), 223, 223*t*, 224*f*
Glottis
    definition of, 162
    of fetal pig, 163, 163*f*, 209, 209*f*
    of frog, 428*f*, 429
    human, 171*f*
Glucagon, 168
Glucose, 30
    in cellular respiration, 87, 87*f*
    in ethanol fermentation, 88
    metabolism of, 214
    photosynthesis and, 95
    structure of, 91
    and yeast fermentation, 91–92
Gluteus maximus muscle, 251, 252*f*, 253*t*
Gluteus medius muscle, 252*f*, 253*t*

Glycerol, 34, 34*f*
    absorption, in small intestine, 181, 181*f*
Glycogen, 30, 214
Goblet cell(s), 147, 147*f*–148*f*, 156, 156*f*
Golgi apparatus, 43, 43*t*
Gonad(s), 199
    of crayfish, 415
    of sea star, 420*f*, 421
Graafian follicle. *See* Vesicular (Graafian) follicle
Gram (g), 11
Grana, in chloroplasts, 95, 95*f*
Grasshopper
    anatomy of, 416–417, 416*f*–417*f*
    and crayfish, comparison of, 417
    metamorphosis, 417–418, 418*f*
Gray fox, 427*f*
Gray matter, of spinal cord, 234, 234*f*–235*f*
Green gland, of crayfish, 415
Green light, in photosynthesis, 100–101
Ground tissue
    of herbaceous stem, 115, 115*f*
    of monocot root, 112, 112*f*
    of plants, 107, 107*t*, 108, 108*f*
Growth plate, 246*f*
Growth rings, of clam, 403, 403*f*
G_1 stage, 58, 58*f*
G_2 stage, 58, 58*f*
Guanine (G), 305
Guard cells, of plant stoma, 119, 119*f*
Gymnosperms, 369, 371, 371*f*
    and angiosperms, comparison of, 382

Hair, of fetal pig, 160
Hair cells, in ear, 240
Hair follicle(s), 157, 157*f*
Hamstring muscle group, 252*f*, 253*t*
Hand(s), bones of, 250
Handplate, in human embryo, 273*f*
Haploid (n) generation, 130*f*
    in plants, 128, 128*f*–129*f*, 357, 357*f*
Hard palate
    of fetal pig, 162, 162*f*
    human, 171*f*
Haversian system(s). *See* Osteon(s)
Hay infusion culture(s)
    effects of pollutants on, 454–455
    microorganisms in, 454, 454*f*
HCl. *See* Hydrochloric acid (HCl)
Head
    chimpanzee and human, comparison of, 326, 327*f*
    of crayfish, 414, 414*f*
    of fetal pig, 160, 161*f*
    of grasshopper, 416, 416*f*
    muscles of, 252*f*, 253*t*
    of squid, 405, 406*f*
Heart, 183, 183*f*, 189*f*, 214*f*
    anatomy of, 184–186, 185*f*
    apex of, 185*f*
    blood flow through, 186
    chambers of, 184, 188
    of clam, 404*f*, 405
    of crayfish, 415
    as double pump, 188
    embryonic development of
        in chick embryo, 264–266, 265*f*–266*f*
        in human, 273*f*
    external anatomy of, 184, 185*f*
    of fetal pig, 166, 167*f*, 169*f*, 210

Micrometer (μm), 10, 10*t*
Microphylls, 362, 362*f*
Micropyle, 130*f*
Microscope(s), 14
    dissecting. *See* Stereomicroscope
    electron. *See* Electron microscope(s)
    light. *See* Light microscope(s)
    parfocal, 20
Microscopic observation(s)
    of *Euglena*, 25, 25*f*
    of human epithelial cells, 23–24, 23*f*
    of onion epidermal cells, 24, 24*f*
    of pond water, 25
Microscopy, 14–25
Microsporangia, 129, 370, 370*f*
    of pines, 375, 375*f*
Microspore(s), 129, 129*f*–130*f*
    of pines, 372*f*, 373
Microtubules, in spindle fibers, 59
Microvilli (sing., microvillus), 144*f*, 146
Midbrain, 231*f*, 232
    in chick embryo, 265*f*
    in human embryo, 273*f*
Middle ear, 240*f*, 240*t*
    embryology of, 325
Middle layer, of blood vessel, 196, 196*f*, 197
Milligram (mg), 11
Milliliter (ml), 12
Millimeter (mm), 10, 10*t*
Millipede(s), 413, 413*f*
Mitochondrion (pl., mitochondria), 43, 43*t*
    in cellular respiration, 87, 87*f*
    of skeletal muscle, 257*f*
Mitosis, 58*f*, 59–62, 60*f*–61*f*
    in animal cells, 59, 60*f*–61*f*
    definition of, 57
    and meiosis, comparison of, 70–71, 70*f*–71*f*
    and meiosis I, comparison of, 70–71, 70*f*–71*f*
    and meiosis II, comparison of, 71, 71*f*
    phases of, 58*f*, 60*f*–61*f*, 61
    in plant cells, 59, 60*f*–61*f*, 62, 128, 128*f*–130*f*, 357, 357*f*
    structures associated with, 58*t*
    in whitefish blastula, 59, 60*f*–61*f*
Mitral valve. *See* Bicuspid valve
μm. *See* Micrometer (μm)
Molecular evidence, of evolution, 323, 331–333, 333*f*
Molecular genetics, 303
Molecule(s), 27
    inorganic, 27
    organic, 27
Molluscs, 401–407
    diversity of, 402, 402*f*
Monocots, 108–109, 109*f*, 377, 377*t*
    flowers of, 127, 127*f*, 127*t*
    herbaceous stem of, 114–115, 115*f*
    roots of, anatomy, 112, 112*f*
    seeds of, 140, 140*f*
Monocyte(s), 151*f*
Monohybrid cross, 276–280, 277*f*–278*f*
Monomer(s), 27
Monosaccharide(s), 27, 30
Monotremes, 427*f*
Morel, 350*f*
Morula, 259, 261*f*
    formation of, 260, 260*f*
Moss(es), 358
    life cycle of, 358, 358*f*

Moth(s)
    metamorphosis, 417, 418*f*
    as pollinators, 132, 133*f*
Motor neuron(s), 155, 155*f*, 234–235, 235*f*
Mouth
    in chick embryo, 266
    of Cnidarians, 388, 388*f*
    of crayfish, 414*f*, 415
    of earthworm, 409, 409*f*
    of fetal pig, 162, 162*f*, 210
    of frog, 428, 428*f*
    human, 174*f*
    of *Hydra*, 389, 389*f*
    of squid, 405
    of vertebrates, 426
Mouthparts, of grasshopper, 416, 416*f*
mRNA. *See* Messenger RNA (mRNA)
M stage, 58, 58*f*
Mucosa, of intestine, 156, 156*f*
Mucus, secretion of, 147, 147*f*
Muscle(s)
    actions of, 245, 245*f*, 253*t*
    antagonistic pairs of, 245, 254
    cardiac. *See* Cardiac muscle
    contraction of, 153
        isometric, 255
        isotonic, 255
    insertion of, 245, 245*f*
    involuntary, 153–154
    naming of, 251
    origin of, 245, 245*f*
    skeletal. *See* Skeletal muscle
    smooth. *See* Smooth muscle
    striated, 153, 153*f*, 256, 257*f*
    types of, 245
    in vertebrates, 426
    voluntary, 153
Muscle fiber(s), 144*f*, 153, 256
    contraction of, mechanism of, 256
    microscopic anatomy of, 256, 257*f*
Muscular foot, of clam, 402, 404*f*, 405
Muscularis, of intestine
    circular, 156, 156*f*
    longitudinal, 156, 156*f*
Muscular system, 245
Muscular tissue, 153–154
    in human body, 144*f*
Musculoskeletal system, definition of, 245
Mushroom(s), 352–353, 352*f*
Mutation(s), 58, 323
MYA, 318
Mycelium, fungal, 350, 350*f*
Myelencephalon, in chick embryo, 266*f*
Myelin sheath, 155*f*, 234
Myofibrils, 256, 257*f*
Myofilament(s), 256, 257*f*
Myosin, 28, 153, 256, 257*f*

Nanometer (nm), 10, 10*t*
Nares (sing., naris), of frog, 427*f*, 428
Nasal bone(s), 248*f*, 250
Nasal pit, in human embryo, 273*f*
Nasopharynx
    definition of, 162
    of fetal pig, 162*f*
    human, 171*f*
Nautilus, 402, 402*f*
Near point, 239
Neck
    of fetal pig, 160, 161*f*

    muscles of, 252*f*, 253*t*
Neck region, of fetal pig, 166, 167*f*
Nectar, 132, 133*f*
Negative control, 2, 28
Negative test, 27
Nematocyst, of *Hydra*, 389*f*, 390
Nematodes, 396–398
Nephridia, of earthworm, 410, 411*f*–412*f*
Nephron(s), 201, 201*f*
    blood circulation around, 223, 224*f*
    blood vessels of, 223, 223*t*, 224*f*
    structure of, 222–223, 222*f*, 224*f*
Nereocystis, 344*f*
Nerve(s), 155. *See also* Cranial nerve(s); Spinal nerve(s)
    of bone, 150, 150*f*
Nerve cord
    dorsal tubular, 423, 423*f*
        of lancelet, 425*f*, 426, 426*f*
    of planarians, 392*f*–393*f*
    ventral, of earthworm, 410, 412, 412*f*
Nervous system, 229, 229*f*
    of clam, 405
    embryology of, 262
    embryonic development of, 264
    of frog, 434, 434*f*
    of planarians, 392, 392*f*
    of roundworms, 396
Nervous tissue, 155, 155*f*
    in human body, 144*f*–145*f*
Neural folds, 262, 262*f*
Neural groove, 261*f*, 262, 262*f*
Neural plate, 261*f*, 262, 262*f*
Neural tube, 261*f*, 262, 264
Neurofibromatosis (NF), 293
Neuroglia, 155, 155*f*
Neuroglial cells, 145*f*
Neuron(s), 145*f*, 155, 155*f*, 229
Neurulation, frog, 261*f*, 262, 262*f*
Neutrophil(s), 151*f*
NF. *See* Neurofibromatosis (NF)
Nipple(s), of fetal pig, 160, 161*f*
Nociceptors. *See* Pain receptors
Nonpolar molecules, 35
Northern leopard frog, 427*f*
Nose, of fetal pig, 161*f*
Nostril(s). *See* Nares (sing., naris)
Notochord, 261*f*, 262, 263*f*, 423, 423*f*
    of lancelet, 425*f*, 426, 426*f*
    presumptive, 261*f*, 262
Nuclear division, 57. *See also* Meiosis; Mitosis
Nuclear envelope
    of eukaryptoc cell, 43
    fragmentation, in prophase, 60*f*
Nucleoid, 42, 42*f*
    bacterial, 336, 337*f*
Nucleolus, 43, 43*t*, 58*t*
    in prophase, 60*f*
    in telophase, 61*f*
Nucleoplasm, of eukaryptoc cell, 43
Nucleus, 58*t*
    cell, 43, 43*t*, 335
    in green algae, 342, 342*f*
    in mitosis, 59
Nut(s), 137
Nutrient(s), absorption, in small intestine, 181, 181*f*
Nymph(s), grasshopper, 417, 418*f*

Observation(s), making, in scientific method, 2, 3*f*

Occipital bone, 248f, 249, 249f
Occipitalis muscle, 252f, 253t
Occipital lobe, 230, 231f, 233f
Oil(s), 34, 177
    emulsification of, 35–36, 35f
Olfactory bulb, 231f
    of frog, 434f
Olfactory nerve, of frog, 434f
Oligochaetes, 408
Omnivores, 444
One-trait crosses, 276–280
Onion
    epidermal cells, microscopic observation
        of, 24, 24f
    root tip cells, mitosis in, 62
    starch, test for, 32
Oocyte(s), 74, 74f
    mammalian, 199
    primary, 73f
    production of, 73, 73f. See also Oogenesis
    secondary, 73, 73f, 74, 74f, 207f
Oogenesis, 73, 73f
Operculum, 427f
    of moss, 358f, 360
Optic chiasma, 231f
Optic lobe, of frog, 434f
Optic nerve, 237f, 237t
Oral cavity. See also Mouth
    definition of, 162
    of fetal pig, 162, 162f
Oral hood, of lancelet, 425f, 426
Orbicularis oculi muscle, 252f, 253t
Orbicularis oris muscle, 252f, 253t
Organ(s), 156–157. See also specific organs
    definition of, 143
Organelles, 42–43, 335
Organism(s), evolutionary relationships of, 317
Organizer(s), in embryonic development, 262
Organ system(s), definition of, 143
Oscillatoria, 340, 340f
Osculum, of sponges, 385, 386f
Osmosis, 49–53, 50f
    definition of, 49
Osmotic pressure, and capillary exchange, 218,
    218f
Ossicles, 240f, 240t
Osteoblast(s), 246f
Osteocyte(s), 145f, 150, 150f, 246f
Osteon(s), 150, 150f, 246f, 247
Otic vesicle, in chick embryo, 266f
Outer ear, 240f, 240t
Outer layer, of blood vessel, 196, 196f, 197
Ova (sing., ovum). See Egg(s)
Ovarian artery, of fetal pig, 194f
Ovarian ligament, 207f
Ovarian vein, of fetal pig, 192f, 195f, 206f
Ovary(ies), 205, 205t, 207f
    of earthworm, 410
    of fetal pig, 205, 206f
    of flower, 126, 126f, 130f, 376, 376f
    of frog, 429, 430f, 431, 433f
    mammalian, 199
    microscopic anatomy of, 74, 74f
    oogenesis in, 73, 73f
    of planarians, 392f
Oviduct(s), 205, 205t, 207f
    of fetal pig, 205, 206f
    of frog, 431, 433f
Ovipositor(s), of grasshopper, 416, 416f–417f
Ovulation, 74f

Ovule(s), 130f
    of flower, 126, 126f, 129, 129f, 376, 376f,
        378f, 380
    of flowering plants, 377, 378f
    of pines, 372f, 373, 375, 375f
    of seed plants, 369, 369f, 370
Oxygen, release of, in photosynthesis, 95, 95f,
    98, 102, 102f

Pacinian corpuscles, 241f
Pain receptors, in skin, 241f
Palatine bone(s), 248f, 250
Pancreas
    endocrine, 168
    exocrine, 168
    of fetal pig, 167f, 168, 169f, 170
    of frog, 430
    human, 171f, 174f
Pancreatic juice, 168, 174f, 177, 177f
Paper chromatography, 96–97, 96f
Papillary muscle, 187f
Paramecium, 347, 347f
Parathyroid gland(s)
    embryology of, 325
Parenchyma, of herbaceous stem, 114f–115f
Parenchyma cells, in plants, 108
Parent cell, 59
Parfocal focusing, 20
Parietal bone(s), 248f, 249, 249f
Parietal lobe, 230, 231f, 233f
Patella, 249f, 251
Patellar reflex. See Knee-jerk reflex
Paternity testing, blood types and, 296–297
Pea, fruit of, 136f
Pearl river redbelly turtle, 427f
Pectoral girdle, 248, 249f, 250
Pectoralis major muscle, 252f, 253t
Pedigree(s)
    analysis, 299–300
    autosomal, 299, 299f
    construction of, 300–301
    symbols used in, 299, 299f
Peduncle, of flower, 126, 126f, 130f
Pellicle, of protozoan, 347f, 348, 348f
Pelvic girdle, 248, 249f, 251, 251f
Pelvis
    chimpanzee and human, comparison of,
        326, 327f
    female, 251
    male, 251
Penis, 202, 202t, 204f
    of fetal pig, 161, 200f, 202, 203f
    of grasshopper, 416
    of planarians, 392, 392f
Pepsin, protein digestion by, 175–176, 175f
Peptide(s), 28
Peptide bond, 27–28, 28f
Perch. See also Vertebrate(s)
    anatomy of, 436f
Pericardial cavity, of fetal pig, 166
Pericardial sac, of clam, 404f, 405
Pericardium, 183f
    of clam, 404f, 405
    of frog, 429
Pericarp, 136–137, 140, 140f
Pericycle, of plant root, 111, 111f–112f
Period(s), of geologic timescale, 318, 319f
Periosteum, 246f, 247
Peripheral nervous system (PNS), 229, 229f,
    234–236

Peristalsis, 156
Peritoneal cavity, 201
Peritoneum, 156
    of fetal pig, 168, 201
Peritubular capillary network, 223, 223t, 224f
Peroneus longus muscle, 252f, 253t
Peroxisome, 43, 43t, 78
Petal(s), 126, 126f, 130f, 376, 376f
pH
    acidic, 54, 54f
    basic, 54, 54f
    and cells, 54–55
    definition of, 54
    and enzyme activity, 82–83
    neutral, 54, 54f
Phalanges (foot), 249f, 251
Phalanges (hand), 249f, 250
Pharyngeal pouches, 325, 423, 423f
Pharynx
    definition of, 162
    of earthworm, 410, 411f
    of fetal pig, 163, 163f, 209, 209f
    human, 171f, 174f
    of lancelet, 425f, 426
    of planarians, 392f–393f, 394
Phenol red, 178
Phenotype
    definition of, 275, 289
    dominant, 289
    genotype and, 275f
    recessive, 289
Phenotypic ratio, 276
Phenylketonuria, 294
Philodina, 399f
Phloem, 107t, 108, 111, 111f–112f
    of herbaceous stem, 114, 114f–115f
    secondary, 117
    of woody stem, 118, 118f
Phospholipid(s), 34
Photomicrograph, 14, 14f
Photoreceptor(s), 236
Photosynthesis, 95–106, 108, 443
    action spectrum for, 100, 100f
    algal, 341
    carbon dioxide uptake in, 102, 102f
    and cellular respiration, 102, 102f
    by cyanobacteria, 339
    equation for, 95
    green light in, 100–101
    gross, 99
    light reactions and Calvin cycle reactions
        in, relationship of, 103
    net, 98
    rate of, 99, 101
    white light in, 98–99, 99f
Photosynthetic pigments, 96–97
    action spectrum for, 100, 100f
pH scale, 54, 54f
Phylogenetic (evolutionary) tree, 384–385,
    384f
    of chordates, 424, 424f
Pigeon. See also Vertebrate(s)
    anatomy of, 438f
Pillbug(s), 1–2, 1f
    external anatomy of, 1, 1f, 4
    motion of, 5
    response to potential foods, 6–7
Pine(s), 371–375
    life cycle of, 372f
Pineal body, 231f

Pyrenoids, 348, 348*f*
   in green algae, 342, 342*f*

Quadriceps femoris muscle group, 252*f*, 253*t*
Quickening, 272

Radial canal(s), of sea star, 420*f*, 421
Radial symmetry, 385, 388, 388*f*, 419
Radicle
   of eudicot embryo, 134, 135*f*, 139–140, 139*f*
   of monocot embryo, 140, 140*f*
Radius, 249*f*, 250
Radula, of squid, 405
Rana pipiens. See Frog
Rat. *See also* Vertebrate(s)
   anatomy of, 441*f*
Ray(s), 427*f*
   of woody stem, 118, 118*f*
Ray-finned fishes, 424, 424*f*, 427*f*
Reactant(s), 77
Receptacle(s)
   of flower, 126, 126*f*, 130*f*, 376, 376*f*
   of *Fucus,* 343
Rectum
   of fetal pig, 170, 205, 206*f*
   human, 174*f*
Rectus abdominis muscle, 251, 252*f*, 253*t*
Red blood cells, 145*f*, 151, 151*f*
   abnormality, in sickle cell disease, 314–315, 314*f*
   in lungs, 219, 219*f*
   normal, 314, 314*f*
   sickle-shaped, 314–315, 314*f*
   tonicity and, 51, 51*f*
Red tide, 346
Reflex(es), definition of, 234–235
Refraction, 236
Renal artery(ies), 189*f*, 201*f*, 214*f*, 222*f*, 224*f*
   of fetal pig, 192*f*, 194, 194*f*
Renal cortex, 201, 201*f*, 222, 222*f*
Renal medulla, 201, 201*f*, 222, 222*f*
Renal pelvis, 201, 201*f*, 222, 222*f*
Renal pyramid(s), 201*f*, 222*f*
Renal vein(s), 189*f*, 201*f*, 214*f*, 222*f*, 224*f*
   of fetal pig, 192*f*, 194, 195*f*
Reproductive system, 199
   of *Ascaris,* 397*f*
   of earthworm, 410, 411*f*
   female, 205, 205*t*
      of fetal pig, 205, 206*f*
         and human, comparison of, 207
      human, 207*f*
         and fetal pig, comparison of, 207
   male, 202–204
      of fetal pig, 202–204, 203*f*
         and human, comparison of, 204, 204*f*
      human, and fetal pig, comparison of, 204, 204*f*
   of planarians, 392, 392*f*
Reptiles, 424, 427*f*
   birds as, 436
   cardiovascular system in, 437*f*
Resolution, of microscope, 15
Respiratory system
   of fetal pig, 209, 209*f*
   of frog, 429
   human, 171*f*
   in vertebrates, comparison of, 439–440
Respiratory volume(s), 219–221

Respirometer, 90, 91*f*
Retina, 237*f*, 237*t*
R*f* (ratio-factor), 97
R group(s), 28, 28*f*
Rh blood groups, 296
Rhizoids, of liverworts, 361, 361*f*
Rhizome(s), 116, 116*f*, 362–363
   of ferns, 364, 364*f*
Rib(s), 248, 249*f*, 250
   of fetal pig, 167*f*
   floating, 250
Rib cage, movement, with breathing, 219, 220*f*
Ribosomal RNA (rRNA), 303, 308
Ribosome(s), 42, 42*f*, 43*t*, 308, 310
   bacterial, 336, 337*f*
   binding sites of, 311, 311*f*
Ring canal, of sea star, 420*f*, 421
Ringworm, 353, 353*f*
RNA (ribonucleic acid), 303
   structure of, 307–308, 307*f*
      and DNA structure, comparison of, 308
Rockweed, 343–344, 344*f*
Rod cells, 236, 237*t*
Root(s), plant, 105, 105*f*
   adventitious, 140*f*
   branch, 106*f*
   diversity of, 113, 113*f*
   eudicot vs monocot, 111–112, 111*f*–112*f*
   major tissues of, 107, 107*t*
   of monocots vs eudicots, 109*f*
   primary, 106*f*, 139*f*–140*f*
   prop, 113, 113*f*, 140*f*
   secondary, 139*f*
Root cap, of plant, 110, 110*f*
Root hair plexus, 241*f*
Root hairs, of plants, 106*f*, 110, 110*f*, 111, 111*f*
Root system, of plants, 106, 106*f*, 110–113
Root tip, of plants, 106, 106*f*, 107, 107*f*
   anatomy of, 110, 110*f*
Rotation, 254, 254*f*
Rotifers, 398–399, 399*f*
Rough endoplasmic reticulum, 43*t*
Round ligament, 207*f*
Round window, 240*f*
Roundworms, 396–398
rRNA. *See* Ribosomal RNA (rRNA)
Ruffini endings, 241*f*

Saccule, 240, 240*f*, 240*t*
Sacrum, 249*f*, 250
Sagittal crest, chimpanzee, 328, 328*f*
St. Martin, Alexis, 173
Salamander(s), 427*f*
Saliva, 174*f*
Salivary gland(s), 174*f*
Samara, 137
Sand dollar(s), 419, 419*f*
Sarcolemma, 257*f*
Sarcomeres, 256, 257*f*
Sarcoplasm, 257*f*
Sarcoplasmic reticulum, 257*f*
Sartorius muscle, 252*f*, 253*t*
Scallop(s), 402, 402*f*
Scanning electron micrograph (SEM), of lymphocyte, 14*f*
Scanning electron microscope, 9, 15
Scapula, 249*f*, 250
Scarlet hood, 350*f*
Scientific method, 2–3, 3*f*
Scientific theory(ies), development of, 3, 3*f*

Scissor-tailed flycatcher, 427*f*
Sclera, 237*f*, 237*t*
Sclerenchyma cells, in plants, 108
Scolex, of tapeworm, 395, 395*f*, 396, 396*f*
Scorpion(s), 413, 413*f*
Scrotal sac, of fetal pig, 161*f*, 202, 203*f*
Scrotum, 204*f*
SDA. *See* Specific dynamic action
Sea anemone, 391, 391*f*
Sea cucumber, 419, 419*f*
Sea lily(ies), 419, 419*f*
Sea squirts, 425, 425*f*
Sea star, 419, 419*f*
   aboral side, 420*f*, 421
   anatomy of, 420–421, 420*f*
   embryonic development of, 260, 260*f*
   oral side, 421
Sea urchin, 419, 419*f*
Seaweed, 343–344, 344*f*
Sebaceous glands, 157, 157*f*
Sebdenia, 345*f*
Sectioning, of tissue, 143
Seed(s), 131, 369. *See also* Germination
   development of, 129, 130*f*
   distribution of, 125–126
   eudicot, 109, 109*f*, 139–140, 139*f*
   of flowering plants, 377, 378*f*
   monocot, 109, 109*f*, 140, 140*f*
   of monocots vs eudicots, 109, 109*f*
   of pines, 374, 374*f*
Seed coat, 130*f*, 131, 135*f*, 139, 139*f*
Seed cone(s), of pines, 372*f*, 373–375, 374*f*–375*f*
Segmented worms, 407
SEM. *See* Scanning electron micrograph (SEM)
Semicircular canals, 240, 240*f*, 240*t*
Seminal receptacles
   of crayfish, 415
   of earthworm, 410, 411*f*
Seminal vesicle(s), 202, 202*t*, 204*f*
   of earthworm, 410, 411*f*
   of fetal pig, 202, 203*f*
Seminiferous tubules, 75, 75*f*
Sensory neuron(s), 234–235, 235*f*
Sensory receptor(s), in skin, 241–242, 241*f*
Sepal(s), 126, 126*f*, 130*f*, 376, 376*f*
Septum (pl., septa)
   cardiac, 184, 187*f*
   of earthworm, 410, 411*f*
   of fungal hypha, 350, 350*f*
Serosa, of intestine, 156, 156*f*
Sessile [term], 385
Sessile filter feeder(s), 386, 402, 402*f*
Setae
   of annelids, 407*f*, 408
   of earthworm, 409, 409*f*, 412*f*
Sex cells. *See* Gamete(s)
Sex chromosomes, 285, 289
   anomalies of, 298, 298*f*
Sexual reproduction
   in black bread mold, 351, 351*f*–352*f*
   in Cnidarians, 388, 388*f*
   genetic variation due to, 67
   in mushrooms, 352, 352*f*
   in sponges, 385
   in vertebrates, 426
   in *Volvox,* 342
Shark(s), 427*f*
Shoot system, of plants, 106, 106*f*
Shoot tip. *See* Terminal bud (shoot tip)
Shoulder, of fetal pig, 160, 161*f*